Immunopsychiatry

Immunopsychiatry

A Clinician's Introduction to the Immune Basis of Mental Disorders

EDITED BY ANTONIO L. TEIXEIRA

AND

MOISES E. BAUER

OXFORD UNIVERSITY PRESS

Oxford University Press is a department of the University of Oxford. It furthers the University's objective of excellence in research, scholarship, and education by publishing worldwide. Oxford is a registered trade mark of Oxford University Press in the UK and certain other countries.

Published in the United States of America by Oxford University Press
198 Madison Avenue, New York, NY 10016, United States of America.

CIP data is on file at the Library of Congress
ISBN 978–0–19–088446–8

9 8 7 6 5 4 3

Printed by Sheridan Books, Inc., United States of America

CONTENTS

PREFACE
FROM PSYCHONEUROIMMUNOLOGY TO IMMUNOPSYCHIATRY: A BRIEF HISTORY

Carmine M. Pariante

The term *immunopsychiatry* is not new, but has only recently attracted the attention required to deserve a book—this book. Until recently, only *psychoneuroimmunology* was used as a term to define this field of research.

The first citation that I could find online in PubMed using the term *immunopsychiatry* is a 1985 review that contrasts immuno-*psychiatry* with immuno-*neurology*; it refers to the fact that antibodies directed against brain epitopes can induce structural and biochemical changes in multiple neuronal activities, not only with neurological (somatic, motor) effects but also with psychiatric (emotional, behavioral) effects (Janković, 1985). Only much more recently, this term has been introduced to indicate the renewed focus of researchers and clinicians on immune abnormalities in psychiatric disorders.

But do we need a new term to define what many of us—and many authors of these book—have already been studying for the last 20 years? I have already referred to the fact that, in my opinion, *psychoneuroimmunology* and *immunopsychiatry* represent two different conceptualizations of the brain–immune communication, symbolized by the different order of the words that compose these names (Pariante, 2015). The term *immunopsychiatry* highlights a new biological and evolutionary framework where our behaviors and emotions are governed by peripheral immune mechanisms, rather than the other way around. As I have said before, "you cannot cure yourself of a fever by meditation, but fever can make you sad and grumpy" (Pariante, 2015).

For those who have entered this research area more recently—and these include some of the younger authors in this book—it is difficult to remember a time when it wasn't like this. When the term *psychoneuroimmunology* first reached a wider audience in the 1980s, the discipline advocated a bidirectional communication between the brain and the immune system on equal terms, and in fact arguably with an emphasis on the notion that psychological and neural phenomena can and do influence the immune system. Even the dozens of studies in the 1980s that examined the relationship between major depression and the immune system were based on a model that saw depression as a mental state that is able to influence the immune system (Stein et al., 1991).

Fast-forward to 2008, and the first review was published that spelled out clearly the notion that the immune system can and does "subjugate the brain," through mechanisms underpinning sickness behavior (Dantzer et al., 2008). The shift had started, bringing the field toward an acceptance of the new notion, central to today's immunopsychiatry concept, that the immune abnormalities are the cause, rather than the consequence, of psychiatric disorders. New research is also pushing the boundaries of animal models, as suggested by a recent review on the "immunopsychiatry of zebrafish" (de Abreu et al., 2018).

So where are we going now? And how does this book add to the debate?

First of all, immunopsychiatry is about *all* psychiatric disorders. While research in depression is possibly ahead of research in other mental disorders, the book is able to examine a vast number of studies looking at immune mechanisms that either cut across multiple psychiatric diagnoses or are specific to individual mental disorders: addiction, bipolar disorder, depression, dementias, eating disorders, post-traumatic stress disorder, and schizophrenia.

But this clinical research must be supported by the biological and molecular understanding of the mechanism underpinning these measurable immune abnormalities. Hence, the choice of also including in this book chapters dedicated to the overall functioning of the immune system (including its pharmacological manipulation), to how the immune system affects the brain both in adulthood and during its development, and to the role of the microbiota-gut-brain axis. The chapter on the immune system "as a sensor," on how the immune system provides information that the brain processes and integrates with other inputs, really conveys the aforementioned new biological and evolutionary framework where the immune system and the brain are endowed with the same dignity.

And where is the future going? What will the next edition of this book, published, let's say, in 2028, look like? I am certain that the next edition of this book will focus on the novel treatments for psychiatric disorders that this research will have generated by then.

The "known unknowns" are already discussed in a chapter of this edition of the book: the potential therapeutic action of adjunctive treatment with anti-inflammatories in a range of psychiatric disorders, which is something that many research groups, including mine, are currently testing. But I am certain that the immunopsychiatry of the future will impact patients' care through what are still presently "unknown unknowns": novel, hitherto undiscovered targets that can either reduce immune activation, or reduce its effects on the brain, in a way that is safe, well-tolerated, and effective in many patients with many different psychiatric disorders, and especially in those not responding to the currently available medications. Then, and only then, we will truly be able to decide whether or not immunopsychiatry deserve its new name.

ACKNOWLEDGMENTS AND CONFLICT OF INTEREST

I have received research funding from pharmaceutical companies interested in the development of anti-inflammatories strategies for psychiatric disorders, such as Johnson & Johnson and Eleusis Ltd., and from the Wellcome Trust Consortium for Neuroimmunology of Mood Disorders and Alzheimer's Disease, which receives contributions from Johnson & Johnson and Lundbeck. My research is funded by the UK Medical Research Council (MR/L014815/1, MR/J002739/1), the UK National Institute for Health Research (NIHR) Biomedical Research Centre for Mental Health at the South London and Maudsley NHS Trust and King's College London, and the charities, Brain and Behavior Research Foundation and the Psychiatry Research Trust.

REFERENCES

Dantzer R, O'Connor JC, Freund GG, Johnson RW, Kelley KW. From inflammation to sickness and depression: when the immune system subjugates the brain. *Nat Rev Neurosci.* 2008;9(1):46–56.

de Abreu MS, Giacomini ACVV, Zanandrea R, et al. Psychoneuroimmunology and immunopsychiatry of zebrafish. *Psychoneuroendocrinology.* 2018 Mar 22;92:1–12.

Janković BD. From immunoneurology to immunopsychiatry: neuromodulating activity of anti-brain antibodies. *Int Rev Neurobiol.* 1985;26:249–314.

Pariante CM. Psychoneuroimmunology or immunopsychiatry? *Lancet Psychiatry.* 2015 Mar;2(3):197–199.

Stein M, Miller AH, Trestman RL. Depression, the immune system, and health and illness. Findings in search of meaning. *Arch Gen Psychiatry.* 1991;48(2):171–177.

CONTRIBUTORS

Samantha Alvarez-Herrera, PhD
Department of Psychoimmunology
National Institute of Psychiatry
Mexico City, Mexico

Izabela G. Barbosa, MD, PhD
Department of Mental Health
Medical School
Universidade Federal de Minas Gerais (UFMG)
Belo Horizonte, Brazil

Moisés E. Bauer, PhD
Laboratory of Stress Immunology
School of Sciences
PUCRS, National Institute of Science and Technology Neuroimmunomodulation (INCT-NIM)
Porto Alegre, Brazil

Isabelle Bauer, PhD
Department of Psychiatry and Behavioral Sciences, McGovern Medical School
University of Texas Health Science Center
Houston, Texas

Bernhard T. Baune, MD, PhD, MPH, FRANZCP
Department of Psychiatry
Melbourne Medical School
University of Melbourne
Melbourne, Australia

Hugo Besedovsky, MD, PhD
Division of Neurophysiology
Research Group Immunophysiology
Institute of Physiology Pathophysiology
Medical Faculty
Philipps University
Marburg, Germany

Alessandra Borsini, PhD
Department of Psychological Medicine
Psychology and Neuroscience
Institute of Psychiatry
King's College London
London

Akif Camkurt, MD
Department of Psychiatry and Behavioral Sciences, McGovern Medical School
University of Texas Health Science Center
Houston, Texas

Eduardo Candelario-Jalil, PhD
Department of Neuroscience
University of Florida
Gainesville, Florida

Karine Clément, MD, PhD
Institute of Cardiometabolism and Nutrition (ICAN)
Assistance Publique-Hôpitaux de Paris
Pitié-Salpêtrière Hospital
Nutrition Department
Paris, France

John F. Cryan, PhD
Department of Anatomy and Neuroscience
APC Microbiome Institute
University College Cork
Cork, Ireland

Vivian Thaise da Silveira
Department of Pharmacology
Institute of Biological Sciences
Universidade Federal de Minas Gerais (UFMG)
Belo Horizonte, Brazil

Antonio Carlos Pinheiro de Oliveira
Department of Pharmacology
Institute of Biological Sciences, UFMG
Belo Horizonte, Brazil

Adriana del Rey, PhD
Division of Neurophysiology
Research Group Immunophysiology
Institute of Physiology Pathophysiology
Medical Faculty
Philipps University
Marburg, Germany

Timothy G. Dinan, MD, PhD
Department of Psychiatry and Neurobehavioural Science
University College Cork
Cork, Ireland

Breno Satler Diniz, MD, PhD
Department of Psychiatry
Faculty of Medicine
University of Toronto
Clinician Scientist
Geriatric Psychiatry Division
Center for Addiction and Mental Health (CAMH)
Toronto, Canada

Adaliene Versiani Matos Ferreira, PhD
Department of Nutrition
Universidade Federal de Minas Gerais (UFMG)
Belo Horizonte, Brazil

Gilliard Lach, PhD
APC Microbiome Institute
University College Cork
Cork, Ireland

Scott D. Lane, PhD
Center for Neurobehavioral Research on Addiction
Department of Psychiatry and Behavioral Sciences, McGovern Medical School
University of Texas Health Science Center
Houston, Texas

Marion Leboyer, MD, PhD
Fondation Fondamental
Department of Psychiatry
Hôpital Henri Mondor
Université Paris-Est-Créteil
INSERM U955, Translational Psychiatry
Créteil, France

Rafael Coelho Magalhães, Occupational Therapist, MSc
Researcher of the Pediatric Branch at the Interdisciplinary
Laboratory of Medical Investigation
Faculty of Medicine
Universidade Federal de Minas Gerais (UFMG)
Belo Horizonte, Brazil

Geneviève Marcelin, PhD
INSERM
Sorbonne Universités
Institut de Cardio-métabolisme et Nutrition (ICAN)
Pitié-Salpêtrière
Paris, France

Laís Bhering Martins, PhD
Department of Nutrition
Universidade Federal de Minas Gerais (UFMG)
Belo Horizonte, Brazil

Nayara Mussi Monteze, PhD
Department of Nutrition
Universidade Federal de Minas Gerais (UFMG)
Belo Horizonte, Brazil

Janaina Matos Moreira, MD, PhD
Department of Pediatrics
School of Medicine, UFMG
Interdisciplinary Laboratory of Medical Investigation
Belo Horizonte, Brazil

José Oliveira, MD, PhD
Centro Hospitalar Psiquiátrico de Lisboa
Champalimaud Clinical Centre
Champalimaud Centre for the Unknown
Lisboa, Portugal

Carmine M. Pariante, MD, PhD, FRCPsych
Professor of Biological Psychiatry
Head of the Stress, Psychiatry and Immunology Laboratory (SPILab)
Institute of Psychiatry, Psychology and Neuroscience
King's College London
London

Lenin Pavón, MD, PhD
Department of Psychoimmunology
National Institute of Psychiatry
Mexico City, Mexico

Antonio Carlos Pinheiro de Oliveira, PhD
Department of Pharmacology
Institute of Biological Sciences, UFMG
Belo Horizonte, Brazil

Natália P. Rocha, PhD
Neuropsychiatry Program
Department of Psychiatry and Behavioral Sciences, McGovern Medical School
University of Texas Health Science Center
Houston, Texas

Haitham Salem, MD, PhD
Department of Psychiatry and Behavioral Sciences, McGovern Medical School
University of Texas Health Science Center
Houston, Texas

Wilson Savino, PhD
Laboratory on Thymus Research
Oswaldo Cruz Institute
Oswaldo Cruz Foundation (Fiocruz)
Rio de Janeiro, Brazil

Sudhakar Selvaraj, MD, PhD
Department of Psychiatry and Behavioral Sciences, McGovern Medical School
University of Texas Health Science Center
Houston, Texas

Ana Cristina Simões e Silva, MD, PhD
Department of Pediatrics
Faculty of Medicine
Coordinator of the Pediatric Branch of the Interdisciplinary
Laboratory of Medical Investigation
Faculty of Medicine
Universidade Federal de Minas Gerais (UFMG)
Belo Horizonte, Brazil

Gaurav Singhal, MTropVSc, BVSc & AH
Psychiatric Neuroscience Lab
Discipline of Psychiatry
University of Adelaide
Adelaide, Australia

Jair C. Soares, MD, PhD
Department of Psychiatry and Behavioral Sciences, McGovern Medical School
University of Texas Health Science Center
Houston, Texas

Laura Stertz, PhD
Department of Psychiatry and Behavioral Sciences, McGovern Medical School
University of Texas Health Science Center at Houston
Houston, Texas

Laure Tabouy, PhD
Fondation Fondamental
Department of Psychiatry
Hôpital Henri Mondor
Université Paris-Est-Créteil
INSERM U955, Translational Psychiatry
Créteil, France

Antônio L. Teixeira, MD, PhD
Professor of Psychiatry
Neuropsychiatry Program & Immunopsychiatry Lab
Department of Psychiatry and Behavioral Sciences, McGovern Medical School
University of Texas Health Science Center
Houston, Texas

Erica Leandro Vieira, PhD
Interdisciplinary Laboratory of Medical Investigation
Medical School
Universidade Federal de Minas Gerais (UFMG)
Belo Horizonte, Brazil

Consuelo Walss-Bass, PhD
Department of Psychiatry and Behavioral Sciences, McGovern Medical School
University of Texas Health Science Center at Houston
Houston, Texas

Andrea Wieck, PhD
Laboratory of Stress Immunology
School of Sciences, PUCRS
Laboratory of Stress Immunology
School of Sciences
Pontifical Catholic University of the Rio Grande do Sul
Porto Alegre, Brazil

Cristian Patrick Zeni, MD, PhD
Department of Psychiatry and Behavioral Sciences, McGovern Medical School
University of Texas Health Science Center
Houston, Texas

Patricia A. Zunszain, PhD
Department of Psychological Medicine
Psychology and Neuroscience
Institute of Psychiatry
King's College London
London

1

Overview of the Immune System

LENIN PAVÓN, SAMANTHA ALVAREZ-HERRERA, AND MOISES E. BAUER ■

THE ORGANIZATION OF THE IMMUNE SYSTEM

The immune system confers immunity through self–nonself discrimination of antigens. It is organized in primary and secondary lymphoid organs, leukocytes that circulate via lymphatics and circulation, and soluble molecules that perform a set of coordinated mechanisms generated against infectious or self-derived antigens (e.g., cancer, autoimmunity). Following antigen removal and returning to homeostasis, the immune system generates a faster response in a future contact with the same agent (i.e., memory). During an immune response, both innate and adaptive mechanisms are engaged, allowing an integrated system of host defense. The principal characteristics of innate and adaptive immunity are shown in Table 1.1.

The innate immunity is phylogenetically the first system of defense against foreign invaders, mainly microorganisms; their cells and soluble components respond immediately after antigen recognition. The major functions of this innate immunity are the prevention of infections via inflammation, structural remodeling after tissue damage, and supporting the development of the adaptive immune responses. The activation of innate immune responses is dependent on the recognition of highly conserved chemical structures that are common to groups of microorganisms, called pathogen-associated molecular patterns (PAMPs), and consists of cellular and soluble components that act in a coordinated manner.[1] The principal components of innate immunity are the physical and chemical barriers, as well as phagocytic cells, with constitutive receptors that recognize the

Table 1.1 Principal Characteristics of Innate and Adaptive Immunity

Characteristic	**Innate Immunity**	**Adaptive Immunity**
Phylogeny	Primitive (found in all multicellular organisms)	Only in vertebrates
Specificity	It can recognize general patterns shared in groups of pathogens but cannot make a fine distinction	Recognizes highly specific microbial and nonmicrobial antigens of a specific pathogen
Diversity	Limited, it is encoded in germinal line	Enormous, the recognition receptors are produced by somatic recombination of genetic segments
Kind of recognition receptors	PRRs from germinal line	BCR and TCR from somatic recombination
Cells involved in the response	Granulocytes, monocytes, macrophages, dendritic cells, NK cells, and specific types of lymphocytes	Professional antigen presenting cells, T cells, B cells, and plasmatic cells
Immunological memory	NO. The response reacts with equal potency upon repeated exposure to the same pathogen	YES. Cells "remember" specific pathogens: upon re-exposure to a pathogen, these cells mount a much faster and more potent second response
Response and potency	Immediate response, but it has a limited and lower potency	Slower response, but it is much more potent
Course	Attempts to immediately destroy the pathogen, and if it can't, it contains the infection until the adaptive immune system acts	Slower to respond; effector cells are generally produced in 1 week and the entire response occurs over 1–2 weeks
Other important components	Physical and chemical barriers, soluble PRRs	Antibodies

PAMPs, release soluble molecules, and are in charge of orchestrating inflammatory reactions.

On the other hand, the adaptive immune responses are mediated by lymphocytes and appear late in evolution. This response has the capacity of distinguishing microbes and molecules with high specificity, which allows an increase in the defensive capacity after successive exposures to the same foreign agent, i.e., "immunological memory." The principal molecular components of adaptive immunity are the cell membrane bound T cell receptors (TCR) and

B-cell receptors (BCR) and the soluble molecules that are released by them, such as cytokines and antibodies. The major functions of the adaptive immune are to eliminate intra- and extracellular foreign microorganisms that innate immunity cannot eliminate, and to boost innate mechanisms such as phagocytosis.

Several cell-surface molecules with a role in immune responses were originally characterized based on their reactivity to monoclonal antibodies. The antibodies produced by laboratories could form a cluster when they could be grouped together because they recognized the same cell-surface molecule. This led to a nomenclature in which a given molecule was assigned a "cluster of differentiation," or CD, number—for example CD3, CD4, and CD28. This knowledge is crucial, for example, in phenotyping immune cell populations, and describing cellular differentiation and activation states, functionality, and regulatory actions (Table 1.2).

In the last decades, compelling evidence demonstrates that the components of the immune system dynamically interact with the nervous and the endocrine systems, via shared communication mechanisms, improving the homeostasis and host's protection. Cytokines, neurotransmitters, and hormones allow this interaction and modulate multiple physiological functions.[2–4] Cells of nervous and immune systems share receptors for these mediators, and modulate in concert the behavior and function of other systems, improving survival.

Table 1.2 Major Leukocyte Subsets Identified by Clusters of Differentiation (CDs)

Cell Type	Markers
Macrophages	CD45+CD14+
Granulocytes	CD45+CD15+
Dendritic cells (DCs)	CD14+/-CD11c+
T helper (Th) cells	CD3+CD4+
Cytotoxic T cells (CTL)	CD3+CD8+
B cells	CD3–CD19+
Natural Killer (NK) cells	CD3–CD56+
NK T cells	CD3+CD56+
Activated markers (T cells)	CD25+ (early), CD28+ (early) CD69+ (early), or HLADR+ (late)
Regulatory T Cell	CD4+CD25+FOXP3+

Note: The CDs are membrane-bound receptors, uniquely expressed by specific cells, and can be identified by flow cytometry.

Physical and Chemical Barriers

The innate immunity has several physical and chemical barriers aimed at avoiding infections and arresting the pathogen at the beginning of the infection.[5] The principal physical mechanisms are represented by the epidermis, ciliated respiratory epithelium, vascular endothelium, and mucosal surfaces; their importance lies in the fact that these barriers separate the organism from the external world through thin layers of cells that cover a wide surface. The chemical components, such as the saliva, gastric secretions, sweat, tears, and antimicrobial agents (e.g., defensins and lysozyme), prevent the progression of infection by damaging the cellular components of infectious agents. The damage or malfunction of these barriers leads to increased susceptibility to infections.

Recognition Sensors of the Innate Immunity

The discrimination of self from nonself (infectious) structures is carried out by a limited number of germ line–encoded pattern recognition receptors (PRRs), expressed on leukocytes, that recognize PAMPs (Table 1.3).[5] The PRRs may also recognize the damage-associated molecular patterns (DAMPs), which are released from damaged tissues during infections or sterile inflammation. Functionally, the PRRs are divided into three classes. The first class is the endocytic PRRs, including mannose receptors and scavenger receptors, which are located on the cell surface of phagocytes bearing specific PAMPs and these receptors may promote the phagocytosis. The second class is represented by the signaling PRRs, such as Toll-like receptors (TLR), which recognize various PAMPs/DAMPs and induce the production of inflammatory cytokines.

Intracellular sensors are also important in the recognition of PAMPs/DAMPs, and include the formation of inflammasomes, such as nucleotide-binding domain and leucine-rich repeat-containing family, pyrin domain containing 3 (NLRP3). The inflammasomes regulate the activation of caspase-1 and induce synthesis of inflammatory mediators (interleukin [IL]-1β) in response to infectious microbes and molecules derived from host proteins.[6] Finally, the secretory PRRs, such as the surfactants, serum amyloid protein, C-reactive protein, and lectins, are present in the circulation.[7] According to their localization, the PRRs are classified into cytoplasmic, membrane, and serum receptors. When the PAMPs or DAMPs are recognized by PRRs, the leukocytes involved in the innate response engage different functions and activate mechanisms that allow the rapid elimination of pathogens or contain the infection until the development of the adaptive response.

Table 1.3 Principal Characteristics of the Pattern Recognition Receptors (PRRs) with Functional Importance in Innate Immunity

PRR Names	Localization	Class	PRR Example	Ligands
Toll-like receptors (TLR)	Membrane and cytoplasmic	Signaling PRR	TLR1–11	Some bacteria and virus molecules like lipopolysaccharides (LPS), peptidoglycans, viral acid nucleic, flagellin, lipoteichoic acid
RIG-like receptors (RLR)	Cytoplasmic	Signaling PRR	RIG-1 (retinoic acid-inducible gene I) and MDA-5 (Melanoma Differentiation-Associated protein 5)	RNA molecules
NOD-like receptors (NLR)	Cytoplasmic	Signaling PRR	CARD (caspase activation and recruitment domain) family: NOD1 and NOD2	Bacterial peptidoglycan
			PYD (Pyrin domain) family: NLRP1 and NLRP3	Flagellin, urate crystals, muramyl dipeptide, bacterial and viral acid nucleic, asbestos
Scavenger receptors	Membrane	Endocytic PRR	CD36	Modified lipoproteins (phosphatidylserine and phosphatidylcholine), polyanionic ligands (LPS), and lipoteichoic acid
C-type lectin receptors (CLR)	Membrane	Endocytic PRR	CD206 and Dectin-1	Mannose, L-fucose, N-acetyl-glucosamine residues on surface of bacteria and fungi
Formyl peptide receptors (FPR)	Membrane	Signaling PRR	FPRL1, FPRL2 and FPRL3	Peptides with N-formyl methionyl
C-type lectins	Serum	Secretory PRR	Colectins: Mannose-binding protein (MBL) and s urfactant proteins SP-A and SP-D Ficolin	Mannose, L-fucose, N-acetyl-glucosamine motifs
Pentraxins	Serum	Secretory PRR	C-reactive protein and serum amyloid P protein	Bacterial phosphatidylcholine and phosphatidylethanolamine
Complement	Serum	Secretory PRR	C3 and C5	Surface of microbial cells

Cells Involved in Innate Response

Innate immunity involves various types of leukocytes and their soluble mediators.[8,9] The innate immunity is mediated primarily by phagocytic cells and antigen-presenting cells (APC) such as granulocytes, macrophages, and dendritic cells.

Polymorphonuclear Granulocytes

The granulocytes, called neutrophils, eosinophils, and basophils, are produced in the bone marrow and are the most abundant leukocytes in the circulation. The neutrophils are professional phagocytic cells present in the blood as well as in tissues as immune-regulatory cells.[8,10] Neutrophils are the first leukocytes that migrate from the blood to bacterial infection sites[11] and recruit other leukocytes by releasing cytokines and other soluble mediators. The eosinophils are involved in the defense against helminths and are involved in the pathogenesis of allergic diseases.[8] The function of the basophils in immunity is mainly connected to their role in allergic diseases where they release large amounts of histamine stored in their granules and rapidly synthesize leukotrienes.[12] These cells are non-phagocytic cells and do not have the capacity to engage microorganisms directly.[8]

Monocytes and Macrophages

These cells originate in the bone marrow, and they work as phagocytic cells, as APC, as well as release cytokines, and they are involved in immune regulation at different levels.[8] The monocytes are distributed in different tissues where they differentiate into tissue-specific macrophages.

Dendritic Cells

They are of myeloid origin and express multiple innate immune receptors. They have highly effective mechanisms to detect and capture antigens and subsequently determine the magnitude and quality of adaptive immune responses, so these cells are the most potent APCs in promoting activation of naïve T cells.[13]

Natural Killer (NK) Cells

The NK cells are of lymphoid origin and involved in the immune vigilance against viral infections and cancer. They are widely distributed in diverse tissues, where they can recognize cellular targets and engage cytotoxicity via the release of perforins and granzymes. They also release cytokines that regulate other cells of the innate and adaptive immune system.[14]

Inflammation

Inflammation is an important response to tissue damage caused by physical, infectious, and psychological factors.[9] Systemic tissue damage induced by increased levels of circulating stress factors, for instance, may contribute to sterile inflammation. The type and range of the inflammatory response (mediated by receptors, cells, and soluble molecules) depend on the type and intensity of the stimulus. Inflammation can be beneficial as an acute (self-limiting) response, or harmful in chronic conditions such as injury or infection.[15] The inflammatory response may facilitate the repair, turnover, and adaptation of tissues. Inflammation is a multistage response mediated by the reactions of leukocytes, a humoral response mediated by the activation of inflammatory molecules present locally and in body fluids, and a hemostatic response. Chronic inflammation consecutively leads to the tissue degeneration or the establishment of other clinical conditions, including metabolic disorders, neuroinflammation, cardiovascular diseases, and major depression.[15]

Sterile inflammation can be elicited by physical [tissue trauma or necrosis], chemical, or metabolic noxious stimuli. In tissue necrosis, normal cellular constituents are released from dying cells as DAMPs, and include molecules such as high-mobility group box-1 (HMGB1) protein, adenosine triphosphate (ATP), mitochondrial DNA, and nuclear DNA fragments that are released from cells and can stimulate TLRs. This leads to the activation of macrophages and other inflammatory cells that express TLRs in tissue, transcriptional activation of cytokine genes, and recruitment of neutrophils and monocytes that can potentially damage tissues. In addition, DAMPs like ATP can activate the Nalp3 inflammasome in APC cells through binding to purinergic receptors, resulting in the activation of caspase-1, which processes pro-IL-1β or pro-IL-18 to the active cytokines.[16] All inflammatory cytokines produced during sterile inflammation require neutrophils and/or monocytes to cause toxicity.

There are three major cytokines contributing to the inflammatory response: IL-1β, tumor necrosis factor (TNF)-α, and IL-6. The IL-6 induces the synthesis of acute-phase proteins (APP) in the liver, while biological functions of IL-1β and TNF-α are similar and depend on the quantity of cytokines produced. When secreted at low concentrations, IL-1 functions as a mediator of local inflammation. It acts on endothelial cells to increase the expression of adhesion molecules. When secreted in larger quantities, the IL-1 enters the bloodstream and exerts systemic effects (Figure 1.1).

Systemic IL-1β shares with TNF the ability to cause fever, induce synthesis of APP, and initiate metabolic wasting. However, there are several differences between IL-1 and TNF. For instance, the IL-1 does not induce the apoptotic death

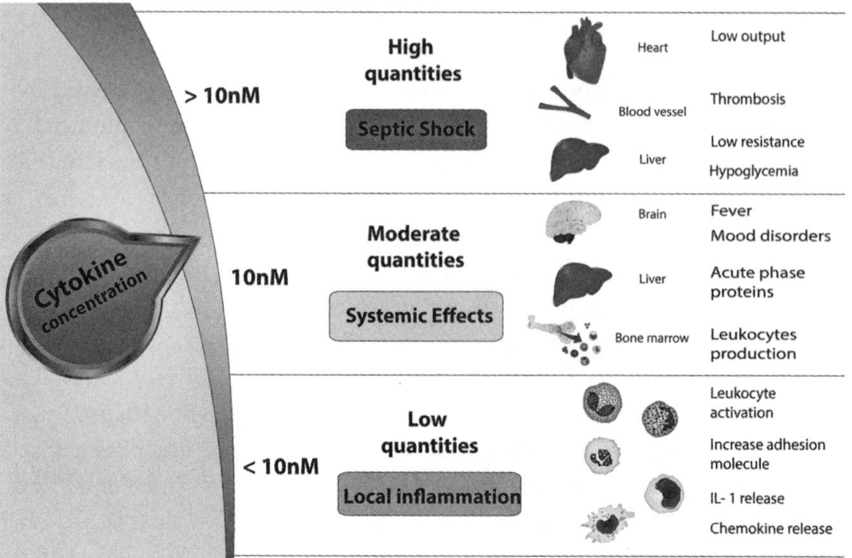

Figure 1.1 *Pro-inflammatory cytokines mediate the inflammatory responses.* The effects on target tissues are related to concentration, ranging from low quantities involved with local inflammation (e.g., edema formation) to moderate-to-high quantities involved with systemic effects.

of cells, and even high systemic levels do not cause the septic shock. The principal physiological function of TNF is to stimulate the recruitment of neutrophils and monocytes to sites of damage or infection and to activate these cells to eradicate microbes. TNF mediates these effects by several actions on vascular endothelial cells (contributing to edema formation) and leukocytes. TNF also has systemic effects: this cytokine acts on the hypothalamus to induce fever and is therefore called an endogenous pyrogen. This effect is mediated by the increased synthesis of prostaglandins by cytokine-stimulated hypothalamic cells. In the liver, the TNF may induce the release of APP, and at muscle and fat cells, induce the wasting named *cachexia*, among other effects. The cytokines have local or systemic effects, depending on their concentration. Their biological effects are regulated by multiple physiological processes, which are essential to maintain homeostasis.

Acute Phase Proteins

APP are secreted by the liver and increase rapidly in circulation in response to an inflammatory insult.[17] Increased levels of IL-6 are necessary to trigger the acute-phase response.[18] The principal functions of APPs are their ability to opsonize bacteria (i.e., increasing phagocytosis), to promote systemic inflammation, and to engage restorative processes. Of note, six biological functions of APPs are recognized: the inhibition of proteinases; blood clotting and fibrinolysis; anti-inflammatory properties; removal of pathogens and foreign materials; modulation of the immune responses; and transport of various materials.[19] Despite their multiple functions, APPs only are part of the initiation and development of the inflammatory reaction.

Complement System

The innate immune response involves the complement system, which functions primarily as a first-line host defense against pathogens. Three pathways have been identified (classical, lectin, and alternative) and more than 30 components and regulators that are widely distributed in our body, and all of them are synthesized and secreted by a number of cells under various stimuli, including cytokines and hormones.[20] To provide instant and effective protection against harmful stimuli, an elaborate network of soluble and cell surface–bound components, pattern-recognition proteins (PRP), proteases, receptors, effectors, and regulators is invoked.

Under normal circumstances, specialized PRPs detect danger-associated patterns on particle surfaces and initiate an enzymatic cascade of complement that leads to the attachment of C4b and C3b opsonins, which form enzyme complexes (termed *convertases*) that cleave the abundant plasma protein C3 into C3b that can again be deposited and form convertases. With continuous amplification, the convertases start cleaving the C5 component, a fragment of which initiates the formation of lytic membrane attack complexes (MAC; C5–C9) that directly destroy or damage susceptible cells (of note, bacteria). The activation of C3 and C5 also liberates potent chemotactic fragments, the anaphylatoxins C3a and C5a, which recruit immune cells to the site of activation and prime them, promoting the inflammation.

Cell Migration

The inflammatory response involves the recruitment of blood leukocytes and the increase in vascular permeability. Successful leukocyte transmigration across postcapillary venules is the culmination of sequential steps mediated by multiple classes of adhesion molecules expressed on the surface of endothelial cells and their interactions with cognate ligands expressed by leukocytes. Of note, the leukocyte recruitment in vivo consists of leukocytes rolling via L-selectin, followed by firm adhesion via β_2 integrins. The vascular endothelium plays an active role in the recruitment of circulating leukocytes. Upon treatment with TNF-α, IL-1β, or LPS (lipopolysaccharide), the normally non-adhesive and non-thrombogenic surface of the endothelium becomes pro-adhesive or "activated."

Inflammatory soluble mediators induce the expression of adhesion proteins in postcapillary venules, which involves the initial attachment of leukocytes and then their slow-velocity rolling (step 1), stable adhesion on endothelial cells (step 2), flattening (step 3), and subsequent crawling on the vascular endothelium, followed by transendothelial cell migration through the vascular endothelium (step 4), and finally uropod elongation to complete the transmigration (step 5). Once stably arrested on the endothelial surface, the leukocytes flatten, probably to reduce their exposure to the shear stress force of flowing blood and collisions with circulating blood cells, and crawl variable distances before initiating transendothelial migration. These events rely on leukocyte β1 and β2 integrins binding to their endothelial-expressed cognate ligands, ICAM-1 (intercellular adhesion molecule 1) and VCAM-1 (Vascular cell adhesion protein 1), respectively. This mechanism allows the leukocytes to reach the areas of infection or damage and perform their biological function.[21]

Phagocytosis

Phagocytosis is an important process that includes the recognition and ingestion of antigens by a plasma membrane–derived vesicle, known as phagosome. Through this mechanism, it is possible to remove microbial pathogens, and importantly, apoptotic cells. Professional phagocytes include monocytes, macrophages, neutrophils, dendritic cells, osteoclasts, and eosinophils. These cells are responsible for processing microorganisms and presenting antigen peptides to T lymphocytes. In addition, fibroblasts, epithelial cells, and endothelial cells can also perform phagocytosis. Although these non-professional phagocytes cannot ingest microorganisms, they are important in eliminating apoptotic bodies.[22,23]

Antigen Presentation

Following antigen trapping in the secondary lymphoid organs, dendritic cells (DCs) and macrophages can process and present antigens to T lymphocytes, via membrane-bound major histocompatibility complex (MHC) molecules. There are two MHC classes involved in antigen presentation, namely MHC class I and class II, which present antigen peptides to CD8+ and CD4+ T cells, respectively. These peptides originate from different sources—intracellular for MHC class I molecules, and exogenous for MHC class II molecules—and are obtained via different pathways.

MHC class I molecules are expressed by all nucleated cells and present protein fragments of cytosolic and nuclear origin at the cell surface. Cytoplasmic proteins are degraded by cytosolic and nuclear proteasomes; the resulting peptides are translocated into the endoplasmic reticulum (ER) by the transporter associated with antigen presentation (TAP) to access MHC class I molecules. In the ER, the MHC class I heterodimer is assembled from a polymorphic heavy chain and a light chain called β2-microglobulin (β2m). A peptide is the intracitoplasmatic third component required for stability, as it inserts itself deep into the MHC class I peptide-binding groove, which accommodates peptides of 8–9 amino acids in size. Without peptides, MHC class I molecules are stabilized by ER chaperone proteins. When peptides bind to MHC class I molecules, the chaperones are released, and fully assembled peptide–MHC class I complexes leave the ER for presentation at the cell surface. This mechanism of antigen presentation to T cells allows protection against tumors, and viral and intracytoplasmic bacterial or parasitic infections.

The antigenic presentation of exogenous antigens in the molecular context of MHC class II molecules is the last step of phagocytosis, and constitutes the onset of the adaptive immunity (Figure 1.2). MHC class II is primarily expressed by professional APCs (of note, DCs). MHC II is a heterodimer formed by one α and one β chain, which are assembled in the ER and associated with the invariant chain (Ii) for stabilization.[24] Vacuoles containing stabilized MHC II bind to a phagosome, allowing an exogenous peptide of up to 30 amino acid (aa) residues join to empty MHC II; once stabilized, they are exported to the surface of the APC for recognition by antigen-specific CD4+ T cells, initiating the adaptive immune response.

THE ADAPTIVE IMMUNE SYSTEM

The adaptive immune system is composed of antigen-specific lymphocytes, generated in the bone marrow (B cells) or in the thymus (T cells), and by serum antibodies. Lymphocytes greatly amplify in number following antigen recognition and activation, differentiating into effector and memory cells. The generation of memory cells provides quicker immune responses to subsequent infections with the same antigen, and constitutes the basis of vaccinations.[1] It produces a larger number of lymphocytes and, in the case of B cells, induces greater levels of antibodies that have a greater affinity for the antigen than the antibody of the primary response. The adaptive immune system is highly integrated with the innate immune cells; for example, receiving important activation cues as well as providing regulatory mechanisms to avoid overshooting of inflammatory responses. Adaptive immune responses are generated in the secondary lymphoid organs, including the lymph nodes, spleen, and mucosa-associated lymphoid tissue.

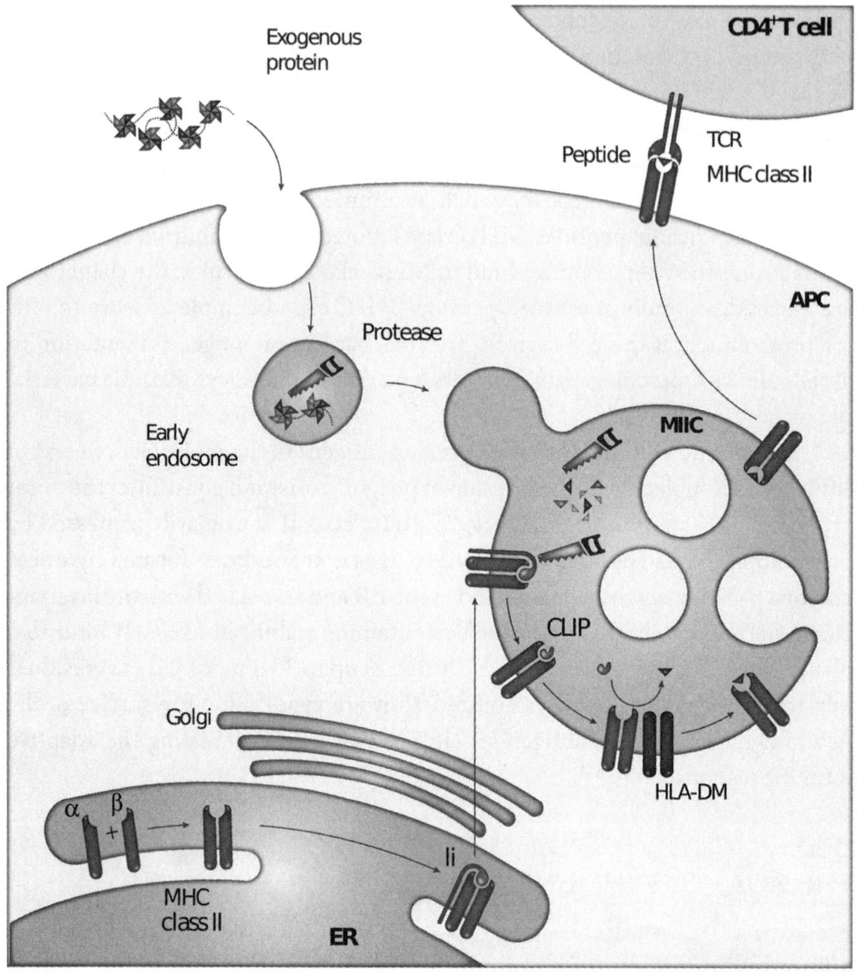

Figure 1.2 *Antigen processing and presentation via major histocompatibility complex (MHC) class II molecules.* Exogenous proteins derived from phagocytosed antigens are enzymatically processed in vesicles (endosomes), generating peptides that are mounted on MHC class II molecules with the help of CLIP (Class II-associated invariant chain peptide). MHC class II loaded with peptide migrates to the cell membrane of antigen presenting cells (APC) where it presents the antigenic peptide to CD4+ T cells. Only CD4+ T cells equipped with T cell receptors (TCR) specific to this peptide will be able to bind and recognize the peptide + MHC class II complex.

It has been estimated that lymphocytes are capable of producing about 10^{15} different antibody variable regions (B cells) and a similar number of T cell–receptor (TCR) variable regions. Remarkably, the vast diversity of the immune repertoire originates from fewer than 400 genes. This extraordinary achievement is performed by nuclear recombination processes that cut, splice, and modify variable-region genes.[25] There are no more than a few thousand lymphocytes specific for each antigen. Since each B cell or T cell is programmed to express only one of the vast number of antigen receptors, all the antigen-receptor molecules on a given lymphocyte have the same specificity. Such clones of lymphocytes are selected to participate in an immune response if they bear a receptor that can bind the relevant antigen, a process known as clonal selection.[26] The antigen-selected cells proliferate, leading to a rapid expansion in the number of B or T cells that can recognize the antigen.

Phases of Adaptive Immune Responses

The adaptive immune responses consist of distinct phases; namely: (a) antigen recognition, (b) lymphocyte activation, (c) proliferation, (d) cellular differentiation, and (e) effector functions (Figure 1.3). The first four phases are mediated in secondary lymphoid organs, such as lymph nodes and spleen, and the last one is mediated in peripheral tissues. For instance, antigen-specific T cells can recognize peptides presented by APCs, via clonal selection mechanisms. This

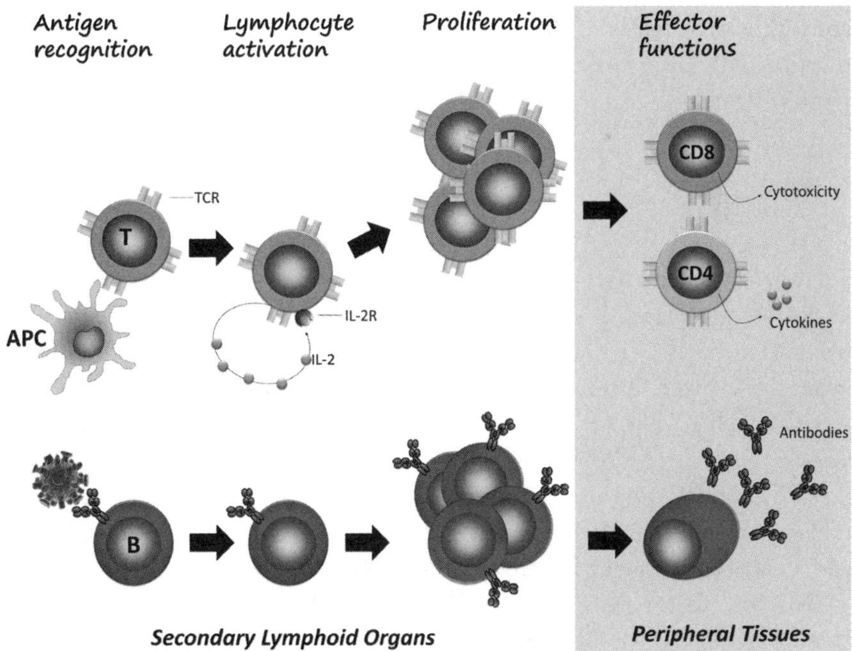

Figure 1.3 *Phases of adaptive immune responses.* The adaptive immune responses consist of distinct phases; namely: (a) antigen recognition, (b) lymphocyte activation, (c) proliferation, (d) cellular differentiation, and (e) effector functions.

process occurs through specific binding of TCR to a peptide presented by MHC molecules in the cell surface of APCs. Following antigen recognition, T lymphocytes promptly activate and express various cell-membrane receptors, including the CD25 (IL-2 receptor), CD69, and CD28. One key consequence of T cell activation is the production of IL-2, which triggers proliferation in an autocrine manner.[28] The result is the clonal expansion antigen-specific T cells, generating a great number of armed T cells necessary for immune responses after the activation of a reduced pool of naïve T cells. Before antigen encounter, the frequency of naïve T cells specific for any antigen is one in 10^5–10^6 lymphocytes. In some infections, the numbers of antigen-specific T cells may increase more than 50,000-fold, and the numbers of specific B cells may increase more than 5,000-fold.[1] This massive lymphocyte expansion enables the adaptive immune response to keep pace with rapidly dividing infections. Swollen lymph glands are the result of this great lymphocyte expansion.

In the effector phase, T cells differentiate into cellular subsets engaged in protective immune responses. Effector B lymphocytes are antibody-secreting plasma cells. Effector T cells include the cytokine-secreting CD4+ helper T cells and CD8+ cytotoxic T cells. Following the elimination of antigens, however, the vast majority of expanded lymphocytes are eliminated by apoptosis—i.e., the contraction phase (homeostasis).[1] The remaining cells constitute the memory T and B cells, which survive for long periods and respond with rapid and enhanced responses to subsequent exposures to the same antigen.

T cell Subpopulations and Effector Functions

There are two major T-cell subpopulations produced in the thymus, CD4+ and CD8+ T cells (Figure 1.4). In accordance with environmental signals provided by antigens or tissues, these cells can be found differentiated into different cellular subtypes with distinct effector functions. CD4+ T cells are mainly cytokine-secreting helper cells, whereas CD8+ T cells are mainly cytotoxic killer cells. CD4+ T cells can be subdivided into major subtypes according to differential cytokine profiles: type 1 (Th1) helper T cells, which secrete IL-2 and interferon (IFN)-γ; type 2 (Th2) helper T cells, which secrete IL-4, IL-5, IL-6, and IL-10; and type 17 (Th17), which secrete IL-17 and IL-22.[29,30]

Of note, there are also several regulatory CD4+ T lymphocytes, including natural CD4+CD25+Foxp3+ regulatory T cells (Tregs),[31] adaptive CD4+CD25+Foxp3+ Tregs,[32] T regulatory 1 (Tr1) cells,[33] and T helper (Th) 3 cells,[34] which are all critically involved in the maintenance of immune homeostasis and the prevention of autoimmune diseases. The mechanisms of action include

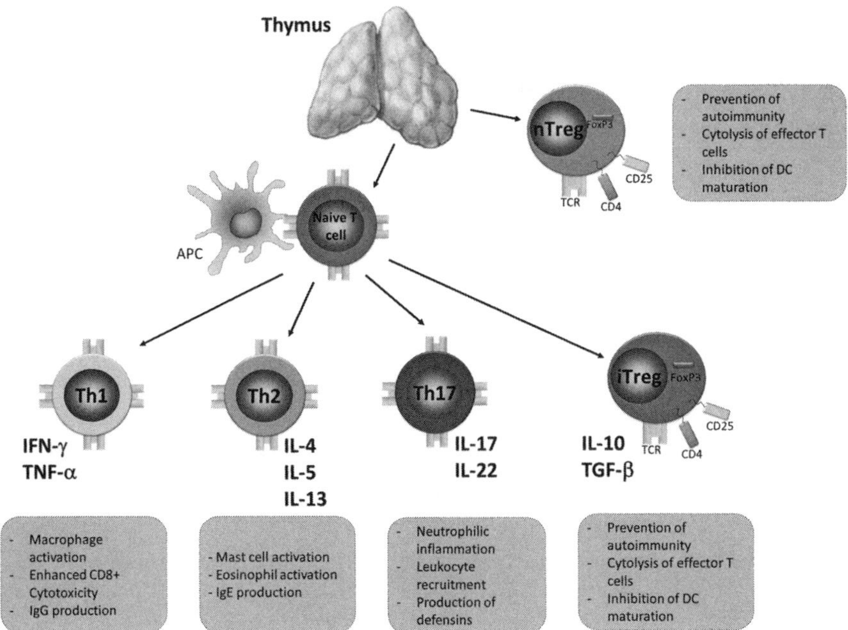

Figure 1.4 *CD4+ T cell subsets and effector functions.* T cells are produced in the thymus and, according to environmental signals in peripheral tissues, these cells can be differentiated into different cellular subtypes with distinct effector functions. The effector functions mediated by CD4+ T cells are largely dependent on their secreted cytokines.

the secretion of suppressive cytokines and molecules (e.g., IL-10, transforming growth factor [TGF]-β), metabolic disruption (e.g., tryptophan depletion), cytolysis of effector CD4+ or CD8+ T cells, and targeting of dendritic cells (e.g., inhibition of DC maturation and co-stimulation blockade).[35]

Cytokines have a pivotal role in influencing the type of immune response needed for optimal protection against particular types of infections, immunity against cancer, as well as allergic and autoimmune responses. For instance, the production of IL-12 dendritic cells stimulates the production of IFN-γ by Th1 cells. IFN-γ is essential for the activation of macrophages, enabling them to kill intracellular organisms (e.g., *Mycobacterium tuberculosis*). In general, the production of Th1 cytokines facilitates cell-mediated immunity, including the activation of macrophages and T cell–mediated cytotoxicity; in contrast, Th2 cytokines help B cells in the production of antibodies.[30]

The CD8+ T cells are chiefly engaged in cytotoxic actions against virally infected cells and tumors. The infected or cancer cell marks itself as a target for the CD8+ T cell by displaying intracellular peptides derived on its surface. These peptides are bound to the peptide-binding regions of class I MHC molecules. CD8+ T cells bind to this peptide–MHC complex and then kill the infected cell or cancer by at least two different pathways. They can insert perforins into the cell membrane of target cells, which produce pores through which granzymes are passed from the cytotoxic T cells into the target cell.[36] At least one of these proteolytic enzymes activates the caspase enzymes that mediate apoptosis in the target cell. Alternatively, cytotoxic T cells can bind the Fas molecule on the target cell using their receptor Fas ligand (FasL), a process that will also activate caspases within the target cell and ultimately induce apoptosis.[37] In addition, cytotoxic CD8+ T cells are also engaged in different actions, including tissue rejection in transplantation, autoimmunity (e.g., eliminating self-reactive pancreatic cells in type I diabetes), and immune regulation.[27]

CD8+CD28–T cells constitute an important subset involved in immune regulation.[38] This cellular subset has been shown to be expanded during healthy aging as well as in stress-related disorders (e.g., bipolar disease), persistent chronic infections (e.g., human immunodeficiency virus [HIV]), rheumatoid arthritis, etc. These cells are resistant to apoptosis, have shortened telomeres, proliferate poorly, and are often referred to as senescent memory T cells. CD8+CD28–T cells play a regulatory role in autoimmune diseases, transplantation, and protection against cancer, and were called CD8+ suppressor cells (Ts) or CD8+ Tregs.[39] Similar to CD4+ Tregs, their mechanisms of action involve the secretion of anti-inflammatory cytokines (IL-10, TGF-β), induction of cytolysis of activated T cells or antigen-presenting cells (APCs), and induction of inhibitory receptors on APCs.[40]

Immunoglobulins and the Humoral Response

Immunoglobulins are expressed on the surface of B cells as well as secreted by plasma cells as antibodies—the humoral response. There are five major classes of immunoglobulins: namely, IgM, IgG, IgA, IgE, and IgD. The IgM and IgG (with four subclasses) are the most prevalent antibodies in the circulation. These glycoproteins are composed of two identical heavy (H) chains and two identical light (L) chains, fixed by cysteine residues. Each molecule contains constant (C) and variable (V) domains. The V domains of heavy and light chains contain regions with extremely variable amino acid sequences, known as the "hypervariable regions," allowing the specific binding to an antigenic epitope, constituted by amino acid residues of the antigen.[1] Therefore, complex antigens consist of a mosaic of individual epitopes, eliciting different antibodies.

Following antigen recognition and activation in secondary lymphoid organs, B cells undergo differentiation into plasma cells, acquiring the ability to secrete high levels of antibodies. In most cases, T cells are crucially important in providing co-stimulatory stimuli (e.g., cytokines) for the B-cell activation and differentiation into plasma cells.[30] The most important T cell derived cytokines required for the B-cell help are IL-4, IL-5, IFN-γ, and TGF-β. Consequently, individuals with immunocompromised T cells (e.g., elderly; HIV-AIDS patients) have impaired antibody responses.

Antibodies have several functions during humoral responses. Antibodies can be directly protective if they inhibit the binding of a microorganism (e.g., viruses) or toxin to the corresponding cellular receptor. This is the basis by which vaccination provides immunity by eliciting neutralizing antibodies.[1] By coating microorganisms with antibodies, known as opsonization, the phagocytosis is greatly enhanced by macrophages or neutrophils. During this process, molecules of the complement system (C1–C9), produced by the liver into the body's circulatory system, also bind to antibodies already fixed to antigens and are activated in cascade by an enzymatic process. This was termed the *classical pathway of complement activation*, and it provides further opsonins (e.g., C3b) and terminates with complement-mediated cellular lysis. IgG and IgE antibodies can mediate antibody-dependent cellular cytotoxicity, an extracellular killing process in which cells bearing receptors for these classes of antibody become linked to antibody-coated target cells or parasites.[41]

REFERENCES

1. Abbas AK, Lichtman AH, Pillai S. *Cellular and Molecular Immunology.* 9th ed. Philadelphia, PA: Elsevier; 2017.
2. Arreola R, Alvarez-Herrera S, Perez-Sanchez G, et al. Immunomodulatory effects mediated by dopamine. *J Immunol Res.* 2016;2016:3160486.
3. Arreola R, Becerril-Villanueva E, Cruz-Fuentes C, et al. Immunomodulatory effects mediated by serotonin. *J Immunol Res.* 2015;2015:354957.
4. Savino W, Mendes-da-Cruz DA, Lepletier A, Dardenne M. Hormonal control of T cell development in health and disease. *Nat Rev Endocrinol.* 2016;12(2):77–89.
5. Mogensen TH. Pathogen recognition and inflammatory signaling in innate immune defenses. *Clin Microbiol Rev.* 2009;22(2):240–273, Table of Contents.
6. Kinoshita T, Imamura R, Kushiyama H, Suda T. NLRP3 mediates NF-kappaB activation and cytokine induction in microbially induced and sterile inflammation. *PLoS One.* 2015;10(3):e0119179.
7. Wu MH, Zhang P, Huang X. Toll-like receptors in innate immunity and infectious diseases. *Front Med China.* 2010;4(4):385–393.
8. Koenderman L, Buurman W, Daha MR. The innate immune response. *Immunol Lett.* 2014;162(2 Pt B):95–102.
9. Pavón-Romero L, Jiménez-Martínez MC, Garcés-Alvarez ME. *Inmunología Molecular, Celular y Traslacional.* Barcelona, Spain: Wolters Kluwer; 2016.
10. Nathan C. Neutrophils and immunity: challenges and opportunities. *Nat Rev Immunol.* 2006;6(3):173–182.

11. Teng TS, Ji AL, Ji XY, Li YZ. Neutrophils and immunity: from bactericidal action to being conquered. *J Immunol Res.* 2017;2017:9671604.
12. Cromheecke JL, Nguyen KT, Huston DP. Emerging role of human basophil biology in health and disease. *Curr Allergy Asthma Rep.* 2014;14(1):408.
13. Boltjes A, van Wijk F. Human dendritic cell functional specialization in steady-state and inflammation. *Front Immunol.* 2014;5:131.
14. Wu Y, Tian Z, Wei H. Developmental and functional control of natural killer cells by cytokines. *Front Immunol.* 2017;8:930.
15. Headland SE, Norling LV. The resolution of inflammation: principles and challenges. *Semin Immunol.* 2015;27(3):149–160.
16. Woolbright BL, Jaeschke H. Sterile inflammation in acute liver injury: myth or mystery? *Expert Rev Gastroenterol Hepatol.* 2015;9(8):1027–1029.
17. Gabay C, Kushner I. Acute-phase proteins and other systemic responses to inflammation. *N Engl J Med.* 1999;340(6):448–454.
18. Uhlar CM, Whitehead AS. The kinetics and magnitude of the synergistic activation of the serum amyloid A promoter by IL-1 beta and IL-6 is determined by the order of cytokine addition. *Scand J Immunol.* 1999;49(4):399–404.
19. Koj A. From the obscure and mysterious acute phase response to Toll-like receptors and the cytokine network. *Curr Immunol Rev.* 2008;4(4):16.
20. Noris M, Remuzzi G. Overview of complement activation and regulation. *Semin Nephrol.* 2013;33(6):479–492.
21. Leick M, Azcutia V, Newton G, Luscinskas FW. Leukocyte recruitment in inflammation: basic concepts and new mechanistic insights based on new models and microscopic imaging technologies. *Cell Tissue Res.* 2014;355(3):647–656.
22. Gordon S. Phagocytosis: an immunobiologic process. *Immunity.* 2016; 44(3):463–475.
23. Flannagan RS, Jaumouille V, Grinstein S. The cell biology of phagocytosis. *Annu Rev Pathol.* 2012;7:61–98.
24. Neefjes J, Jongsma ML, Paul P, Bakke O. Towards a systems understanding of MHC class I and MHC class II antigen presentation. *Nat Rev Immunol.* 2011; 11(12):823–836.
25. Tonegawa S. Somatic generation of antibody diversity. *Nature.* 1983;302(5909): 575–581.
26. Burnet FM. A modification of Jerne's theory of antibody production using the concept of clonal selection. *Aust J Sci.* 1957;20:67–69.
27. Delves PJ, Roitt IM. The immune system. First of two parts. *N Engl J Med.* 2000;343(1):37–49.
28. Huppa JB, Davis MM. The interdisciplinary science of T cell recognition. *Adv Immunol.* 2013;119:1–50.
29. Littman DR, Rudensky AY. Th17 and regulatory T cells in mediating and restraining inflammation. *Cell.* 2010;140(6):845–858.
30. Annunziato F, Romagnani S. Heterogeneity of human effector CD4+ T cells. *Arthritis Res Ther.* 2009;11(6):257.

31. Sakaguchi S, Sakaguchi N, Asano M, Itoh M, Toda M. Immunologic self-tolerance maintained by activated T cells expressing IL-2 receptor alpha-chains (CD25). Breakdown of a single mechanism of self-tolerance causes various autoimmune diseases. *J Immunol.* 1995;155(3):1151–1164.
32. Akbar AN, Vukmanovic-Stejic M, Taams LS, Macallan DC. The dynamic co-evolution of memory and regulatory CD4+ T cells in the periphery. *Nat Rev Immunol.* 2007;7(3):231–237.
33. Roncarolo MG, Gregori S, Battaglia M, Bacchetta R, Fleischhauer K, Levings MK. Interleukin-10-secreting type 1 regulatory T cells in rodents and humans. *Immunol Rev.* 2006;212:28–50.
34. Weiner HL. Oral tolerance: immune mechanisms and the generation of Th3-type TGF-beta-secreting regulatory cells. *Microbes Infect.* 2001;3(11):947–954.
35. Plitas G, Rudensky AY. Regulatory T cells: differentiation and function. *Cancer Immunol Res.* 2016;4(9):721–725.
36. Podack ER. Functional significance of two cytolytic pathways of cytotoxic T lymphocytes. *J Leukoc Biol.* 1995;57(4):548–552.
37. Wong P, Pamer EG. CD8 T cell responses to infectious pathogens. *Annu Rev Immunol.* 2003;21:29–70.
38. Strioga M, Pasukoniene V, Characiejus D. CD8+ CD28– and CD8+ CD57+ T cells and their role in health and disease. *Immunology.* 2011;134(1):17–32.
39. Gershon RK, Kondo K. Cell interactions in the induction of tolerance: the role of thymic lymphocytes. *Immunology.* 1970;18(5):723–737.
40. Suzuki M, Konya C, Goronzy JJ, Weyand CM. Inhibitory CD8+ T cells in autoimmune disease. *Hum Immunol.* 2008;69(11):781–789.
41. Delves PJ, Roitt IM. The immune system. Second of two parts. *N Engl J Med.* 2000;343(2):108–117.

2

Immunoneuropharmacology

VIVIAN THAISE DA SILVEIRA,
EDUARDO CANDELARIO-JALIL, AND
ANTONIO CARLOS PINHEIRO DE OLIVEIRA

INTRODUCTION

Inflammation has traditionally been described as a defensive process involving the cardinal signs of pain, heat, redness, swelling, and loss of function, in a homeostatic adaptation to exogenous and endogenous challenges.[1] However, since the 1970s, an increasing number of studies has shown that immune molecules could interact with neurons and neuronal circuits. In this context, the brain–immune interactions are actually seen as an essential component of the evolutionary survival strategy. Moreover, a self-limiting inflammatory response is essential for the maintenance of homeostasis, playing important roles in learning, memory, neuroplasticity, and neurogenesis in a physiological orchestrated form. However, if the immune response persists, a chronic activation of inflammatory mechanisms can become maladaptive, negatively influencing the neurotransmitter systems, as well as the behavioral, emotional, and cognitive processes that contribute to the pathophysiology of mental disorders in vulnerable individuals.[2–4]

As shown in several systematic reviews and meta-analyses, there is a strong evidence for an imbalance in the levels of inflammatory mediators and their receptors in blood, cerebrospinal fluid, and neuroanatomical areas of the brains of patients with psychiatric disorders. These data strongly support the hypothesis of inflammation as a common mechanism in several endophenotypes of mental illnesses.[5–9]

The observation that altered levels of chemical mediators of inflammation have a crucial role in the pathophysiology of chronic diseases was considered by a special issue of the *Science* journal as one of substantial scientific insights in the first decade of the 21st century.[4,10] Albeit far from being conclusive, the brain–immune interactions occur especially through inflammatory mediators such as cytokines, acute phase proteins, and adhesion molecules, molecules that act as mediators in the cross-talk between the peripheral immune system and the central nervous system (CNS). To date, the most frequently studied immune molecules in psychiatric disorders have been tumor-necrosis factor alpha (TNF-α), interleukin (IL)-1β, IL-4, IL-6, IL-10, and interferon (IFN)-γ.[5–9]

It has not been established yet whether the imbalance of these inflammatory mediators occurs as a cause or an adaptive consequence of chronic pathophysiological processes seen in psychiatric disorders. In addition, the exact neurobiological processes and biochemical mechanisms/cellular pathways by which the neuro–immune interaction occurs have not been completely defined, and how they influence neurons/neuronal circuits in both physiological and pathological states. These neurobiological processes include regulation of the synthesis, release, reuptake, and metabolism of multiple neurotransmitters, modulation of oxidative stress and neurotropic support.[6] There are four major potential molecular signaling pathways by which this can happen:

1. The activation of the tryptophan-degrading enzyme indoleamine 2,3-dioxygenase (IDO) along the kynurenine pathway;
2. Through disruption of the enzyme tyrosine hydroxylase tetrahydrobiopterin (BH4);
3. Through p38 mitogen-activated protein kinase (p38 MAP) signaling pathway;[2,3,11] and finally,
4. By the increase in the metabolism of the arachidonic acid cascade.[12]

The activation of glial cells by inflammatory mediators, particularly microglia and astrocytes, can be regarded as an additional mechanism with direct and indirect effects on all of these signaling pathways.

In this chapter, we will discuss how immune molecules can influence neurons and neural circuits, eventually leading to cognitive and behavioral impairment, as well as pharmacological strategies in inflammatory signaling pathways in the CNS. The pursuit of understanding these signaling pathways has become an area of intensive investigation since the 1990s. A better understanding of pathophysiological mechanisms by which inflammation and neurons are linked is fundamental for the development of novel preventive and therapeutic strategies for neuropsychiatric disorders.[11]

The Neuro–Immune Interaction

In the periphery, inflammatory mediators are secreted by macrophages, lymphocytes, and many parenchymal cells. However, despite the existence of the blood–brain barrier (BBB), inflammatory mediators can access the brain and interact with neurons and neuronal circuits. Additionally, inflammatory mediators are produced by various cell types in the brain, such as neurons, astrocytes, and activated microglia. These inflammatory mediators, in turn, attract peripheral cells to the brain and interact with neurons and neuronal circuits in a vicious inflammatory cycle (for more details, see Chapter 3).

Inflammatory Signaling Pathways in the CNS, and Pharmacological Strategies

The most explored mechanism underlying the influence of cytokines on neurons/neuronal circuits is related to their ability to alter the synthesis, release, and reuptake of multiple neurotransmitters. Abundant data from experimental and clinical studies that strongly support the immune hypothesis of psychiatric disorders, especially the "inflammatory hypothesis of depression" include:

1. Enhanced inflammatory response in a variety of psychiatric disorders;
2. The observation that cytokine immunotherapy in animal models or IFN-alpha therapy in patients with hepatitis C or cancer can lead to depressive-like symptoms such as fatigue, decreased concentration, anorexia, anxiety, anhedonia, and behavioral alterations;[13,14] and
3. The aforementioned symptoms can be significantly improvement by anti-cytokines therapy,[13,15] or prevented by immune-regulatory effects of antidepressants.[16,17]

There are four major putative mechanisms involved in the neuro–immune interaction, detailed in the following sections.

Activation of the Tryptophan-Degrading Enzyme IDO

Since the 20th century, great attention has been directed to the capacity of inflammatory mediators to regulate the tryptophan-degrading enzyme named indoleamine 2,3-dioxygenase (IDO), the first enzyme of the metabolic route called the kynurenine pathway (KP). The up- and/or downregulation of KP is related to the imbalance of the production of neuroactive KP metabolites collectively known as kynurenines. Different studies have shown that kynurenines play an important role in the pathophysiology of psychiatric disorders such as schizophrenia, bipolar disorder, and major depression.[18]

A causative interaction between IDO, inflammation, and mental illness was tested by O'Connor et al. in 2009. They found that peripheral administration of lipopolysaccharide (LPS) in mouse elevates the activity of IDO, leading to depressive-like behavior (increased duration of immobility in the forced-swim and tail suspension tests). Blocking IDO activation indirectly with minocycline, as well directly with the IDO antagonist 1-methyltryptophan, prevented depressive like-behaviors with the normalization of the levels of kynurenine and tryptophan in the plasma and brain.[19]

Although the essential amino acid tryptophan is a major substrate for the generation of serotonin (5-hydroxytryptamine or 5-HT), approximately 95% of all dietary tryptophan is catabolized in mammalians by KP.[20] In this context, three KP enzymes are responsible for oxidative cleavage of the dietary tryptophan: the tryptophan 2,3-dioxygenase (TDO), the indoleamine 2,3-dioxygenase 1 (IDO1), and the most recently discovered, indoleamine 2,3-dioxygenase 2 (IDO2). While TDO is expressed only in the liver and is activated by corticosteroid stress hormones, IDO2 is expressed in the liver, kidney, and epididymis. IDO1 is expressed in a variety of extra-hepatic tissues, including endothelial cells, macrophages, dendritic cells, astrocytes, and microglia. Importantly, IDO1 is up- or downregulated by pro-inflammatory or anti-inflammatory cytokines, respectively, under both normal and pathological situations. Several lines of evidence suggest that IDO-1–mediated tryptophan metabolism plays a critical role in the immunotherapy-induced depletion of brain monoamines.[18,21]

The activity of IDO1 is upregulated by several cytokines. TFN-α and IFN-γ are the main inducers of IDO1 activity. IFN-γ acts synergistically with TNF-α and other cytokines, such as other IFNs (IFNα and IFNβ), IL-1, IL-6, and TLR (Toll-like receptors), inducing a potent stimulus for transcription of the IDO1 gene. The binding of TNF-α, IL-1, and IL-6 to receptors activates relevant inflammatory-related intracellular signaling pathways such as NF-κB (nuclear factor kappa B) pathway, STAT1 (signal transducer and activator of transcription 1), IRF (interferon regulatory factor), and MAPK (mitogen-activated protein kinase), upregulating the expression of IFN-γ receptors and increasing cell sensitivity to IFN-γ.[22,23]

Physiologically, IDO activation leads to local depletion of L-tryptophan, and, as a consequence, prevents pathogen growth, as first demonstrated in 1986 by Byrne and collaborators.[24] Thus, IDO activation has protective effects in controlling intracellular parasites, being recognized as a defense molecule to deal with infections. However, the increased IDO activity induced by dysregulated inflammatory process degrades tryptophan into the bioactive metabolite NFK (N-formyl Kynurenine) through an oxidative cleavage, leading to insufficient levels of tryptophan.[25,26] At a very low concentration, tryptophan is not converted into serotonin in the brain, negatively impacting serotoninergic neurotransmission, leading to the serotonin deficit typically found in depression.[18,26]

In addition, NFK is quickly hydrolyzed to L-Kynurenine (KYN) by formidase. KYN is easily transported across the BBB into the brain. In the brain, KYN serves as a substrate for three different KP enzymes found in activated glial cells: kynurenine amino-transferase (KATs), kynureninase (KYNU), and kynurenine 3-monooxygenase (KMO), producing, respectively, the neuroactive compounds KYNA (kynurenic acid), AA (anthranilic acid), and 3-HK (3-hydroxykynurenine). These compounds are produced in three different branches of the KP (Figure 2.1).

The KMO branch, considered the major route of the KP in physiological conditions due to the high affinity of KYN to KMO enzyme, is initiated with the conversion of KYN to 3-HK by KMO. 3-HK is then metabolized by KATs, yielding the inert product xanthurenic acid, and KYNU, yielding 3-hydroxyanthranilic acid (3-HAA). 3-HAA can also be generated by the metabolism of AA by

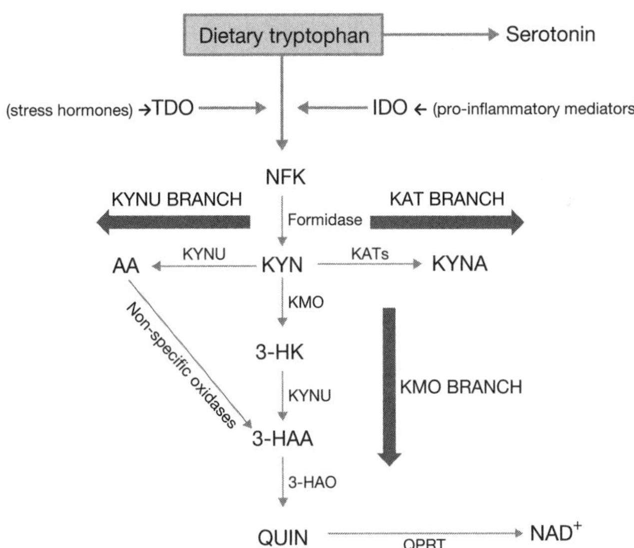

Figure 2.1 *Kynurenine pathway (KP) metabolic route.* A diagram of tryptophan metabolism in mammalian cells.

Abbreviations: TDO, tryptophan 2,3-dioxygenase; IDO, indoleamine 2,3-dioxygenase; NFK, L-formylkynurenine; KYN, L-kynurenine; KYNU, kynureninase; AA, anthranilic acid; KATs, kynurenine amino-transferases; KYNA, kynurenic acid; KMO, kynurenine 3-mononxygenase; 3-HK, 3-hidroxykynurenine; 3-HAA, 3-hydroxy anthranilic acid; 3- HAO, 3- hydroxyanthranilic acid oxygenase; QA, quinolinic acid; QPRQ, quinolinate phosphoribosyltransferase; NAD^+, nicotinamide adenine nucleotide.

non-specific oxidases. As KMO exhibits very high affinity for KYN, the production of 3-HAA by the KYNU branch only occurs when KYN concentration is high.

3-HAA is catalyzed by 3-hydroxyanthranilic acid oxygenase, producing the instable intermediate 3-amino-2-carboxy muconic semialdehyde which, by a non-enzymatic cyclization, is converted into QUIN (quinolinic acid). Finally, the enzyme quinolinic acid phosphoribosyl transferase degrades QUIN, producing the end product nicotinamide adenine nucleotide (NAD+),[25–27] a fundamental coenzyme mediator in several biochemical processes, including apoptosis, cellular respiration and energy production, genome stability, DNA repair, and transcriptional regulation.[28] The fundamental steps of the KP in mammalian cells are depicted in Figure 2.1.

The levels of kynurenines in the periphery tend to be much higher than CNS levels, and under physiological conditions, most kynurenines have renal excretion. However, L-tryptophan, L-KYN, and 3-HK can cross the BBB via an amino acid transporter. In the brain, these substrates are metabolized by KP enzymes, which are preferentially expressed in infiltrating macrophages, astrocytes, and microglia (Figure 2.2). L-KYN has been described as a neuroprotective compound in physiological concentrations with antitumor and antioxidant properties. Nevertheless, in inflammatory states, when IDO is upregulated,

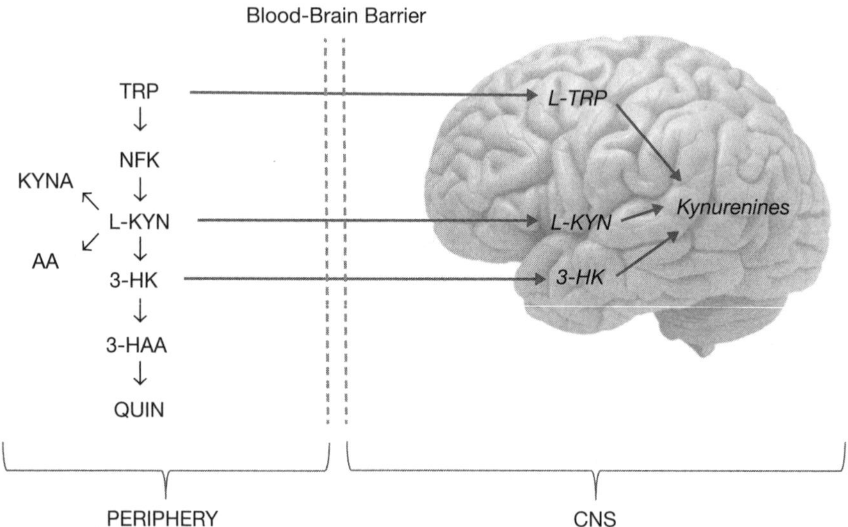

Figure 2.2 *Dynamics of peripheral and central kynurenines metabolism.*

Abbreviations: TRP, tryptophan; NFK, L-formylkynurenine; KYN, L-kynurenine; AA, anthranilic acid; KYNA, kynurenic acid; 3-HK, 3-hidroxykynurenine; 3-HAA, 3-hydroxy anthranilic acid; QA, quinolinic acid.

some neuroactive KP metabolites can have neurotoxic effects.[18,25] This observation was first provided in 1978, when Lapin and colleagues noted convulsions in mice after intracerebroventricular injection of QUIN.[29] Actually, the roles of kynurenines continue to be explored, and newer studies not only confirmed these data, but also indicated that KYNA, 3-HK, and 3-HAA metabolites may determine an imbalance in neurotransmitter systems with pathological consequences (see further).

Kynurenic Acid (KYNA)

Originally found in canine urine, KYNA is the end metabolite formed in the KAT branch of the KP (Figure 2.1). Metabolized especially in astrocytes, this astrocyte-derived molecule is a non-competitive antagonist of α7-nicotinic acetylcholine receptor (α7nAChR) and competitive antagonist of N-methyl-D-aspartate receptors (NMDAR). Due to the blockade of NMDAR, KYNA is considered a physiological anti-excitotoxic and anticonvulsant compound. KYNA is also considered an endogenous antioxidant due to its ability to scavenge reactive oxygen species such as hydroxyl radicals, superoxide anions, and peroxynitrite.[18]

Despite neuroprotective properties in physiological concentrations, fluctuations in glial production with abnormal accumulation of KYNA can induce NMDAR hypoactivation, with low glutamate release and consequent cognitive dysfunction. In addition, the suppression α7nAChR activity is strongly associated with deficits of auditory suppress gating, one of the most established biological markers of schizophrenia.[18] The imbalance of KYNA levels is also found to affect gamma-aminobutyric acid (GABA) neurotransmission. Inhibition of KYNA production increases extracellular GABA levels, while increased KYNA decreases GABA levels in a dose-dependent manner. This effect that can be prevented by an α7nAChR agonist,[30,31] strongly suggesting that it is mediated by α7nAChR antagonism. KYNA also inversely regulates dopamine levels in the brain.[32,33]

As a competitive antagonist of NMDAR, bidirectional effects of KYNA in glutamatergic system are found. Fluctuations in endogenous production of KYNA in the prefrontal cortex of rats modulate extracellular glutamate.[34,35] KYNA stimulates the release of glutamate in a dose-dependent manner, leading to increased levels of glutamate in the synaptic cleft and, hence, glutamate excitotoxicity. This effect is found to be potentiated by the activation of NMDAR by QUIN, as discussed later.

3-Hydroxykynurenine (3-HK)

Precursor of 3-HAA in KP, 3-HK is a highly reactive kynurenine compound metabolized in the KMO branch in both periphery and brain (Figure 2.2). A body of literature has suggested that 3-HK and 3-HAA are free-radical generators, i.e., these metabolites could generate the toxic free radicals superoxide and hydrogen peroxide, initiating a cascade of intracellular events that results in oxidative damage and cell death.[36–39] However, these redox properties of 3-HK and 3-HAA are controversial. Some studies have supported antioxidant and anti-inflammatory effects,[40,41] dual action,[42] as well as redox modulatory activity of these KP metabolites.[43] Despite controversies, the pro- or antioxidant properties of 3-hydroxykynurenine appear to be dependent on the local redox conditions.[44]

On the other hand, Reyes-Ocampo (2015) reported mitochondrial dysfunction induced by these kynurenines, neurotoxic effects that occur through a mechanism that is non-dependent on reactive oxygen species production.[45] Future studies are needed to elucidate not only mechanisms of action of theses KP metabolites in the brain, but their imbalanced impact on neurotransmitter systems and neuropsychiatric disorders. Importantly, in KP, 3-HAA is readily converted to QUIN.

Quinolinic Acid (QUIN)

One of the first descriptions of QUIN was done by L. Henderson, an author who extensively observed in the 1940s the metabolism of this molecule by humans and other mammalians.[46–49] Metabolized in microglia and infiltrating macrophages in CNS, QUIN is an endogenous neuroactive metabolite in the KMO branch of KP. QUIN is found in nanomolar concentrations under physiological conditions. In high concentrations, as induced by aberrant IDO activation, QUIN exhibits neuroexcitatory and convulsant properties.

While the exact signaling pathways activated by QUIN remain elusive, it is well documented that QUIN is an endogenous glutamate analog, inducing neurotoxic effects through two main mechanisms. Increasing evidence has shown that QUIN inhibits glutamate reuptake and degradation by astrocytes, increasing extracellular glutamate levels, and, as a consequence, excitotoxicity that may be related to convulsion and neuronal death.[50] Another relevant target of QUIN is NMDAR. As an agonist at the NMDAR, QUIN specifically binds to NMDARs NR1 and NR2 subunits, inducing the release of glutamate, especially from astrocytes.[51]

In addition, through a mechanism independent of NMDAR activation, QUIN leads to disruption of neuronal cytoskeletal homeostasis,[51,52] progressive mitochondrial dysfunction,[53] and cytotoxic lipid peroxidation.[54,55] Lipid peroxidation is mediated through QUIN interaction with Fe^{2+}, forming QUIN-Fe^{2+} complexes that decrease antioxidant enzymes with the generation of reactive oxygen species in a concentration-dependent manner.[51,54,56] Accordingly, QUIN is an endogenous neurotoxin with multiple neuronal and glial targets. These different insults can induce apoptotic cell death and critical alterations of brain homeostasis (Table 2.1).

Table 2.1 Putative Links Between Neuroactive Metabolites of KP, Its Effect in CNS, and Clinical Impact on Psychiatry

KP Neuroactive Metabolites	Brain Metabolism by KP Enzymes Present Especially In:	Effect	Effect Mediated By:	Clinical Impact
Kynurenic acid (KYNA)	Astrocytes	Bi-directional effect in glutamatergic system, with anticonvulsant, neuroprotection, or excitotoxicity	Competitive antagonist of NMDAR	Cognitive impairment; deficits of auditory suppress gating in schizophrenia
		Inversely affect GABA and dopamine levels in brain	Non-competitive antagonist of α7nAChR	
		Neuroprotection	Scavenge ROS	
3-Hydroxykynurenine (3-HK) and 3-hydroxyanthranilic acid (3-HAA)	?	Oxidative stress, mitochondrial dysfunction, apoptosis, pro- or antioxidant effect	Local redox conditions	Neurodegeneration or neuroprotection
Quinolinic acid (QUIN)	Microglia and infiltrating macrophages	Oxidative stress, excitotoxicity, apoptosis, neuronal death	Free radical generation, overstimulation of NMDA, inhibition of glutamate reuptake, and degradation by astrocytes	Convulsion, neuronal death, neurodegeneration

Abbreviations: α7nAChR = α7-nicotinic acetylcholine receptor; GABA = gamma-aminobutyric acid; NMDAR = N-methyl-D-aspartate receptors; ROS = reactive oxygen species.

In conclusion, inflammatory conditions are characterized by dysregulated KP, negatively influencing brain serotonin levels, and also generating neurotoxic metabolites. Taken together, these data make IDO an attractive molecular target, especially in patients with psychiatric disorders that are resistant to drug therapy and that exhibit an inflammatory phenotype.

Indeed, patients with depression and bipolar disorder present upregulation of the KMO branch and/or downregulation of the KAT branch, characterized by increased circulating levels of QUIN and/or decreased levels of KYNA metabolites. It has also been demonstrated that patients with schizophrenia present upregulation of the KAT branch and/or down regulation of the KMO branch, characterized by increased levels of KYNA and/or decreased levels of QUIN metabolites in the cerebrospinal fluid. Although it remains unclear whether KP metabolite imbalance is a causal factor for these psychiatric disorders, there is no doubt that KP metabolites play a critical role in their pathophysiology.[18]

Given that the levels of kynurenine metabolites are controlled by enzymes, the design and synthesis of new pharmacological agents have been of particular interest to medicinal chemistry. In this regard, KP enzymes' inhibition, particularly IDO1, KMO, and KAT inhibitors, are promising therapeutic strategies. These KP enzymes also have an essential physiological role, and the development of precise and accurate monitoring techniques to access the normalization of KP metabolites is a current challenge.[18,57]

Taking into account the significant imbalance of kynurenines with consequent alteration of the glutamatergic system and the potential neurotoxic properties of glutamate, NMDAR antagonism has also been considered a promising pharmacological strategy. A recent meta-analysis of randomized controlled trials (RCTs) found that memantine, a non-competitive NMDAR antagonist, appears to be an efficacious and safe treatment, improving neurocognitive performance and negative symptoms of schizophrenia.[58]

Inactivation and/or Depletion of Tetrahydrobiopterin (BH4)

A second mechanism by which inflammatory mediators can influence the biosynthesis of biogenic amines is through the inactivation and/or depletion of tetrahydrobiopterin (BH4).[59] BH4 is a critical co-factor of the mono-oxygenases tryptophan hydroxylase, phenylalanine hydroxylase, and tyrosine hydroxylase, which are enzymes necessary for the biosynthesis of monoamines. The BH4-dependent enzyme tryptophan hydroxylase converts the essential amino acid tryptophan to serotonin, influencing serotoninergic neurotransmission. In addition, the BH4-dependent enzyme phenylalanine hydroxylase converts phenylalanine into tyrosine. Subsequently, tyrosine is converted by the BH4-dependent enzyme tyrosine hydroxylase to levodopa (L-DOPA), the intermediate product in the biosynthesis of the dopamine. With dopamine as a substrate (precursor), adrenaline and noradrenaline are formed. Therefore, BH4 is required as a coenzyme for several enzymes responsible for neurotransmitter synthesis,[60] as shown in Figure 2.3.

BH4 disruption by inflammatory mediators limits the biosynthesis of serotonin and dopamine (and also epinephrine and norepinephrine), which are catecholamines largely documented as involved in diverse brain functions, mood states, and behavioral responses.

It has been demonstrated that IFN-α and IL-6 can disrupt BH4 through two mechanisms. First, BH4 is a critical co-factor of the enzyme nitric oxide synthases (NOS). The BH4-dependent NOS convert the amino acid arginine to nitric oxide (NO). Inflammatory mediators can overstimulate NOS to produce NO, increasing the utilization of BH4. In this process, BH4 is oxidized to dihydrobiopterin (BH2) in a reversible reaction. BH4 can be regenerated through pathways involving folic acid, L-methylfolate and S-adenosylmethionine (SAMe).[61] Interestingly, the NO formed by NOS after, for instance, the injection of IFN-α is a able to cross the BBB and suppress BH4 and dopamine, an effect reversed by N(G)-monomethyl L-arginine, an inhibitor of NO synthase.[59] Second, BH4 is a relatively unstable compound and extremely sensitive to non-enzymatic oxidation induced by oxidative stress. Accordingly, inflammatory-related increase of oxidative stress may be responsible for the irreversible degradation of BH4 to dihydroxyanthopterin (XPH2) (Figure 2.4).[62]

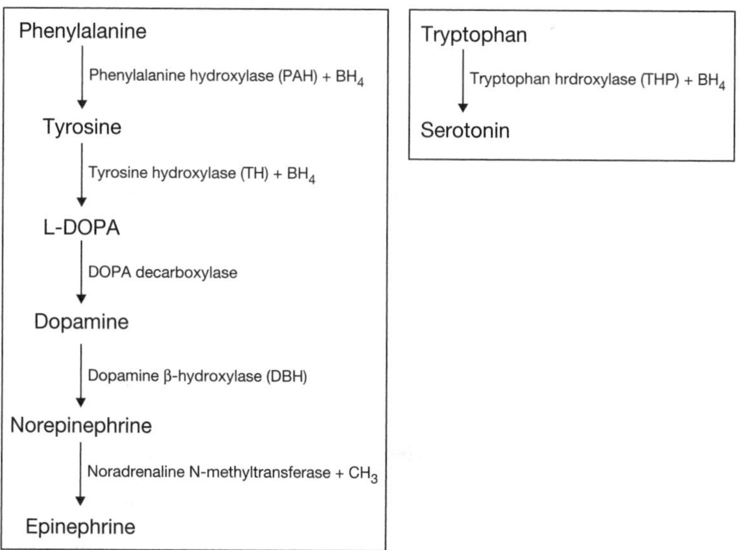

Figure 2.3 *Monoamines biosynthesis.* Biochemical conversion of phenylalanine and tryptophan to form dopamine, norepinephrine, epinephrine, and serotonin. Reactions occurs with PAH, TH, THP, and its co-factor tetrahydrobiopterin (BH4).

Pro-inflammatory cytokines negatively influence the synthesis of neurotransmitters, with decrease not only of serotonin, but also dopamine, epinephrine, and norepinephrine synthesis, by inactivation and/or depletion of BH4. Pharmacological strategies to restore or maintain the availability of BH4 include administration of BH4 itself, folic acid, L-methylfolate, and SAMe.

BH4 is a co-factor for phenylalanine hydroxylase, an enzyme that catalyzes phenylalanine to form tyrosine. Deficiency of BH4 can lead to excessive phenylalanine in blood (hyperphenylalaninemia), causing phenylketonuria (PKU), a genetically inherited metabolic disorder characterized by progressive impairment of brain functions. Sapropterin dihydrochloride (Kuvan/phenoptin) is a synthetic form of BH4 used for PKU treatment. Despite the fact that Koch and colleagues (2002) had reported that patients with hyperphenylalaninemia (due to a mutation on the phenylalanine hydroxylase gene) can develop depression and panic disorder, and that these conditions improved after one year of BH4 therapy,[63] the non-dietary administration of BH4 has not been investigated in the context of psychiatric disorders.

Some studies have investigated the potential of folic acid, L-methylfolate, and SAMe in depression based on the observation of low blood levels of these molecules in patients with depression. In a recent meta-analysis of RCTs and non-randomized comparative studies, Schefft et al. (2017) evaluated their efficacy in

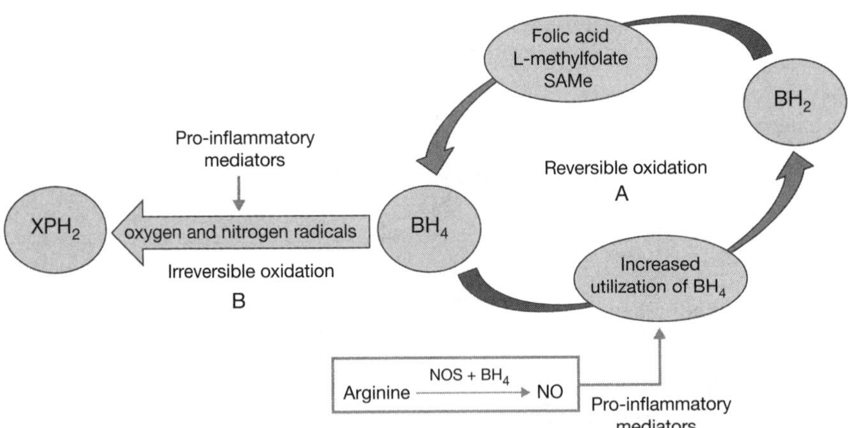

Figure 2.4 *Effects of pro-inflammatory cytokines on BH_4 oxidative loss.* **A.** BH_4 is also a co-factor for the nitric oxide synthases (NOS), responsible for conversion of arginine to NO. Pro-inflammatory mediators can overstimulate the BH4-dependent NOS to produce NO, increasing the utilization of BH_4 and favoring its oxidation to BH_2. BH_4 can be again regenerated in a reversible reaction through pathways supported by the substances folic acid, L-methylfolate, and S-adenosyl-methionine (SAMe). **B.** Pro-inflammatory mediators lead to formation of oxygen and nitrogen radicals, contributing to irreversible oxidation of tetrahydrobiopterin (BH^4) to dihydroxyanthopterin (XPH^2).

the treatment of major depression, concluding that the available data do not support the efficacy of folic acid and/or are too limited to draw firm conclusions in relation to SAMe.[64] Importantly, the International Society for Nutritional Psychiatry Research published a position statement about the robust relationship between nutrition and physical and mental health, and the importance of well-designed studies to advance the field.[65]

Activation of the p38 Mitogen-Activated Protein Kinase (p38 MAPK)

A third mechanism by which inflammatory mediators can influence brain homeostasis is through activation of the mitogen-activated protein kinase isoform p38 (p38 MAPK). This is one of the best-characterized kinases that play a role in inflammatory processes. This important class of MAPK has 38 KDa and is largely expressed in brain regions implicated in learning and memory, and it is activated in response to several stimuli such as stress, growth factors, and cytokines. For its activation, p38 MAPK is phosphorylated at its tyrosine residues, inducing conformational changes, and facilitating substrate binding.[66] The cascades regulated by p38 MAPK coordinate pro- and anti-inflammatory gene expression. The activation of p38 MAPK signaling can activate the gene expression of inflammatory molecules, including IL-1β, IL-6, IL-8, IL-12p40, TNF-α, and IFN-γ, which, in turn, are able to induce catalytic activation of a p38 MAPK signaling cascade in a continuous cycle (Figure 2.5).[66,67]

The activation of p38 MAPK signaling increases the expression and activity of reuptake pumps (transporters) of serotonin transporter and dopamine transporter in astrocytes and neurons in a dose- and time-dependent manner, with consequent decrease of metabolism and bioavailability of these biogenic amines at the synapse.[3] In accordance with diverse in vitro and in vivo data on the relationship between the aberrant activation of p38 MAPK by pro-inflammatory cytokines and dysfunctional monoamine metabolism,[68–70] Sanchez and colleagues observed an interesting association between p38 MAPK activation and serotoninergic dysfunction in a non-human primate model. In this study, increased p38 MAPK signaling was positively correlated with a decrease in the serotonin metabolite 5-hydroxyindoleacetic in the cerebrospinal fluid (CSF) of 17 female rhesus monkeys who rejected their offspring during infancy. Some of these females were abused/rejected as infants by their own mothers, and later became abusive mothers themselves. This study first demonstrated that p38 MAPK activation (manifested by increased peripheral blood monocytes staining positive for phosphorylated p38) may lead to permanent serotoninergic dysfunction, which, in turn, is extensively related to behavioral morbidity and maternal rejection.[71] Taken together, these data suggest that p38 MAPK signaling cascade may display a therapeutic role in clinical psychiatry.

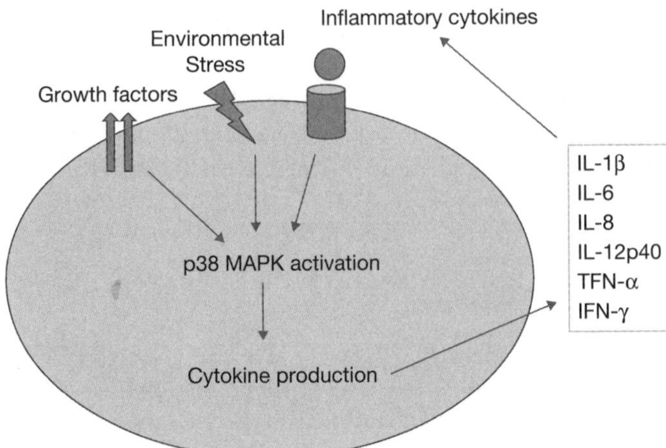

Figure 2.5 *The p38 MAPK activation.* The p38 MAPK is activated in response to several stimuli such as environmental stress, growth factors, and cytokines. In turn, p38 MAPK signaling activates the gene expression of IL-1β, IL-6, IL-8, IL-12p40, TNF-α, and IFN-γ. These inflammatory cytokines induce catalytic activation of p38 MAPK, in a continuous cycle.

To date, four isoforms of p38 have been identified (p38α, p38β, p38γ, and p38δ). The first of them, the p38α isoform, has been the most investigated isoform as a pharmacological target due to its central role in modulation of cytokine production. Since the mid-1990s, several inhibitors of p38 MAPK have been investigated as a pharmacological strategy in a wide variety of preclinical studies involving neurological diseases with an inflammatory component, with promising results in animal models of seizures, stroke, neuropathic pain, Alzheimer's disease, and Parkinson's disease. However, clinical studies with p38 MAPK inhibitor were marked by potential toxicity for the human liver and a wide range of side effects, interrupting the human trials and the development of several molecules considered potential p38 MAPK inhibitors.[72]

Of particular interest for psychiatry, Losmaspimod (GW856553X) is a selective inhibitor of both p38α and p38β isoforms of p38 MAPK that revealed anti-inflammatory properties in several preclinical studies. Safety and tolerability for healthy subjects were demonstrated in phase I studies, but in phase II, Losmaspimod was not effective in patients with major depressive disorder with symptoms of abulia and psychomotor retardation.[73] Phase III clinical trials of p38 MAPK inhibitors have not been performed yet. The efficacy of p38 MAPK inhibitors in a wide variety of chronic diseases with an inflammatory component in animal models from different species,[74] associated with some positive clinical trials, maintains the interest of the pharmaceutical industries in the design and synthesis of p38 inhibitors.[72]

The activation of astrocytes and microglia has direct and indirect influence in all the signaling pathways mentioned here. Therefore, these glial cells are potential targets for neuropsychopharmacological approaches. Minocycline, an antibiotic with anti-inflammatory properties that blocks microglial and astrocyte activation, has been extensively tested in preclinical and clinical studies in psychiatric conditions. The potential of minocycline to inhibit p38 MAPK and NF-κB seems to be related to its efficacy in these conditions. In a meta-analysis performed by Li et al. (2013), minocycline revealed neuroprotective effects, with anti-inflammatory, antiapoptotic, and antioxidant activities in the brains of rodents.[75] Consistent with these data, several systematic reviews and meta-analyses of RCTs have shown that minocycline has acceptable tolerability and side effects, and improves positive and especially negative symptoms of schizophrenia.[76–78]

Increased Metabolism of Arachidonic Acid Cascade

A fourth mechanism by which inflammatory mediators can influence neurons and neuronal circuits is through increased metabolism of arachidonic acid (AA). The polyunsaturated fatty acid AA comprises a significant fraction of the essential fatty acids in the brain. AA and their metabolites are second messengers with central roles in signal transduction of multiple processes, such as membrane excitability, gene transcription, neuroinflammatory response, synaptic plasticity, spatial learning, and sleep, among others.[79,80]

AA incorporated in the phospholipid membrane is mobilized/released from synaptic membrane through activation of phospholipase A_2 (PLA_2). The activation of PLA_2 can occur through two independent mechanisms: (1) by calcium entry via NMDA activation or via activation of G-protein coupled receptors, such as dopaminergic D2 receptors, serotoninergic $5\text{-}HT_{2A/2C}$ receptors, muscarinic $M_{1,3,5}$, bradykinin B2, and adrenergic β2 receptors; or (2) via coupling to TNF-α and IL-1β receptors on astrocytes. The major fraction of AA released is recycled back and reincorporated into the membrane, whereas a smaller portion (about 4%) is metabolized. AA can be metabolized by auto-oxidation (producing hydroperoxy acids), and can serve as substrate to cytochrome P450 enzymes (producing epoxides), lipoxygenases (producing leukotrienes), or cyclooxygenases 1 and 2 (producing prostaglandins and thromboxanes). By a mechanism of synthesis not completely understood, AA can also be metabolized to anandamide, an endogenous, non-oxygenated cannabinoid. The AA-related metabolites hydroperoxy acids, epoxides, leukotrienes, prostaglandins, and thromboxanes are oxygenated metabolites collectively called eicosanoids[79] (see Figure 2.6).

Neurotransmitters and inflammatory mediators modulate the turnover and the levels of AA in the CNS. In turn, AA can be metabolized through several signaling pathways, producing neuroactive metabolites that can influence neurotransmission. For instance, both AA and their metabolites act as negative regulators of the serotoninergic neurotransmission.[79]

Dysregulation of the immune system is associated with increased prostaglandin E2 (PGE2) production and increased cyclooxygenase-2 (COX-2)

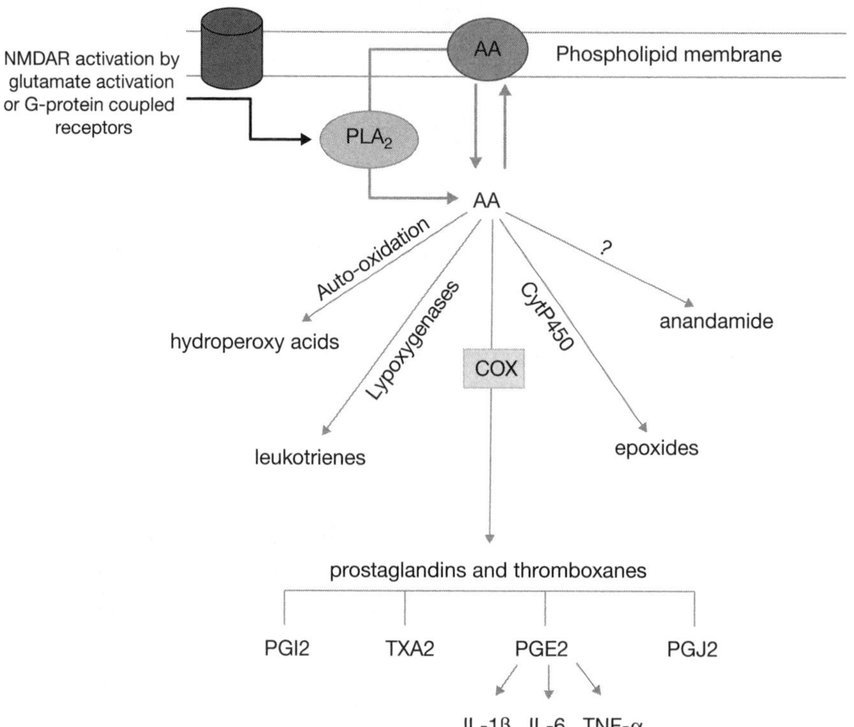

Figure 2.6 *Diagram illustrating arachidonic acid cascade.* PLA2 is stimulated by activation of NMDA or G-protein coupled receptors. AA is mobilized by PLA2 from phospholipid membrane and metabolized through several pathways and enzymes. AA can be metabolized by the COX enzyme to produce PGI2, TXA2, PGE2, and PGJ2. PGE2 stimulates the production of IL-1β, IL-6, and TNF-α.

Abbreviations: AA, arachidonic acid; PLA2, phospholipase A2; NMDA, N-methyl-D-aspartate; CytP450, cytochrome P450; COX, cyclooxygenase; PG, prostaglandin; TX, thromboxane.

expression in schizophrenia and depression. PGE2, in turn, induces the production of IL-1β, IL-6, and TNF-α (Figure 2.6), frequently found elevated in patients with these two conditions.[81,82] Altered AA cascade enzymes are also upregulated in postmortem brains of bipolar disorder patients.[82] Postprandial metabolism of polyunsaturated fatty acids and eicosanoids was found dysregulated in patients with anorexia nervosa.[83] Studies connecting AA cascade and other psychiatric disorders are still limited, with heterogeneous methodology, and small sample sizes.

Nonsteroidal anti-inflammatory drugs (NSAIDs), such as COX-2 inhibitors, particularly celecoxib, have been evaluated as potential adjuvant treatments in clinical psychiatry. Na et al. (2014) performed a systematic review with a meta-analysis of RCTs comparing adjunctive celecoxib with placebo in the treatment of patients with major depressive disorder. Despite not being statistically significant ($p = 0.09$), the adjunctive celecoxib group showed better remission of depressive symptoms than placebo.[84] In a recent meta-analysis of double-blind RCTs, Zheng et al. (2017) reported that treatment with adjunctive celecoxib presents significant efficacy and tolerability in schizophrenia treatment, improving positive and negative symptoms in the first episode of psychosis, but not in chronic patients.[85] Meta-analysis of adjunctive anti-inflammatory agents in the treatment of bipolar disorder was also performed, indicating that a variety of anti-inflammatory agents, including NSAIDs, have a safe and moderate antidepressant effect.[86]

The inhibition of cytokines and chemokines using biological compounds, such as soluble cytokine receptors (e.g., IL-1ra, Etanercept) and anti-inflammatory cytokines (e.g., IL-10), as well as pharmacological inhibitors of TNF-α-converting enzyme (TACE inhibitors), chemokine receptor antagonists (e.g., maraviroc, plerixafor) have been considered as promising therapeutic strategies. The use of these compounds in clinical psychiatry may benefit from the low drug interaction rate between them and psychotropic agents. On the other hand, changes of immune compounds with physiological roles may increase the risk of infections and/or malignancies.[61]

Our knowledge on the mechanisms of interaction between the immune and nervous systems has greatly expanded. The schematic representation of the four major neuro–immune interaction pathways is shown in Figure 2.7. Despite the clear advances, several questions remain to be explored, and one of the most challenging obstacles is the absence of valid biomarkers in psychiatry. Immune biomarkers able to predict the development and/or progression of specific psychiatric disorders would not only support a more precise diagnosis, but could also could tailor treatment.

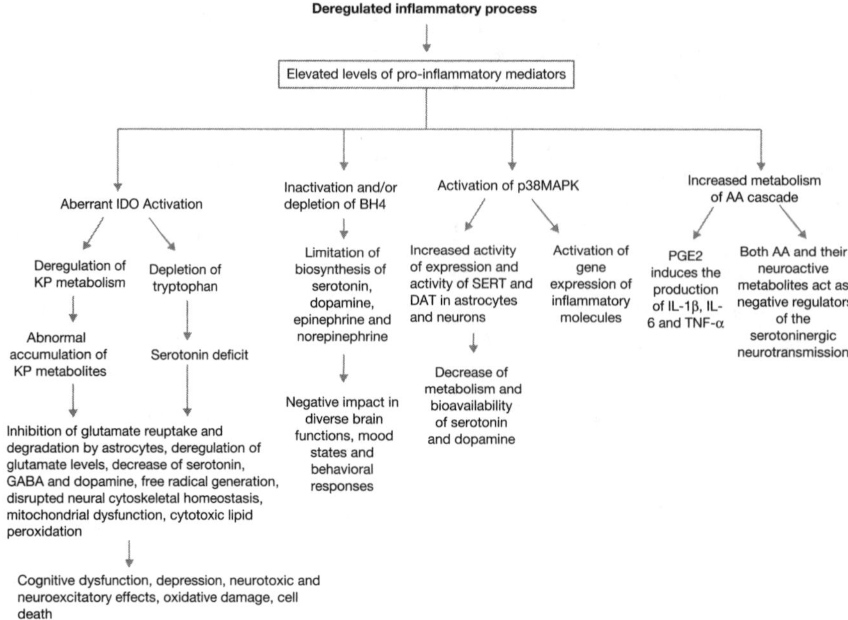

Figure 2.7 *Schematic representation of the neuro–immune interaction.*

Abbreviations: IDO, indoleamine 2,3-dioxygenase; KP, kynurenine pathway; BH4, tetrahydrobiopterin; p38MAPK, p38 mitogen-activated protein kinase; SERT, serotonin transporter; DAT, dopamine transporter (DAT); PGE2, prostaglandin E2; AA, arachidonic acid.

It is important to acknowledge that, although represented individually, immune signaling pathways work simultaneously and in an integrated way. Thus, the complexity of these interactions should not be underestimated (Figure 2.8).

CONCLUSION

Inflammatory and/or immune mediators interact with neurons, modulating multiple signaling pathways. These effects are influenced by different factors, such as concentration and interaction with other mediators. The available data make these inflammatory signaling pathways as well as cytokines attractive pharmacological targets. Future studies may contribute not only to elucidating the mechanisms underlying neuroimmune interactions, but also to providing new insights into neuropsychopharmacology and, ultimately, to better and/or more precise therapeutic strategies.

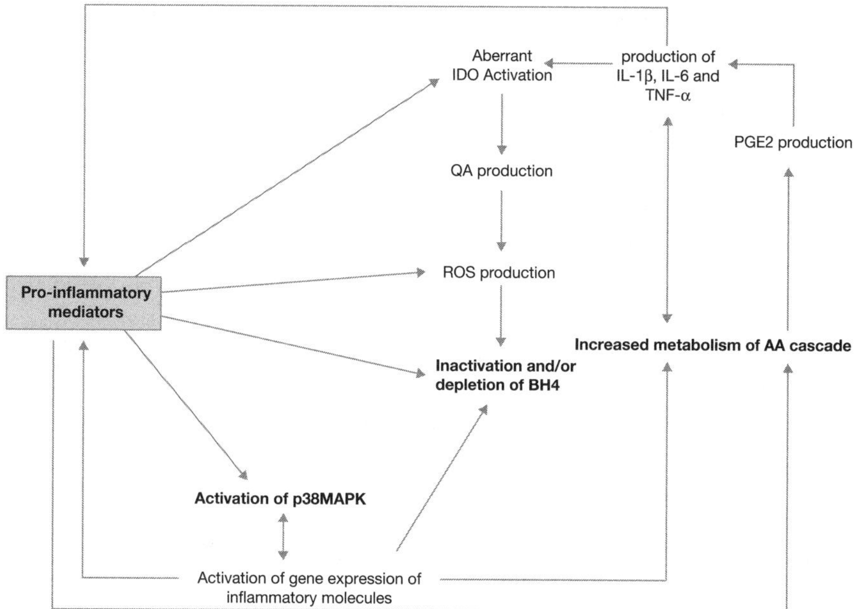

Figure 2.8 *Diagram illustrating the mutual interaction of inflammatory signaling pathways.* Pro-inflammatory mediators lead to aberrant indoleamine 2,3-dioxygenase (IDO) activation. IDO activation results in formation of the kynurenine pathway metabolite quinolinic acid (QUIN). QA, in turn, leads to the generation of reactive oxygen species (ROS) responsible for irreversible oxidation of tetrahydrobiopterin (BH4). In addition, p38 mitogen-activated protein kinase (p38MAPK) is activated in response to pro-inflammatory mediators. The activation of this signaling pathway is able to activate gene expression of inflammatory molecules, which act in all inflammatory pathways here represented and lead to increased metabolism of arachidonic acid (AA) cascade. In AA cascade, the metabolite prostaglandin n E2 (PGE2) is formed, inducing the production of more pro-inflammatory mediators, IL-1β, IL-6, and TNF-α, in a continuous cycle of all system.

REFERENCES

1. Ader, R. (2000). On the development of psychoneuroimmunology. *Eur J Pharmacol, 405*(1–3), 167–176.
2. Miller, A. H. (2009). Norman Cousins Lecture. Mechanisms of cytokine-induced behavioral changes: psychoneuroimmunology at the translational interface. *Brain Behav Immun, 23*(2), 149–158.
3. Miller, A. H., Haroon, E., Raison, C. L., & Felger, J. C. (2013). Cytokine targets in the brain: impact on neurotransmitters and neurocircuits. *Depress Anxiety, 30*(4), 297–306.
4. Couzin-Frankel, J. (2010). Inflammation bares a dark side. *Science, 330*(6011), 1621.

5. Mitchell, R. H., & Goldstein, B. I. (2014). Inflammation in children and adolescents with neuropsychiatric disorders: a systematic review. *J Am Acad Child Adolesc Psychiatry, 53*(3), 274–296.
6. Stuart, M. J., Singhal, G., & Baune, B. T. (2015). Systematic review of the neurobiological relevance of chemokines to psychiatric disorders. *Front Cell Neurosci, 9*, 357.
7. Goldsmith, D. R., Rapaport, M. H., & Miller, B. J. (2016). A meta-analysis of blood cytokine network alterations in psychiatric patients: comparisons between schizophrenia, bipolar disorder and depression. *Mol Psychiatry, 21*(12), 1696–1709.
8. Rajabally, Y. A., Seri, S., & Cavanna, A. E. (2016). Neuropsychiatric manifestations in inflammatory neuropathies: a systematic review. *Muscle Nerve, 54*(1), 1–8.
9. Wang, A. K., & Miller, B. J. (2017). Meta-analysis of cerebrospinal fluid cytokine and tryptophan catabolite alterations in psychiatric patients: comparisons between schizophrenia, bipolar disorder, and depression. *Schizophr Bull, 44*(1), 75–83.
10. *Science*, Vol. 330, issue 6011, pp. 1621, 2010. Table of Contents. Special Issue: Insights of the decade (2010, December). Inflammation bares a dark side. Retrieved October 11, 2017, from the web site: http://science.sciencemag.org/content/330/6011
11. D'Mello, C., & Swain, M. G. (2017). Immune-to-brain communication pathways in inflammation-associated sickness and depression. *Curr Top Behav Neurosci, 31*, 73–94.
12. Sublette, M. E., Russ, M. J., & Smith, G. S. (2004). Evidence for a role of the arachidonic acid cascade in affective disorders: a review. *Bipolar Disord, 6*(2), 95–105.
13. Ma, K., Zhang, H., & Baloch, Z. (2016). Pathogenetic and therapeutic applications of tumor necrosis factor-α (TNF-α) in major depressive disorder: a systematic review. *Int J Mol Sci, 17*(5), 1–21.
14. Abbott, R., Whear, R., Nikolaou, V., et al. (2015). Tumour necrosis factor-α inhibitor therapy in chronic physical illness: a systematic review and meta-analysis of the effect on depression and anxiety. *J Psychosom Res, 79*(3), 175–184.
15. Kappelmann, N., Lewis, G., Dantzer, R., Jones, P. B., & Khandaker, G. M. (2018). Antidepressant activity of anti-cytokine treatment: a systematic review and meta-analysis of clinical trials of chronic inflammatory conditions. *Mol Psychiatry, 23*(2), 335–343.
16. Szuster-Ciesielska, A., Tustanowska-Stachura, A., Słotwińska, M., Marmurowska-Michałowska, H., & Kandefer-Szerszeń, M. (2003). In vitro immunoregulatory effects of antidepressants in healthy volunteers. *Pol J Pharmacol, 55*(3), 353–362.
17. Hannestad, J., DellaGioia, N., & Bloch, M. (2011). The effect of antidepressant medication treatment on serum levels of inflammatory cytokines: a meta-analysis. *Neuropsychopharmacology, 36*(12), 2452–2459.
18. Fujigaki, H., Yamamoto, Y., & Saito, K. (2017). L-tryptophan-kynurenine pathway enzymes are therapeutic target for neuropsychiatric diseases: focus on cell type differences. *Neuropharmacology, 112*(Pt B), 264–274.
19. O'Connor, J. C., Lawson, M. A., André, C., et al. (2009). Lipopolysaccharide-induced depressive-like behavior is mediated by indoleamine 2,3-dioxygenase activation in mice. *Mol Psychiatry, 14*(5), 511–522.
20. Musajo, L., & Benassi, C. A. (1964). Aspects of disorders of the kynurenine pathway of tryptophan metabolism in man. *Adv Clin Chem, 7*, 63–135.

21. Murakami, Y., Ishibashi, T., Tomita, E., et al. (2016). Depressive symptoms as a side effect of interferon-α therapy induced by induction of indoleamine 2,3-dioxygenase 1. *Sci Rep, 6*, 29920.
22. Robinson, C. M., Hale, P. T., & Carlin, J. M. (2006). NF-kappa B activation contributes to indoleamine dioxygenase transcriptional synergy induced by IFN-gamma and tumor necrosis factor-alpha. *Cytokine, 35*(1–2), 53–61.
23. Udina, M., Moreno-España, J., Navinés, R., et al. (2013). Serotonin and interleukin-6: the role of genetic polymorphisms in IFN-induced neuropsychiatric symptoms. *Psychoneuroendocrinology, 38*(9), 1803–1813.
24. Byrne, G. I., Lehmann, L. K., & Landry, G. J. (1986). Induction of tryptophan catabolism is the mechanism for gamma-interferon-mediated inhibition of intracellular *Chlamydia psittaci* replication in T24 cells. *Infect Immun, 53*(2), 347–351.
25. Schwarcz, R., & Pellicciari, R. (2002). Manipulation of brain kynurenines: glial targets, neuronal effects, and clinical opportunities. *J Pharmacol Exp Ther, 303*(1), 1–10.
26. Parrott, J. M., & O'Connor, J. C. (2015). Kynurenine 3-monooxygenase: an influential mediator of neuropathology. *Front Psychiatry, 6*, 116.
27. Bryleva, E. Y., & Brundin, L. (2017). Kynurenine pathway metabolites and suicidality. *Neuropharmacology, 112*(Pt B), 324–330.
28. Ying, W. (2007). NAD+ and NADH in brain functions, brain diseases and brain aging. *Front Biosci, 12*, 1863–1888.
29. Lapin, I. P. (1978). Stimulant and convulsive effects of kynurenines injected into brain ventricles in mice. *J Neural Transm, 42*(1), 37–43.
30. Beggiato, S., Antonelli, T., Tomasini, M. C., et al. (2013). Kynurenic acid, by targeting α7 nicotinic acetylcholine receptors, modulates extracellular GABA levels in the rat striatum in vivo. *Eur J Neurosci, 37*(9), 1470–1477.
31. Beggiato, S., Tanganelli, S., Fuxe, K., Antonelli, T., Schwarcz, R., & Ferraro, L. (2014). Endogenous kynurenic acid regulates extracellular GABA levels in the rat prefrontal cortex. *Neuropharmacology, 82*, 11–18.
32. Nilsson, L. K., Linderholm, K. R., & Erhardt, S. (2006). Subchronic treatment with kynurenine and probenecid: effects on prepulse inhibition and firing of midbrain dopamine neurons. *J Neural Transm (Vienna), 113*(5), 557–571.
33. Linderholm, K. R., Alm, M. T., Larsson, M. K., et al. (2016). Inhibition of kynurenine aminotransferase II reduces activity of midbrain dopamine neurons. *Neuropharmacology, 102*, 42–47.
34. Konradsson-Geuken, A., Wu, H. Q., Gash, C. R., et al. (2010). Cortical kynurenic acid bi-directionally modulates prefrontal glutamate levels as assessed by microdialysis and rapid electrochemistry. *Neuroscience, 169*(4), 1848–1859.
35. Wu, H. Q., Pereira, E. F., Bruno, J. P., Pellicciari, R., Albuquerque, E. X., & Schwarcz, R. (2010). The astrocyte-derived alpha7 nicotinic receptor antagonist kynurenic acid controls extracellular glutamate levels in the prefrontal cortex. *J Mol Neurosci, 40*(1–2), 204–210.
36. Kuznezova, L. E. (1969). Mutagenic effect of 3-hydroxykynurenine and 3-hydroxyanthranilic acid. *Nature, 222*(5192), 484–485.
37. Okuda, S., Nishiyama, N., Saito, H., & Katsuki, H. (1996). Hydrogen peroxide-mediated neuronal cell death induced by an endogenous neurotoxin, 3-hydroxykynurenine. *Proc Natl Acad Sci U S A, 93*(22), 12553–12558.

38. Goldstein, L. E., Leopold, M. C., Huang, X., et al. (2000). 3-Hydroxykynurenine and 3-hydroxyanthranilic acid generate hydrogen peroxide and promote alpha-crystalline cross-linking by metal ion reduction. *Biochemistry, 39*(24), 7266–7275.
39. Ramírez-Ortega, D., Ramiro-Salazar, A., González-Esquivel, D., Ríos, C., Pineda, B., & Pérez de la Cruz, V. (2017). 3-Hydroxykynurenine and 3-hydroxyanthranilic acid enhance the toxicity induced by copper in rat astrocyte culture. *Oxid Med Cell Longev, 2017*, 2371895.
40. Krause, D., Suh, H. S., Tarassishin, L., et al. (2011). The tryptophan metabolite 3-hydroxyanthranilic acid plays anti-inflammatory and neuroprotective roles during inflammation: role of hemeoxygenase-1. *Am J Pathol, 179*(3), 1360–1372.
41. Leipnitz, G., Schumacher, C., Dalcin, K. B., et al. (2007). In vitro evidence for an antioxidant role of 3-hydroxykynurenine and 3-hydroxyanthranilic acid in the brain. *Neurochem Int, 50*(1), 83–94.
42. Colín-González, A. L., Maldonado, P. D., & Santamaría, A. (2013). 3-Hydroxykynurenine: an intriguing molecule exerting dual actions in the central nervous system. *Neurotoxicology, 34*, 189–204.
43. Colín-González, A. L., Maya-López, M., Pedraza-Chaverrí, J., Ali, S. F., Chavarría, A., & Santamaría, A. (2014). The Janus faces of 3-hydroxykynurenine: dual redox modulatory activity and lack of neurotoxicity in the rat striatum. *Brain Res, 1589*, 1–14.
44. Darlington, L. G., Forrest, C. M., Mackay, G. M., et al. (2010). On the biological importance of the 3-hydroxyanthranilic acid: anthranilic acid ratio. *Int J Tryptophan Res, 3*, 51–59.
45. Reyes-Ocampo, J., Ramírez-Ortega, D., Cervantes, G. I., et al. (2015). Mitochondrial dysfunction related to cell damage induced by 3-hydroxykynurenine and 3-hydroxyanthranilic acid: non-dependent-effect of early reactive oxygen species production. *Neurotoxicology, 50*, 81–91.
46. Henderson, L. M. (1949). Quinolinic acid excretion by the rat receiving tryptophan. *J Biol Chem, 178*(2), 1005.
47. Henderson, L. M., & Hirsch, H. M. (1949). Quinolinic acid metabolism: urinary excretion by the rat following tryptophan and 3-hydroxyanthranilic acid administration. *J Biol Chem, 181*(2), 667–675.
48. Henderson, L. M., & Ramasarma, G. B. (1949). Quinolinic acid metabolism; formation from 3-hydroxyanthranilic acid by rat liver preparations. *J Biol Chem, 181*(2), 687–692.
49. Henderson, L. M., Ramasarma, G. B., & Johnson, B. C. (1949). Quinolinic acid metabolism, urinary excretion by man and other mammals as affected by the ingestion of tryptophan. *J Biol Chem, 181*(2), 731–738.
50. Tavares, R. G., Tasca, C. I., Santos, C. E., et al. (2002). Quinolinic acid stimulates synaptosomal glutamate release and inhibits glutamate uptake into astrocytes. *Neurochem Int, 40*(7), 621–627.
51. Pierozan, P., Zamoner, A., Soska, A. K., et al. (2010). Acute intrastriatal administration of quinolinic acid provokes hyperphosphorylation of cytoskeletal intermediate filament proteins in astrocytes and neurons of rats. *Exp Neurol, 224*(1), 188–196.
52. Pierozan, P., & Pessoa-Pureur, R. (2017). Cytoskeleton as a target of quinolinic acid neurotoxicity: insight from animal models. *Mol Neurobiol, 55*(5), 4362–4372.

53. Luis-García, E. R., Limón-Pacheco, J. H., Serrano-García, N., et al. (2017). Sulforaphane prevents quinolinic acid-induced mitochondrial dysfunction in rat striatum. *J Biochem Mol Toxicol, 31*(2), 1–7.
54. Stípek, S., Stastný, F., Pláteník, J., Crkovská, J., & Zima, T. (1997). The effect of quinolinate on rat brain lipid peroxidation is dependent on iron. *Neurochem Int, 30*(2), 233–237.
55. St'astný, F., Lisý, V., Mares, V., Lisá, V., Balcar, V. J., & Santamaría, A. (2004). Quinolinic acid induces NMDA receptor-mediated lipid peroxidation in rat brain microvessels. *Redox Rep, 9*(4), 229–233.
56. Braidy, N., Grant, R., Adams, S., Brew, B. J., & Guillemin, G. J. (2009). Mechanism for quinolinic acid cytotoxicity in human astrocytes and neurons. *Neurotox Res, 16*(1), 77–86.
57. Schwarcz, R., & Stone, T. W. (2017). The kynurenine pathway and the brain: challenges, controversies and promises. *Neuropharmacology, 112*(Pt B), 237–247.
58. Zheng, W., Li, X. H., Yang, X. H., et al. (2017). Adjunctive memantine for schizophrenia: a meta-analysis of randomized, double-blind, placebo-controlled trials. *Psychol Med, 48*(1), 1–10.
59. Kitagami, T., Yamada, K., Miura, H., Hashimoto, R., Nabeshima, T., & Ohta, T. (2003). Mechanism of systemically injected interferon-alpha impeding monoamine biosynthesis in rats: role of nitric oxide as a signal crossing the blood-brain barrier. *Brain Res, 978*(1–2), 104–114.
60. Sperner-Unterweger, B., Kohl, C., & Fuchs, D. (2014). Immune changes and neurotransmitters: possible interactions in depression? *Prog Neuropsychopharmacol Biol Psychiatry, 48*, 268–276.
61. Haroon, E., Raison, C. L., & Miller, A. H. (2012). Psychoneuroimmunology meets neuropsychopharmacology: translational implications of the impact of inflammation on behavior. *Neuropsychopharmacology, 37*(1), 137–162.
62. Förstermann, U., & Sessa, W. C. (2012). Nitric oxide synthases: regulation and function. *Eur Heart J, 33*(7), 829–837; 837a–837d.
63. Koch, R., Guttler, F., & Blau, N. (2002). Mental illness in mild PKU responds to biopterin. *Mol Genet Metab, 75*(3), 284–286.
64. Schefft, C., Kilarski, L. L., Bschor, T., & Köhler, S. (2017). Efficacy of adding nutritional supplements in unipolar depression: a systematic review and meta-analysis. *Eur Neuropsychopharmacol, 27*(11), 1090–1109.
65. Sarris, J., Logan, A. C., Akbaraly, T. N., et al. (2015). International Society for Nutritional Psychiatry Research consensus position statement: nutritional medicine in modern psychiatry. *World Psychiatry, 14*(3), 370–371.
66. Cuadrado, A., & Nebreda, A. R. (2010a). Mechanisms and functions of p38 MAPK signalling. *Biochem J, 429*(3), 403–417.
67. Mayer, R. J., & Callahan J. F. (2006). P38 MAP kinase inhibitors: a future therapy for inflammatory diseases. *Drug Disc Today Ther Strateg*, 3(1), 49–54.
68. Samuvel, D. J., Jayanthi, L. D., Bhat, N. R., & Ramamoorthy, S. (2005). A role for p38 mitogen-activated protein kinase in the regulation of the serotonin transporter: evidence for distinct cellular mechanisms involved in transporter surface expression. *J Neurosci, 25*(1), 29–41.

69. Zhu, C. B., Lindler, K. M., Owens, A. W., Daws, L. C., Blakely, R. D., & Hewlett, W. A. (2010). Interleukin-1 receptor activation by systemic lipopolysaccharide induces behavioral despair linked to MAPK regulation of CNS serotonin transporters. *Neuropsychopharmacology, 35*(13), 2510–2520.
70. Baganz, N. L., Lindler, K. M., Zhu, C. B., et al. (2015). A requirement of serotonergic p38α mitogen-activated protein kinase for peripheral immune system activation of CNS serotonin uptake and serotonin-linked behaviors. *Transl Psychiatry, 5*, e671.
71. Sanchez, M. M., Alagbe, O., Felger, J. C., et al. (2007). Activated p38 MAPK is associated with decreased CSF 5-HIAA and increased maternal rejection during infancy in rhesus monkeys. *Mol Psychiatry, 12*(10), 895–897.
72. Yasuda, S., Sugiura, H., Tanaka, H., Takigami, S., & Yamagata, K. (2011). P38 MAP kinase inhibitors as potential therapeutic drugs for neural diseases. *Cent Nerv Syst Agents Med Chem, 11*(1), 45–59.
73. Inamdar, A., Merlo-Pich, E., Gee, M., et al. (2014). Evaluation of antidepressant properties of the p38 MAP kinase inhibitor Losmapimod (GW856553) in major depressive disorder: results from two randomised, placebo-controlled, double-blind, multicentre studies using a Bayesian approach. *J Psychopharmacol, 28*(6), 570–581.
74. Fehr, S., Unger, A., Schaeffeler, E., et al. (2015). Impact of p38 MAP kinase inhibitors on LPS-induced release of TNF-α in whole blood and primary cells from different species. *Cell Physiol Biochem, 36*(6), 2237–2249.
75. Li, C., Yuan, K., & Schluesener, H. (2013). Impact of minocycline on neurodegenerative diseases in rodents: a meta-analysis. *Rev Neurosci, 24*(5), 553–562.
76. Oya, K., Kishi, T., & Iwata, N. (2014). Efficacy and tolerability of minocycline augmentation therapy in schizophrenia: a systematic review and meta-analysis of randomized controlled trials. *Hum Psychopharmacol, 29*(5), 483–491.
77. Solmi, M., Veronese, N., Thapa, N., et al. (2017). Systematic review and meta-analysis of the efficacy and safety of minocycline in schizophrenia. *CNS Spectr, 22*(5), 415–426.
78. Xiang, Y. Q., Zheng, W., Wang, S. B., et al. (2017). Adjunctive minocycline for schizophrenia: a meta-analysis of randomized controlled trials. *Eur Neuropsychopharmacol, 27*(1), 8–18.
79. Basselin, M., Ramadan, E., & Rapoport, S. I. (2012). Imaging brain signal transduction and metabolism via arachidonic and docosahexaenoic acid in animals and humans. *Brain Res Bull, 87*(2-3), 154–171.
80. Müller, N., & Schwarz, M. J. (2008). COX-2 inhibition in schizophrenia and major depression. *Curr Pharm Des, 14*(14), 1452–1465.
81. Tang, B., Capitao, C., Dean, B., & Thomas, E. A. (2012). Differential age- and disease-related effects on the expression of genes related to the arachidonic acid signaling pathway in schizophrenia. *Psychiatry Res, 196*(2-3), 201–206.
82. Kim, H. W., Rapoport, S. I., & Rao, J. S. (2011). Altered arachidonic acid cascade enzymes in postmortem brain from bipolar disorder patients. *Mol Psychiatry, 16*(4), 419–428.
83. Shih, P. B. (2017). Integrating multi-omics biomarkers and postprandial metabolism to develop personalized treatment for anorexia nervosa. *Prostaglandins Other Lipid Mediat, 132*, 69–76.

84. Na, K. S., Lee, K. J., Lee, J. S., Cho, Y. S., & Jung, H. Y. (2014). Efficacy of adjunctive celecoxib treatment for patients with major depressive disorder: a meta-analysis. *Prog Neuropsychopharmacol Biol Psychiatry, 48*, 79–85.
85. Zheng, W., Cai, D. B., Yang, X. H., et al. (2017). Adjunctive celecoxib for schizophrenia: a meta-analysis of randomized, double-blind, placebo-controlled trials. *J Psychiatr Res, 92*, 139–146.
86. Rosenblat, J. D., Kakar, R., Berk, M., et al. (2016). Anti-inflammatory agents in the treatment of bipolar depression: a systematic review and meta-analysis. *Bipolar Disord, 18*(2), 89–101.

3

Immune Mechanisms Affecting the Functioning of the Central Nervous System (CNS)

MOISES E. BAUER, NATÁLIA P. ROCHA, WILSON SAVINO, AND ANTONIO L. TEIXEIRA ■

INTRODUCTION

The immune system is generally conceived to work independently from other physiological systems, and as primarily engaged in mounting immune responses against microorganisms. Accordingly, the immune system was thought to work only in pathological conditions (i.e., immunity) as well as in limiting the microbiota growth—as shown by immunoglobulin (Ig) A-mediated control of bacterial growth in the gut. A contemporary view considers the immune system as acting in an integrated way with other physiological systems, especially the nervous system, in the sense of maintaining homeostasis or body balance. In this chapter, we will discuss how cells and proteins of the immune system influence the brain functions, such as behavior, emotion, and cognition.

Historically, Hugo Besedovsky and Adriana del Rey were the first to show that the activated immune system was able to release a soluble product that increased the firing rate of hypothalamic neurons.[1,2] To date, we know that cytokines and leukocytes are indeed major effectors of the immune-to-brain communication, and this cross-talk is established by three independent pathways: the humoral, neural, and cellular (leukocyte) routes (Figure 3.1). The trafficking of lymphoid cells to the brain, including the meningeal space, has become an area of special interest with the recent description of a brain lymphatic system that heretofore

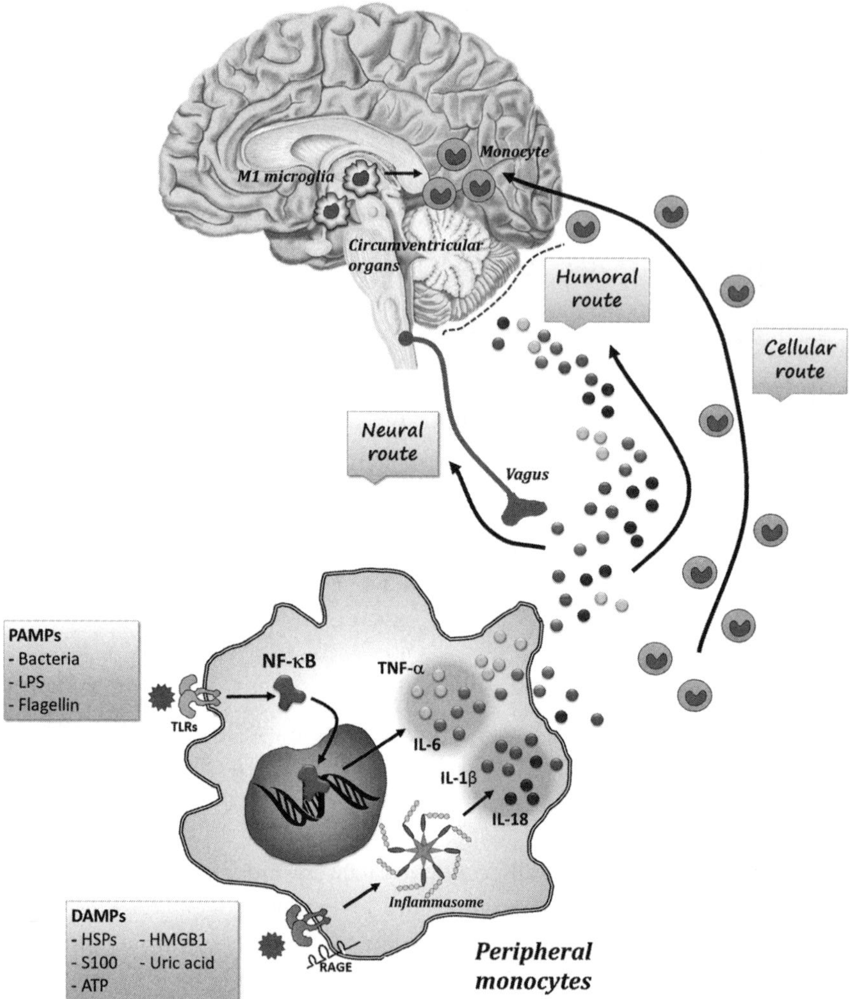

Figure 3.1 *Major pathways of immune-to-brain communication.* Three independent pathways mediate the immune-to-brain communication: the humoral, neural, and leukocyte routes. Infections provide pivotal signaling to circulating monocytes, including pathogen-associated molecular patterns (PAMPs) or damage-associated molecular patterns (DAMPs). Both PAMPs and DAMPs (sterile injury) engage inflammatory signaling pathways, such as nuclear factor-κB (NF-κB) and inflammasomes. In consequence, the pro-inflammatory cytokines (TNF-α, IL-1β, IL-6, and IL-18) are readily secreted and enter the circulation. Circulating cytokines can reach the brain through various mechanisms, including **(i)** the active transport into the brain, crossing the brain–blood barrier (BBB) through leaky areas in the circumventricular organs (humoral route); or **(ii)** through the activation of neural pathways such as the vagus nerve. The leukocytic route is another mechanism of immune-to-brain communication, and is mediated by the migration of circulating leukocytes to the brain borders. Leukocytes are present in small numbers in brain circumventricular organs and the choroid plexus. Under healthy conditions, these peripheral immune cells support neuronal function and scan the brain for pathogens or tissue damage. Adapted from[8].

had gone unrecognized.[3] Due to the destructive effects of inflammation upon the (CNS) under pathological conditions, the brain has long been considered protected from the action of the immune system, i.e., it is "an immune privileged site." Recent research, however, points out that immune cells migrate to the CNS at steady state and cross-talk to neurons and glial cells. This newly described immune-to-brain cellular (leukocyte) pathway has been described in crucial physiological processes, interfering with behavior, emotion, and cognition.

THE HUMORAL PATHWAY OF IMMUNE-TO-BRAIN COMMUNICATION

The sickness behavior is the prototype of immune-to-brain communication. Following an infection, the organism rapidly starts a set of changes to enhance host survival. This set includes the sickness behavior, as shown by physiological (e.g., fever, alterations in blood composition), behavioral (e.g., decreased locomotion, food and water intake), and hormonal (e.g., secretion of hormones from the hypothalamic-pituitary-adrenal [HPA] axis) changes. Sickness behavior may maximize chances of surviving infections by diverting energetic resources to the immune system and minimizing contact with other pathogens and predators. The expression "sickness behavior" is also employed in the context of an existing, chronic low-grade inflammation as observed in different psychiatric disorders, including major depression, bipolar disorder, and schizophrenia.[4]

Circulating cytokines can trigger the "sickness behavior" by means of several immune-to-brain pathways:

1. Pro-inflammatory cytokines produced during infections or tissue damage may sensitize afferent sensory nerves (i.e., vagus nerve in visceral infections and the trigeminal nerve during cranial infections);
2. Macrophages in the circumventricular organs and the choroid plexus (CP) produce cytokines that can enter the brain by volume diffusion;
3. Circulating cytokines may gain access into the brain via saturable transporters at the blood–brain barrier (BBB); and
4. The activation of interleukin (IL)-1 receptors expressed by perivascular macrophages and endothelial cells results in the production of prostaglandin E2 in the brain.[5]

The "sickness behavior" can be induced by administration of pro-inflammatory cytokines and lipopolysaccharide (LPS). IL-1β and TNF-α are the main cytokines involved in triggering the sickness behavior. In rodents, both systemic and intracerebral administration of IL-1β or TNF-α resulted in signs of sickness behavior (i.e., decreased motor activity, social withdrawal, reduced food and water intake, and impaired cognition), in time- and dose-dependent manners. Furthermore, IL-6–deficient mice have attenuated signs of LPS-induced sickness behavior as compared with wild-type animals, indicating that IL-6 expression in the brain may contribute to the expression of other cytokines (including IL-1β and TNF-α) following immune challenges.[5] The nuclear factor (NF)-κB, an intracellular signaling molecule, was described as an essential mediator of the immune-to-brain communication. In rodents, the central blockade of NF-κB inhibited IL-1β- and LPS-induced behavioral changes.[6]

In line with these experimental studies, clinical data revealed that intravenous injection of LPS from *Salmonella abortus* in humans resulted in a transient increase in the levels of anxiety and depressive mood, and a decrease in verbal and nonverbal memory functions. The administration of this endotoxin was also followed by an increase in circulating levels of TNF-α, soluble TNF receptors, IL-6, and cortisol.[7] The clinical use of cytokines was also associated with sickness behavior in humans. For example, up to 50% of patients receiving chronic interferon (IFN)-α therapy for the treatment of hepatitis C or cancer developed clinical depression. The IFN-α-induced depressive syndrome is responsive to antidepressants and clinically overlaps with primary major depression. Furthermore, the administration of peripheral cytokines results in the amplification of neuroinflammation in the CNS. Patients with hepatitis C receiving peripheral administration of IFN-α presented greater concentrations of IFN-α in the cerebrospinal fluid (CSF), which correlated with increased CSF levels of IL-6 and the chemokine CCL2/MCP-1.[6]

What is the evolutionary role of sickness behavior? This has been discussed in light of the "pathogen host defense hypothesis of depression." According to this hypothesis, humans inherited a genomic bias towards inflammation and the associated set of depressive symptoms. This pattern of response, which may be currently seen as negative or detrimental, enhanced host survival and increased the likelihood of reproduction in the highly dangerous and pathogenic environments, such as the African savanna, where humans evolved.[8]

HOW DO CYTOKINES MODULATE BEHAVIOR?

The mechanisms underlying the "sickness behavior" involve changes in brain monoamine levels, excitotoxicity (i.e., increased glutamate levels), and reduced brain plasticity (Figure 3.2). The depletion of brain serotonin levels seems to be an important mechanism by which cytokines may influence behavior. Both experimental studies and clinical trials have demonstrated that the decrease in synaptic availability of serotonin was highly correlated with the development

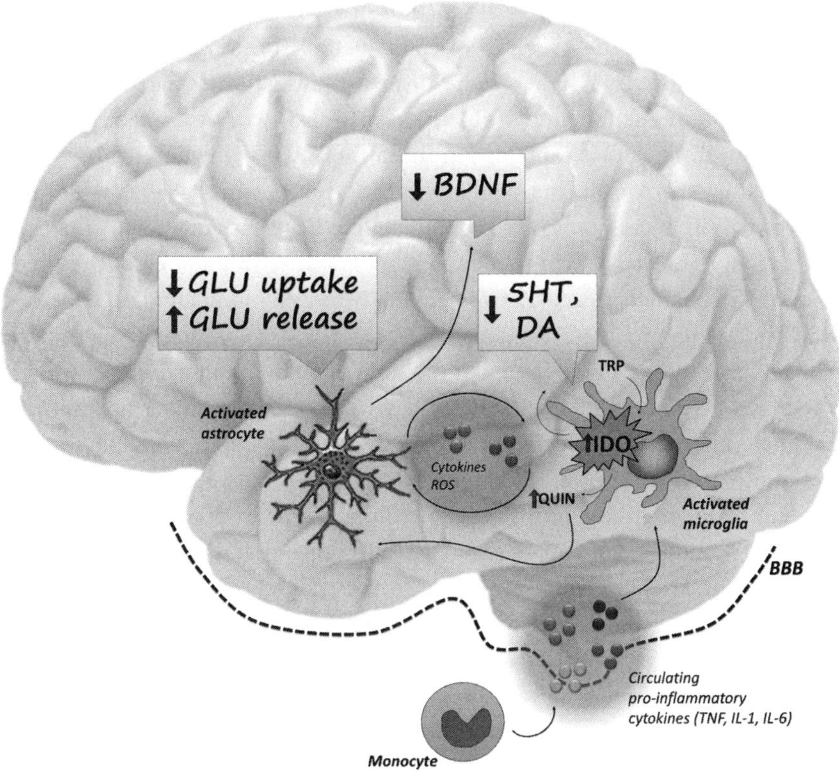

Figure 3.2 *Cytokine effects on depressive-like behavior.* Peripheral pro-inflammatory cytokines (TNF-α, IL-1, IL-6) can reach the CNS and readily activate microglia, which in turn amplify inflammation in the brain by activating astrocytes through secretion of pro-inflammatory cytokines and reactive oxygen species (ROS). Several astrocytic functions are impaired by excessive cytokine signaling, leading to downregulation of glutamate (GLU) transporters, impairing GLU reuptake and increasing GLU release. This is known as excitotoxicity. Circulating inflammatory cytokines activate the enzyme indoleamine 2,3-dioxygenase (IDO), which breaks down tryptophan (TRP), the precursor of serotonin (5-HT) and dopamine (DA), into quinolinic acid (QUIN), a potent NMDA agonist that stimulates GLU release. The binding of GLU to extrasynaptic NMDA receptors on neurons leads to decreased production of brain-derived neurotrophic factor (BDNF), impairing neuroplasticity.

of cytokine-induced depressive symptoms. Accordingly, polymorphisms in the IL-6 and serotonin transporter genes predicted depression during IFN-α treatment for hepatitis C.[9] Previous studies investigating the effects of IFN-α upon the indoleamine 2,3-dioxygenase (IDO) provided further evidence that the serotonin pathway is involved in IFN-α-induced depression. IDO is involved in breaking down tryptophan into kynurenine. Tryptophan is an essential amino acid (which must be obtained through the diet) and the primary precursor of serotonin. Usually, only a small portion of tryptophan is used for the synthesis of serotonin. The great majority of dietary tryptophan (>95%) is degraded in the liver through the kynurenine pathway by tryptophan dioxygenase. Tryptophan degradation into kynurenine can also occur extrahepatically by the enzyme IDO. IDO is expressed in the brain, and is highly inducible by proinflammatory cytokines. Under inflammation, tryptophan availability for serotonin synthesis decreases, while kynurenine levels increase due to an enhanced IDO activity.[5] Moreover, kynurenine easily cross the BBB and enter the brain, where it will be metabolized by two cellular pathways: (1) microglia, generating 3-hydroxykynurenine (3-HK) and quinolinic acid (QA); and (2) astrocytes, producing kynurenic acid (KA). 3-Hydroxykynurenine is an oxidative stressor, whereas QA is an N-methyl-D-aspartate (NMDA) receptor agonist, stimulating glutamate release and blocking glutamate reuptake by astrocytes.[8] Quinolinic acid is also associated with lipid peroxidation and oxidative stress. These activities, in combination, can lead to excitotoxicity and neurodegeneration. In contrast to QA, the KA can reduce glutamate and dopamine release, which in turn can contribute to cognitive dysfunction.[8] Increased levels of QA have been found in microglia in the brain of suicide victims who suffered from depression.[10]

In addition to pro-inflammatory cytokines, IDO can be activated by multiple inflammatory signaling pathways, including STAT1a, interferon regulatory factor (IRF)-1, NF-κB, and p38 mitogen-activated protein kinase (MAPK).[6] Importantly, blocking NMDA glutamate receptors with ketamine or inhibiting IDO activity protects mice from LPS- or stress-induced depressive-like behavior.[11,12] Accordingly, drugs that block IDO would be also useful for patients with depression and increased inflammation. In addition, recent studies have reported the rapid antidepressant effect of ketamine in humans.[13]

Cytokines may also interfere in the dopamine synthesis. Intramuscular injection of IFN-α in rats resulted in a significant decrease in the levels of tetrahydrobiopterin (BH_4) and dopamine in the amygdala and raphe areas. BH_4 is an important cofactor for tyrosine hydroxylase, the rate-limiting enzyme in dopamine synthesis.[6] The impact of cytokines on the brain dopaminergic pathways leads to a state of decreased motivation or anhedonia, a core symptom of depression.

In a second vein, the binding of glutamate to extrasynaptic NMDA receptors leads to decreased production of brain-derived neurotrophic factor (BDNF), impairing neuroplasticity.[14] BDNF is a neurotrophin involved with neurogenesis, essential for an antidepressant response, and it has been shown to be reduced by TNF-α and IL-1β and their downstream signaling pathways, including NF-κB in stress-induced animal models of depression.[15]

THE NEURAL PATHWAY OF IMMUNE-TO-BRAIN COMMUNICATION

Sensory neurons throughout the body, and in proximity to immune cells, including in the lymphoid organs,[16] are capable of carrying afferent immune-related signals to the brain via the spinal cord (sympathetic) and vagus nerve (parasympathetic).[17] The afferent pathway of this neural communication includes the vagal stimulation by pro-inflammatory cytokines, enabling a (not conscious) representation in the CNS of a peripheral inflammatory signal.[18] In this sense, the immune system would be acting as a "sensory organ," a concept originally proposed by J. Edwin Blalock during the 1980s.[19] Afferent vagus neurons end primarily in the nucleus *tractus solitarius* in the brainstem medulla. This afferent signal is then communicated through neural contacts to other brainstem nuclei, the hypothalamus, and forebrain regions associated with integration of visceral sensory information as well as coordination of autonomic function and behavioral responses.[20]

Most of the neural signaling from the brain to the lymphoid organs is mediated via efferent noradrenergic innervation by the sympathetic nervous system (SNS). The efferent vagus does not directly communicate with lymphoid organs.[16] However, it is known that cholinergic stimulation via acetylcholine (ACh) secretion suppresses excessive inflammation in the liver, heart, pancreas, and gastrointestinal tract.[20] The efferent arm of vagal communication, termed the "inflammatory reflex,"[21] can also send signals from the efferent vagus to the splenic nerve responsible for secreting norepinephrine (NE) and inducing the release of ACh by a subset of splenic T cells, with important anti-inflammatory actions.[22] For example, in nude mice, which lack T cells, the vagal stimulation cannot refrain inflammatory responses. However, transfer of ACh-producing T cells, which repopulate the spleen in nude mice, restores the integrity of this anti-inflammatory neural circuit.[22]

The vagal efferent pathway, the inflammatory reflex, is responsible for the attenuation of TNF-α levels during septic shock. Accordingly, a range of sickness responses is abolished by cutting the vagus nerve: including fever, decreased food-motivated behavior, increased sleep, decreased activity, decreased social interaction, changes in brain activity, and release of stress hormones.[23] ACh interacts with nicotinic receptors (α7) expressed in several leukocytes, and the intracellular pathways of this regulation have already been elucidated.[24] The vagal efferent pathway may thus work as a brake on inflammatory responses.

THE LEUKOCYTE PATHWAY OF THE IMMUNE-TO-BRAIN COMMUNICATION: MICROGLIA AND MACROPHAGES

The leukocyte pathway is a novel axis of immune-to-brain communication and includes infiltrating leukocytes (e.g., granulocytes, dendritic cells, T cells) present in small numbers in brain vasculature, choroid plexus (CP), and meninges (Figure 3.3). The CP, an epithelial tissue located within the brain ventricles, is responsible for filtering the blood, producing the CSF with low-protein content. In addition, recent evidence indicates that CP is an important selective gateway for immune-cell trafficking to the CNS.[25] Under physiological conditions, these peripheral immune cells support neuronal functions and scan the brain for pathogens or tissue damage.[26]

Under healthy conditions, leukocytes are not found in the cerebral parenchyma. Microglia are resident cells of the CNS with macrophagic activity, representing up to 10% of all CNS cells, and are crucially involved in brain homeostasis. During development, microglial cells migrate very early from the yolk sac into the brain (embryonic day 9.5) and before the differentiation of other resident nervous system cells. Of key note, microglia participate in the development of neural circuits, maintenance of synapses, and neurogenesis. For instance, microglia help the brain in eliminating the excess of neurons during development, a phenomenon known as *neuronal pruning*.[27] Indeed, microglia have been implicated in a range of developmental processes, including the regulation of cell number and spatial patterning of brain cells, myelination, and formation and refinement of neural circuits.[28] However, after challenge by a biological insult or psychological stress, the resting microglia are transformed and activated, undergoing a series of changes, notably in shape, increased proliferation, and production of inflammatory mediators. The activated microglia are then promptly recruited to injured sites where they will phagocyte debris and unwelcome dying cells. Once activated, microglia function like peripheral macrophages.[29]

Activated microglial cells have been associated with the neuroinflammation reported in neurodegenerative and psychiatric disorders, and they promote behavioral changes by modulating key stress-responsive brain areas, including the prefrontal cortex, hypothalamus, amygdala, and hippocampus (Figure 3.1).[30]

Noradrenergic signaling plays a pivotal role in microglia activation. Pretreatment with propranolol (i.e., β-adrenergic receptor antagonist), known to impair anxiety-like behavior, prevents stress-induced neuronal activation and microglia activation.[31] Conversely, the administration of isoproterenol, a β-adrenergic receptor agonist, results in the induction of both peripheral and central (i.e., hippocampus) production of pro-inflammatory cytokines.[32] Treatment with minocycline, an antibiotic known to limit microglia response, prevents or

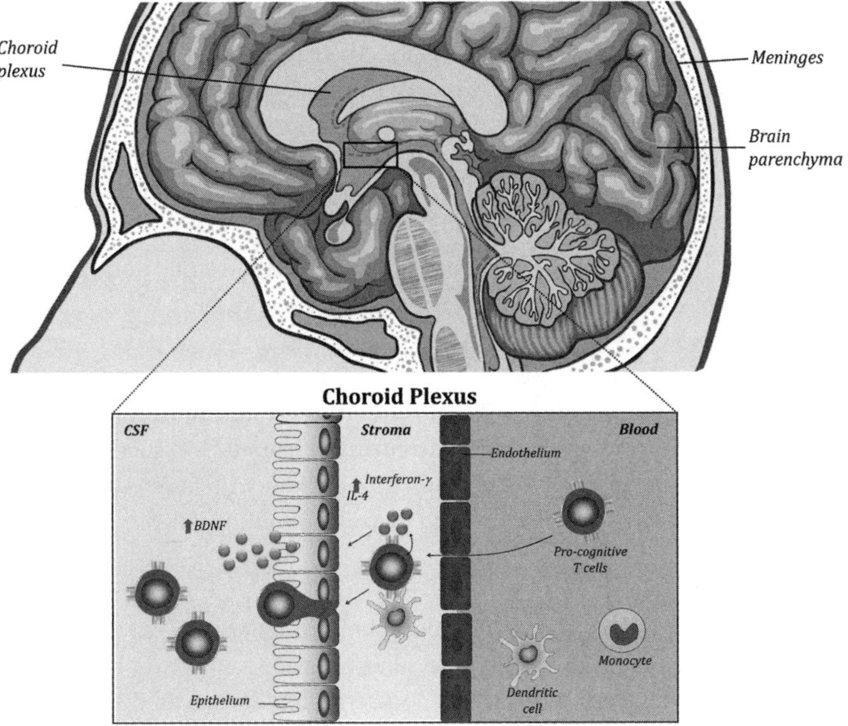

Figure 3.3 *The choroid plexus as major gateway in immune-to-brain communication.* This figure depicts the blood-cerebrospinal barrier at the choroid plexus (CP). The CP is located in all four ventricles and is constituted of fenestrated capillaries surrounded by a monolayer of epithelial cells. In physiological conditions, the CP stroma (a space between the endothelium of blood vessels and the epithelial surface) is populated by several immune cells, including T cells (i.e., CD4+ effector and memory T lymphocytes), monocytes, and dendritic cells. The CP controls the trafficking of leukocytes to the CNS and maintains the production of neurotrophic factors, including brain-derived neurotrophic factor (BDNF), by the CP epithelium. The T cell derived cytokines interferon (IFN)-γ and IL-4 are the major local BDNF inducers.

attenuates stress-induced microglia activation, pro-inflammatory cytokine production inside the brain (especially IL-1β), and persistent neuronal activation. In addition, the same treatment diminishes anxiety- and depressive-like behaviors and cognitive deficits following stress.[33]

Following stress-induced microglia activation, the produced cytokines signal to increase neuroendocrine outflow, resulting in a cycle of stress-related responses and further microglia activation. IL-1β is thought to be the key cytokine in inducing stress-related responses. The release of IL-1β in the hypothalamus activates the HPA axis by inducing the secretion of corticotrophin-releasing hormone (CRH) from the paraventricular nucleus. CRH, in turn, induces the secretion of adrenocorticotropin (ACTH) from the pituitary, which in turn stimulates the secretion of cortisol from the adrenal cortex. In addition, IL-1β is capable of directly inducing ACTH and cortisol secretion from the pituitary and adrenal, respectively.[34]

The brain-to-immune system communication is potentiated by the SNS. As discussed previously, sympathetic nerve fibers innervate lymphoid tissues, and SNS activation results in NE release into immune organs, including the bone marrow. As peripheral immune cells express NE receptors, they undergo functional alterations after sympathetic activation. Under chronic stress, the prolonged or repeated sympathetic activation results in increased production and release of myeloid cells by the bone marrow. The enhanced cycling of myeloid cells culminates in a phenotype of less mature and more inflammatory circulating monocytes (M1 phenotype), which will move throughout the tissues where they become effector cells. These inflammatory monocytes traffic even to the brain, where they differentiate into brain macrophages, promoting inflammation.[33]

The recruitment of blood monocytes to the brain has been extensively described in classical inflammatory conditions, such as neurotrauma and neuroinfection. Recently, studies demonstrated that monocytes' trafficking into the brain is required for inducing anxiety-like behavior after neurophatic pain,[35] sickness behavior due to inflammation,[36] and cognitive decline following peripheral surgery.[37] Peripheral monocyte migration to the brain was also demonstrated following psychological stress.[38] This cellular trafficking to the brain was associated with prolonged responses after stress. Using the repeated social-defeat murine model of stress, a study showed that monocyte infiltration into the brain is necessary for the development of stress-induced, prolonged, anxiety-like behavior.[38] A postmortem study has also recently shown that more peripheral monocytes are being recruited into the brains of depressed patients than in those of healthy controls.[39]

THE ROLE OF T CELLS IN THE IMMUNE-TO-BRAIN COMMUNICATION

Recent data indicate that circulating T cells may also play an important role in neuroprotection and resilience against development of neuropsychiatric disorders. The relative concentration of leukocytes found circulating in ventricular CSF is low—approximately 1,000-fold less as compared to blood.[40] More than 90% of the leukocytes in the CSF of healthy brain are T cells, half of which are CD4+ T helper cells. However, it should be noted that, in the healthy brain, T cells do no penetrate the BBB and are rarely found in brain parenchyma. From a mechanistic point of view, it seems likely that T cells enter the CNS via the

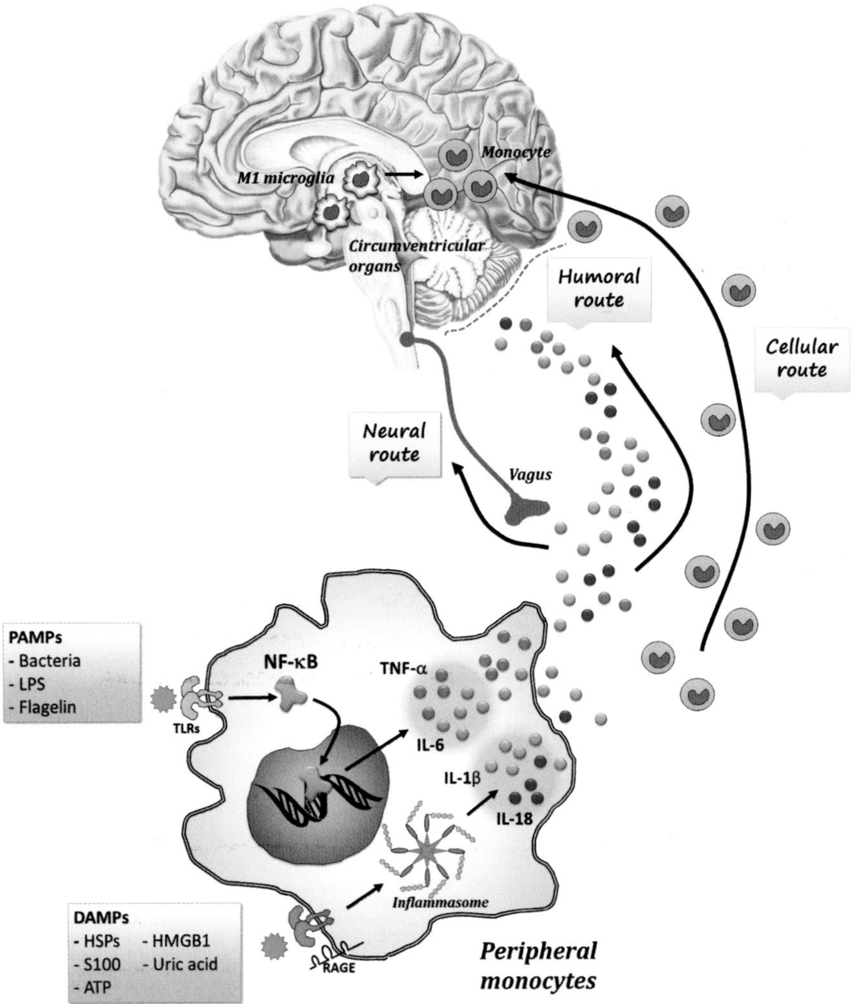

Figure 3.1 *Major pathways of immune-to-brain communication.* Three independent pathways mediate the immune-to-brain communication: the humoral, neural, and leukocyte routes. Infections provide pivotal signaling to circulating monocytes, including pathogen-associated molecular patterns (PAMPs) or damage-associated molecular patterns (DAMPs). Both PAMPs and DAMPs (sterile injury) engage inflammatory signaling pathways, such as nuclear factor-κB (NF-κB) and inflammasomes. In consequence, the pro-inflammatory cytokines (TNF-α, IL-1β, IL-6, and IL-18) are readily secreted and enter the circulation. Circulating cytokines can reach the brain through various mechanisms, including **(i)** the active transport into the brain, crossing the brain–blood barrier (BBB) through leaky areas in the circumventricular organs (humoral route); or **(ii)** through the activation of neural pathways such as the vagus nerve. The leukocytic route is another mechanism of immune-to-brain communication, and is mediated by the migration of circulating leukocytes to the brain borders. Leukocytes are present in small numbers in brain circumventricular organs and the choroid plexus. Under healthy conditions, these peripheral immune cells support neuronal function and scan the brain for pathogens or tissue damage. Adapted from [8].

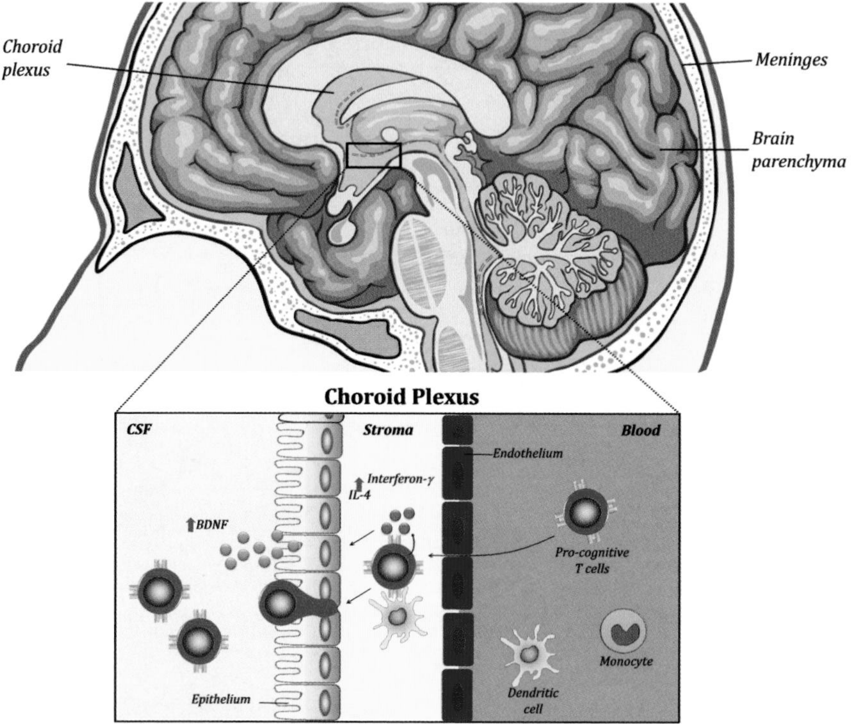

Figure 3.3 *The choroid plexus as major gateway in immune-to-brain communication.* This figure depicts the blood-cerebrospinal barrier at the choroid plexus (CP). The CP is located in all four ventricles and is constituted of fenestrated capillaries surrounded by a monolayer of epithelial cells. In physiological conditions, the CP stroma (a space between the endothelium of blood vessels and the epithelial surface) is populated by several immune cells, including T cells (i.e., CD4+ effector and memory T lymphocytes), monocytes, and dendritic cells. The CP controls the trafficking of leukocytes to the CNS and maintains the production of neurotrophic factors, including brain-derived neurotrophic factor (BDNF), by the CP epithelium. The T cell derived cytokines interferon (IFN)-γ and IL-4 are the major local BDNF inducers.

fenestrated blood capillaries in the CP located in the ventricles.[41] More specifically, the CD4+ T cell subset found in the CNS comprises very few naïve cells,[41] whereas most of the CSF T cells are effector-memory T cells expressing receptors that allow homing to inflamed tissues.[42] Importantly, effector-memory T cells have immediate effector functions without the need for further differentiation.

Previous studies indicated that peripheral T cells play a key role in the maintenance of brain plasticity, as shown by CNS-specific T cells required for supporting hippocampal-dependent memory and neurogenesis.[43] Mice deficient in T cells had reduced spatial learning and memory functions, as well as reduced proliferation of neural progenitor cells and neuronal differentiation, leading to decreased neurogenesis in the adult brain. These cognitive and plastic deficits were specific for CD4+ T cells, because adoptive transfer of CD4+ T cells (but not CD8+) into deficient mice restored proliferation of hippocampal neurons.[43] Furthermore, transgenic mice overexpressing a T cell receptor to myelin basic protein (i.e., T_{MBP} mice) had higher levels of adult neurogenesis relative to the wild type.[43] Considering there are no T cells in the healthy brain parenchyma, the question was how peripheral T cells stimulated cognitive functions, and where this cross-talk might take place. It has been shown that pro-cognitive T cells specific to CNS antigens are located at the brain borders, especially at the meninges[44] and at the CP.[45] In the absence of T cells, meningeal myeloid cells acquire a pro-inflammatory phenotype (M1) with secretion of TNF, IL-1β, and IL-12, cytokines that negatively affect brain function.[5] After learning and memory tasks in mice, the activated T cells found in their meningeal spaces expressed high levels of IL-4[46] and maintained meningeal myeloid cells in an M2 anti-inflammatory state. In addition to maintaining the M2 phenotype of meningeal myeloid cells, IL-4 may also directly mediate an improvement in learning behavior via the upregulation of BDNF expression by neural cells.[46] Actually, BDNF is a key neurotrophic factor supporting neuronal survival and brain plasticity, being also involved with memory and neurogenesis. Decreased levels of BDNF have been in found in psychiatric disorders, such as major depression, and negatively correlated with memory performance.[47]

Effector T cells have important roles in resilience to psychosocial stress. T cell-deficient mice (SCID) are more likely to develop behavioral changes characteristic of a post-traumatic stress disorder–like syndrome than wild-type animals. Reconstitution of SCID mice with T cells isolated from wild-type donors attenuated the overactive stress response. Furthermore, when the T cells were boosted with MBP vaccination in wild-type mice, the long-term pathological response to stress was further diminished.[48] These studies indicate that T cells can attenuate stress responses,[49] and that inhibition of T-cell function by stress and depression may have profound consequences for key immunological elements of healthy brain activity. In this respect, it has been speculated that boosting memory T cells specific for brain antigens might be a novel treatment for depression. Indeed, previous data showed that boosting effector T cells, by reducing Treg-mediated suppression of effector T cells, is an important mechanism for protection of CNS tissue from acute insult and from psychological stress.[50,51]

CNS-specific autoreactive T cells also play a protective role in neurodegenerative conditions. Several studies have demonstrated the protective role of CNS-specific T cells in neurodegenerative diseases, including Alzheimer's disease (AD) and Parkinson's disease.[49,52] These findings have led to the concept of "protective autoimmunity," which is in sharp contrast to the current dogma that autoreactive lymphocytes are pathological in nature. In an experimental model of AD, a transient depletion of Treg cells was followed by amyloid-β plaque clearance, mitigation of the neuroinflammatory response, and reversal of cognitive decline.[53] It has been shown that transient Treg depletion affects the brain's CP and was associated with subsequent recruitment of macrophages and Tregs to cerebral sites of plaque pathology. These findings indicate that targeting Treg-mediated systemic immunosuppression can be beneficial for treating AD. It has also been shown that the immune checkpoint blockade against the programmed death-1 (PD-1) pathway elicited an IFN-γ–dependent systemic immune response, followed by the recruitment of monocyte-derived macrophages to the brain.[54] When induced in mice with established pathology, this immunological response led to the clearance of cerebral amyloid-β (Aβ) plaques and improved cognitive performance. Furthermore, the depletion of Treg cells by administering anti-CD25 antibodies increased anxiety behavior, suggesting that Treg cells are anxiolytic.[55] Taken together, these findings indicate that systemic adaptive immunity should be boosted, rather than suppressed, to drive immune-mediated cascades needed for brain repair and mood behavior.

The regulation of cytokine networks may be of great clinical interest for brain homeostasis. Indeed, a type I/type II interferon balance was shown to be pivotal in the regulation of brain physiology and pathology.[54] Type I interferon consists of a family of pleiotropic cytokines (e.g., IFN-α and IFN-β) produced mainly by innate immune cells and are involved in the induction of antiviral state. Type II interferon (i.e., IFN-γ) is produced chiefly by CD4+ Th1 cells and potentiates several cell-mediated adaptive functions, including cell proliferation, differentiation, and the activation of macrophages. It has been shown that production of IFN-γ by CD4+ T cells at the CP is a key mechanism involved in leukocyte trafficking to the CSF. The lack of IFN-γ-signaling resulted in reduced CSF leukocyte counts, and was correlated with premature cognitive decline in mice during adulthood.[56] In addition, low IFN-γ content at the CP in neurodegenerative diseases, such as Parkinson's and AD, resulted in poorer immune support needed for repair.[54] During human aging and in a mouse model of AD, the insufficient IFN-γ signaling at the CP occurred in parallel with local elevation of type

I IFN (i.e., IFN-α and IFN-β) produced by the CP epithelium.[56] Blockade of this cytokine-induced signaling within the CNS in aged mice led to partial restoration of brain homeostasis, including attenuated age-related inflammation in the hippocampus.[56] IFN-I was shown to affect negatively the epithelial expression of BDNF and IGF-1 at the CP. Since BNDF and IGF-1 are neurotrophic factors supporting neuronal differentiation, growth, survival, and synaptogenesis,[56] these findings indicate that blocking IFN-I signaling in the aged CNS/CP may have a positive effect on the brain aging and AD, and that this effect is mediated, at least in part, by restoration of IFN-γ-dependent activity of the CP.

Finally, the meningeal immunity also seems to be important for social behavior. SCID mice or mice deficient in IFN-γ exhibited severe social deficits and hyper-connectivity of frontocortical brain regions.[57] It was shown that inhibitory neurons respond to IFN-γ and increase GABAergic (γ-aminobutyric-acid) transmission in projection neurons, suggesting that IFN-γ is a molecular link between meningeal immunity and neural circuits recruited for social behavior. Taken together, these data suggest that social deficits observed in several neurological and psychiatric disorders (e.g., autism, frontotemporal dementia, schizophrenia) may result from impaired circuitry homeostasis derived from impaired adaptive immunity.

CONCLUDING REMARKS

The immune-to-brain communication is pivotal in establishing brain homeostasis, as well as in the onset and progression of neuropsychiatric disorders. Humoral, neural, and leukocyte pathways convey immune cells and cytokines to the CNS, especially at the brain borders, where they regulate cognition, emotion, and behavior. Under physiological conditions, leukocytes are not found in the cerebral parenchyma. However, chronic exposure to inflammatory mediators may disrupt brain barriers and allow the transfer of immune cells to the brain parenchyma, modulating the microglial phenotype, reactivity, and driving neuroinflammation. After challenge by a biological insult or psychological stress, the resting microglia are activated, undergoing phenotypical changes and secreting pro-inflammatory cytokines. The activated microglia are promptly recruited to injured sites where they help in phagocytosing debris and unwelcome dying cells. This activation profile is reparative, although its homeostatic control is necessary to avoid brain pathology. Chronic inflammation may lead to increased differentiation of activated microglia in the brain, contributing to impaired cognition and depressive-like behavioral symptoms.

Recent research also indicates novel physiological roles for adaptive immunity, especially T cells, during health and disease. The adaptive immune system affects hippocampal neurogenesis, cognition, mood, and social behavior, and it may prevent the development of neuropsychiatric disorders. The CNS-specific T cells, which have previously been considered as key inducers of pathology, should now be considered as a beneficial support system for the CNS (i.e., protective autoimmunity).

In conclusion, the immune system plays an important role in regulating brain processes at the steady state. The immune changes reported in neuropsychiatric diseases cannot be thus understood as simple epiphenomena, but rather as interconnected with the brain's physiology and playing roles ranging from etiology to disease progression.

REFERENCES

1. Besedovsky, H., Sorkin, E., Felix, D., & Haas, H. (1977). Hypothalamic changes during the immune response. *Eur J Immunol*, *7*(5), 323–325.
2. Besedovsky, H., del Rey, A., Sorkin, E., Da Prada, M., Burri, R., & Honegger, C. (1983). The immune response evokes changes in brain noradrenergic neurons. *Science*, *221*(4610), 564–566.
3. Louveau, A., Smirnov, I., Keyes, T. J., et al. (2015). Structural and functional features of central nervous system lymphatic vessels. *Nature*, *523*(7560), 337–341.
4. Pessoa Rocha, N., Reis, H. J., Vanden Berghe, P., & Cirillo, C. (2014). Depression and cognitive impairment in Parkinson's disease: a role for inflammation and immunomodulation? *Neuroimmunomodulation*, *21*(2–3), 88–94.
5. Dantzer, R., O'Connor, J. C., Freund, G. G., Johnson, R. W., & Kelley, K. W. (2008). From inflammation to sickness and depression: when the immune system subjugates the brain. *Nat Rev Neurosci*, *9*(1), 46–56.
6. Miller, A. H., Maletic, V., & Raison, C. L. (2009). Inflammation and its discontents: the role of cytokines in the pathophysiology of major depression. *Biol Psychiatry*, *65*(9), 732–741.
7. Reichenberg, A., Yirmiya, R., Schuld, A., et al. (2001). Cytokine-associated emotional and cognitive disturbances in humans. *Arch Gen Psychiatry*, *58*(5), 445–452.
8. Miller, A. H., & Raison, C. L. (2016). The role of inflammation in depression: from evolutionary imperative to modern treatment target. *Nat Rev Immunol*, *16*(1), 22–34.
9. Miller, A. H. (2009). Norman Cousins Lecture. Mechanisms of cytokine-induced behavioral changes: psychoneuroimmunology at the translational interface. *Brain Behav Immun*, *23*(2), 149–158.
10. Steiner, J., Walter, M., Gos, T., et al. (2011). Severe depression is associated with increased microglial quinolinic acid in subregions of the anterior cingulate gyrus: evidence for an immune-modulated glutamatergic neurotransmission? *J Neuroinflammation*, *8*, 94.
11. Walker, A. K., Budac, D. P., Bisulco, S., et al. (2013). NMDA receptor blockade by ketamine abrogates lipopolysaccharide-induced depressive-like behavior in C57BL/6J mice. *Neuropsychopharmacology*, *38*(9), 1609–1616.

12. O'Connor, J. C., Lawson, M. A., Andre, C., et al. (2009). Lipopolysaccharide-induced depressive-like behavior is mediated by indoleamine 2,3-dioxygenase activation in mice. *Mol Psychiatry, 14*(5), 511–522.
13. Newport, D. J., Carpenter, L. L., McDonald, W. M., et al. (2015). Ketamine and other NMDA antagonists: early clinical trials and possible mechanisms in depression. *Am J Psychiatry, 172*(10), 950–966.
14. Hardingham, G. E., Fukunaga, Y., & Bading, H. (2002). Extrasynaptic NMDARs oppose synaptic NMDARs by triggering CREB shut-off and cell death pathways. *Nat Neurosci, 5*(5), 405–414.
15. Koo, J. W., Russo, S. J., Ferguson, D., Nestler, E. J., & Duman, R. S. (2010). Nuclear factor-kappaB is a critical mediator of stress-impaired neurogenesis and depressive behavior. *Proc Natl Acad Sci U S A, 107*(6), 2669–2674.
16. Nance, D. M., & Sanders, V. M. (2007). Autonomic innervation and regulation of the immune system (1987–2007). *Brain Behav Immun, 21*(6), 736–745.
17. Herkenham, M., & Kigar, S. L. (2017). Contributions of the adaptive immune system to mood regulation: mechanisms and pathways of neuroimmune interactions. *Prog Neuropsychopharmacol Biol Psychiatry, 79*(Pt A), 49–57.
18. Goehler, L. E., Gaykema, R. P., Nguyen, K. T., et al. (1999). Interleukin-1beta in immune cells of the abdominal vagus nerve: a link between the immune and nervous systems? *J Neurosci, 19*(7), 2799–2806.
19. Blalock, J. E. (1984). The immune system as a sensory organ. *J Immunol, 132*(3), 1067–1070.
20. Pavlov, V. A., & Tracey, K. J. (2012). The vagus nerve and the inflammatory reflex—linking immunity and metabolism. *Nat Rev Endocrinol, 8*(12), 743–754.
21. Tracey, K. J. (2002). The inflammatory reflex. *Nature, 420*(6917), 853–859.
22. Rosas-Ballina, M., Olofsson, P. S., Ochani, M., et al. (2011). Acetylcholine-synthesizing T cells relay neural signals in a vagus nerve circuit. *Science, 334*(6052), 98–101.
23. Watkins, L. R., & Maier, S. F. (2000). The pain of being sick: implications of immune-to-brain communication for understanding pain. *Annu Rev Psychol, 51*, 29–57.
24. Wang, H., Yu, M., Ochani, M., et al. (2003). Nicotinic acetylcholine receptor alpha7 subunit is an essential regulator of inflammation. *Nature, 421*(6921), 384–388.
25. Marin, I. A., & Kipnis, J. (2017). Central nervous system: (immunological) ivory tower or not? *Neuropsychopharmacology, 42*(1), 28–35.
26. Schwartz, M., Kipnis, J., Rivest, S., & Prat, A. (2013). How do immune cells support and shape the brain in health, disease, and aging? *J Neurosci, 33*(45), 17587–17596.
27. Paolicelli, R. C., Bolasco, G., Pagani, F., et al. (2011). Synaptic pruning by microglia is necessary for normal brain development. *Science, 333*(6048), 1456–1458.
28. Frost, J. L., & Schafer, D. P. (2016). Microglia: architects of the developing nervous system. *Trends Cell Biol, 26*(8), 587–597.
29. Bilimoria, P. M., & Stevens, B. (2015). Microglia function during brain development: new insights from animal models. *Brain Res, 1617*, 7–17.
30. Norden, D. M., Muccigrosso, M. M., & Godbout, J. P. (2015). Microglial priming and enhanced reactivity to secondary insult in aging, and traumatic CNS injury, and neurodegenerative disease. *Neuropharmacology, 96*(Pt A), 29–41.

31. Wohleb, E. S., Hanke, M. L., Corona, A. W., et al. (2011). Beta-adrenergic receptor antagonism prevents anxiety-like behavior and microglial reactivity induced by repeated social defeat. *J Neurosci, 31*(17), 6277–6288.
32. Johnson, J. D., Campisi, J., Sharkey, C. M., et al. (2005). Catecholamines mediate stress-induced increases in peripheral and central inflammatory cytokines. *Neuroscience, 135*(4), 1295–1307.
33. Wohleb, E. S., McKim, D. B., Sheridan, J. F., & Godbout, J. P. (2014). Monocyte trafficking to the brain with stress and inflammation: a novel axis of immune-to-brain communication that influences mood and behavior. *Front Neurosci, 8*, 447.
34. Goshen, I., & Yirmiya, R. (2009). Interleukin-1 (IL-1): a central regulator of stress responses. *Front Neuroendocrinol, 30*(1), 30–45.
35. Sawada, A., Niiyama, Y., Ataka, K., Nagaishi, K., Yamakage, M., & Fujimiya, M. (2014). Suppression of bone marrow-derived microglia in the amygdala improves anxiety-like behavior induced by chronic partial sciatic nerve ligation in mice. *Pain, 155*(9), 1762–1772.
36. D'Mello, C., Riazi, K., Le, T., et al. (2013). P-selectin-mediated monocyte-cerebral endothelium adhesive interactions link peripheral organ inflammation to sickness behaviors. *J Neurosci, 33*(37), 14878–14888.
37. Degos, V., Vacas, S., Han, Z., et al. (2013). Depletion of bone marrow-derived macrophages perturbs the innate immune response to surgery and reduces postoperative memory dysfunction. *Anesthesiology, 118*(3), 527–536.
38. Wohleb, E. S., McKim, D. B., Shea, D. T., et al. (2014). Re-establishment of anxiety in stress-sensitized mice is caused by monocyte trafficking from the spleen to the brain. *Biol Psychiatry, 75*(12), 970–981.
39. Torres-Platas, S. G., Cruceanu, C., Chen, G. G., Turecki, G., & Mechawar, N. (2014). Evidence for increased microglial priming and macrophage recruitment in the dorsal anterior cingulate white matter of depressed suicides. *Brain Behav Immun, 42*, 50–59.
40. Seabrook, T. J., Johnston, M., & Hay, J. B. (1998). Cerebral spinal fluid lymphocytes are part of the normal recirculating lymphocyte pool. *J Neuroimmunol, 91*(1–2), 100–107.
41. Prinz, M., & Priller, J. (2017). The role of peripheral immune cells in the CNS in steady state and disease. *Nat Neurosci, 20*(2), 136–144.
42. Kivisakk, P., Tucky, B., Wei, T., Campbell, J. J., & Ransohoff, R. M. (2006). Human cerebrospinal fluid contains CD4+ memory T cells expressing gut- or skin-specific trafficking determinants: relevance for immunotherapy. *BMC Immunol, 7*, 14.
43. Ziv, Y., Ron, N., Butovsky, O., et al. (2006). Immune cells contribute to the maintenance of neurogenesis and spatial learning abilities in adulthood. *Nat Neurosci, 9*(2), 268–275.
44. Kipnis, J., Gadani, S., & Derecki, N. C. (2012). Pro-cognitive properties of T cells. *Nat Rev Immunol, 12*(9), 663–669.
45. Baruch, K., Ron-Harel, N., Gal, H., et al. (2013). CNS-specific immunity at the choroid plexus shifts toward destructive Th2 inflammation in brain aging. *Proc Natl Acad Sci U S A, 110*(6), 2264–2269.
46. Derecki, N. C., Cardani, A. N., Yang, C. H., et al. (2010). Regulation of learning and memory by meningeal immunity: a key role for IL-4. *J Exp Med, 207*(5), 1067–1080.

47. Grassi-Oliveira, R., Stein, L. M., Lopes, R. P., Teixeira, A. L., & Bauer, M. E. (2008). Low plasma brain-derived neurotrophic factor and childhood physical neglect are associated with verbal memory impairment in major depression—a preliminary report. *Biol Psychiatry, 64*(4), 281–285.
48. Lewitus, G. M., Cohen, H., & Schwartz, M. (2008). Reducing post-traumatic anxiety by immunization. *Brain Behav Immun, 22*(7), 1108–1114.
49. Lewitus, G. M., & Schwartz, M. (2009). Behavioral immunization: immunity to self-antigens contributes to psychological stress resilience. *Mol Psychiatry, 14*(5), 532–536.
50. Cohen, H., Ziv, Y., Cardon, M., et al. (2006). Maladaptation to mental stress mitigated by the adaptive immune system via depletion of naturally occurring regulatory CD4+CD25+ cells. *J Neurobiol, 66*(6), 552–563.
51. Kipnis, J., Mizrahi, T., Hauben, E., Shaked, I., Shevach, E., & Schwartz, M. (2002). Neuroprotective autoimmunity: naturally occurring CD4+CD25+ regulatory T cells suppress the ability to withstand injury to the central nervous system. *Proc Natl Acad Sci U S A, 99*(24), 15620–15625.
52. Avidan, H., Kipnis, J., Butovsky, O., Caspi, R. R., & Schwartz, M. (2004). Vaccination with autoantigen protects against aggregated beta-amyloid and glutamate toxicity by controlling microglia: effect of CD4+CD25+ T cells. *Eur J Immunol, 34*(12), 3434–3445.
53. Baruch, K., Rosenzweig, N., Kertser, A., et al. (2015). Breaking immune tolerance by targeting Foxp3 (+) regulatory T cells mitigates Alzheimer's disease pathology. *Nat Commun, 6*, 7967.
54. Baruch, K., Deczkowska, A., Rosenzweig, N., et al. (2016). PD-1 immune checkpoint blockade reduces pathology and improves memory in mouse models of Alzheimer's disease. *Nat Med, 22*(2), 135–137.
55. Kim, S. J., Lee, H., Lee, G., et al. (2012). CD4+CD25+ regulatory T cell depletion modulates anxiety and depression-like behaviors in mice. *PLoS One, 7*(7), e42054.
56. Baruch, K., Deczkowska, A., David, E., et al. (2014). Aging: aging-induced type I interferon response at the choroid plexus negatively affects brain function. *Science, 346*(6205), 89–93.
57. Filiano, A. J., Xu, Y., Tustison, N. J., et al. (2016). Unexpected role of interferon-gamma in regulating neuronal connectivity and social behaviour. *Nature, 535*(7612), 425–429.

4

Immune Mechanisms and Central Nervous System (CNS) Development

ANA CRISTINA SIMÕES E SILVA, JANAINA MATOS MOREIRA, RAFAEL COELHO MAGALHÃES, AND CRISTIAN PATRICK ZENI ■

INTRODUCTION

Embryo and fetus growth require time and an orchestrated process comprising multiple mechanisms and factors that influence healthy development during the gestation period.[1,2] At this stage, the placenta provides the embryo/fetus with growth factors needed for normal body and brain development.[3] Immune cells also play a crucial role in regulating the maternal–fetal interface during pregnancy.[3]

Changes in the morphological and histopathological structure of the placenta and the inflammatory response secondary, for instance, to prenatal infection may result in altered placental function.[4] When challenged with inflammatory stimuli, the placental barrier function is weakened, allowing immune cells to penetrate in fetus brain parenchyma.[5] Concurrently, increased microglial and astrocyte activation, impairment in axonal formation, and macrophage infiltration in the fetal brain are related to the elevation of cytokine levels in the placenta and the fetal membranes, which are risk factors for impaired future neurodevelopment.[4]

Several mechanisms related to the immune system, including the release of cytokines and activation of T cells, may contribute to neuroinflammation, which, in turn, potentially interferes with CNS development.[5] According to this theory, systemic and CNS inflammation alter neuronal and microglial cells' proliferation and differentiation, and may also increase cell death rate via neuronal damage, astrogliosis, and oligodendrocytes loss.[6,7] The possible mechanisms by which inflammation promotes early brain lesion are:

1. reduced blood-flow to the CNS, which reduces oxygen and glucose availability;
2. BBB rupture;
3. leukocyte infiltration into the CNS;
4. increased cytokines and chemokines released in the cerebral parenchyma;
5. mitochondrial dysfunction and energy failure;
6. increased calcium influx, neurotoxins release, oxygen and nitric oxygen free radicals formation; and
7. brain edema.

All these mechanisms have been associated with neuronal and glial cells apoptosis.[8,9] Thus, neuroinflammation is thought to be one of the main factors involved in neurodevelopmental impairment.[5]

In this chapter, we will discuss the role of immune processes in CNS development as well as maternal immune system alterations affecting fetal CNS development, and the link between immune system changes, abnormal CNS development, and neuropsychiatric diseases affecting young populations (children and adolescents).

ROLE OF THE IMMUNE SYSTEM IN THE PHYSIOLOGICAL CNS DEVELOPMENT

Brain development involves a complex sequence of plastic changes occurring over lifetime.[1] CNS development depends on age, variations in metabolic rate, blood flow, neurotransmitter activity, and capacity to tolerate oxidative stress. We will present the data showing that the same developmental mechanisms that promote plastic changes may expose the organism to adverse outcomes, once the products and residuals of the development trigger responses from the organism in a loop system.[10]

The nervous and immune systems have a bidirectional communication facilitated by the brain lymphatic system and the BBB.[11,12] The first one connects peripheral lymphatic tissues to the CNS,[2] and the second regulates endothelial passage of immune cells and associated factors.[11,12]

The brain harbors resident immune cells that protect it against infection and injury, while they also support neurons in remodeling circuit connectivity and plasticity.[11,13] Cytokines, chemokines, and other bioactive molecules exert several functions related to the peripheral immune system. These molecules are also present in the developing brain and recently have been associated with neurodevelopmental outcomes.[11]

Inflammation is a physiological response that mediates the clearance of cellular debris, releases neurotrophic factors, and eludes or minimizes threats.[14–17] By means of microglial activity, inflammation plays a crucial role in CNS protection and restoration.[14–17] Upon stimulation, microglia are activated and assume different morphological states (amoeboid). Microglia also release bioactive molecules, contributing to the activation inflammatory responses, thus combating foreign invasion at damaged sites in order to reestablish CNS homeostasis.[18] Microglia are also essential key components in synaptic plasticity and neurogenesis.[12,13]

Microglial development starts from erythromyeloid progenitors in the yolk sac and continues in the brain during early embryonic stages.[13,19] These cells migrate and spread throughout the CNS, and morphologically evolve to a ramified state, which correlates with brain maturation.[20] Microglia exert many important functions during neurodevelopment, including homeostasis and control of inflammatory conditions within the CNS,[21,22] contributing to the maintenance of CNS functions at a steady state through dynamic interactions with astrocytes and neurons.[21] Considering these dynamic interactions, microglial activation can influence synaptic plasticity and ultimately impact higher functions.[21]

Microglial cells also have multifaceted roles beyond the classical immune response, including guidance of neuronal proliferation, differentiation, and regulation of the formation or elimination of synapses.[13,21] Additionally, these cells control oligodendrocyte differentiation, immune quiescence, clearance of cellular debris, and uptake of glutamate and other neurotransmitters.[19] In this way, microglia regulate the spatial development of neuronal tracts and appear to depend on several chemical signals.[13,21] The actions of microglial cells occur mainly through the release of specific mediators, such as cytokines and growth factors.[23]

The maturation of neurons, astrocytes, and oligodendrocytes occurs concomitantly with microglia development.[20] The maturation process is responsible for leading to dynamic changes in the CNS.[13] In summary, immune system cells, notably microglial cells, and immune mediators such as cytokines and chemokines are critical for normal CNS development.[20]

MATERNAL IMMUNE SYSTEM ALTERATIONS AND COMPROMISED FETAL CNS DEVELOPMENT

Experimental and clinical studies support a role for alterations in maternal immune system on impairment in fetal CNS development.

Evidence from Animal Models

The CNS is characterized by a specific structure and function.[10,11,24] Despite a unique system of brain barriers and autonomous immune system, the CNS is very susceptible to infection or inflammation.[24] Abnormal activation of the immune system during gestational period and early life can trigger exacerbated inflammation with increased release of pro-inflammatory molecules.[25]

The gestational period is critical to CNS development.[1] The placenta provides the embryo/fetus an adequate amount of nutrients and growth factors needed for normal body and brain development.[3,26] Changes in placental function or even placenta dysfunction have been associated with many antenatal conditions, including pre-eclampsia and maternal infection.[3,4,15,20] These conditions are considered risk factors for *in-utero* inflammation, including increased cytokine levels and subsequent CNS damage.[3,4,15,20]

Inflammation and infection during pregnancy and the perinatal period may severely affect CNS development, with short-term and long-term structural and functional consequences.[27,28] Changes in synaptic transmission obtained in adult animals submitted to inflammatory stimuli are similar to those observed in prenatal or perinatal inflammation.[25]

Experimental studies have shown that peripheral infection or administration of lipopolysaccharides (LPS) can intensively stimulate innate immune response, not only in peripheral organs, but also in the CNS of rodents.[25,29] In mice models of systemic inflammation, peripheral circulating exosomes contributed to neuroinflammation.[29] These exosomes cross the BBB, reaching the brain parenchymal cells, and increase neuroinflammation and exosomal uptake by the neural cells.[29] Increased expression of inflammatory microRNAs in serum-derived exosomes may play a role in regulating the CNS immune response and in inducing microgliosis and astrogliosis.[29]

Cellular and molecular mechanisms of inflammation that are related to brain injury involve pro-inflammatory cytokine and chemokine signaling,[1,25] and immune-mediated disruption of the BBB.[12] Inflammatory response triggered by infectious pathogens activates IL-1R-like receptors (ILRs)[25] and Toll-like receptors (TLR), resulting in activation of nuclear factor kappa-B (NF-κB)[30] with production and release of pro-inflammatory cytokines.[1,12] These molecules can cross the damaged BBB into the brain where they in turn activate microglial cells, the primary defense mechanism in the CNS.[1,12,24]

Signals released by microglia recruit hematopoietic cells, such as monocytes and macrophages, to exert their biological function after injury takes place.[31,32] Evidence from inflammation models reveals that activated microglia trigger a cascade of events that ultimately lead to an increase in pro-inflammatory cytokines, such as IFN-γ IL-1β, IL-6, and TNF.[1,25] These cytokines stimulate the immune system to perpetuate the inflammatory response and to exert additional functions on inflammatory control in the CNS.[33]

The increased levels of immune mediators in the injured brain may contribute to cell damage by means of the release of NO, reactive oxidative species (ROS), and glutamate from immune cells, including brain-resident microglia and infiltrating leukocytes.[14] An impaired intrinsic antioxidant defense is also observed.[14] Excitotoxicity is another common mechanism of secondary brain injury.[31] However, while neurons are highly vulnerable to excitotoxicity, astrocytes play key roles in the reuptake of glutamate from synapses, preventing excessive extracellular glutamate accumulation.[31] It has been demonstrated that the blockade of glutamate transporters (EAAT1 and EAAT2) on astrocytes by administering antisense oligonucleotides in rats resulted in increased extracellular glutamate concentration, leading to excitotoxicity and neuronal damage.[31]

In experimental conditions that resulted in neurotoxicity with leukocyte infiltration into the brain parenchyma, the infiltrated peripheral myeloid cells expressed higher levels of pro-inflammatory cytokines than resident glial cells.[22] These responses may represent a more severe type of CNS inflammation and brain injury,[22] resulting in glial scar.[24,31] Glial scar is a post-injury repair response and consists largely of astrocytes, along with microglia, endothelial cells, fibroblasts, and extracellular matrix.[24,31] Knock-out mice studies have demonstrated a role of several mediators, including IFN-γ, Fibroblast growth factor (FGF), Glial Fibrillary Acidic Protein (GFAP), transforming growth factor (TGF)-β1, TGF-2, and fibrinogen, in proper glial scar formation.[31] Glial scar produces a physical barrier to prevent toxic molecules from leaking out into healthy tissue and to prevent access of invading cells after injury.[31]

As the inflammatory response may exert effects at the periphery and CNS, leading to brain tissue damage,[32] animal models may serve as reliable and reproducible systems for elucidation of the molecular processes involved in inflammation and the consequences for synapse structure and CNS function.[25,34] Table 4.1 summarizes findings of animal models that investigated the role of inflammation on neurological outcome.

Table 4.1 Findings Provided by Animal Models Investigating the Role of Inflammation on Neurological Outcome

Author	Models	Laboratory Technique	Inflammatory Markers	Neurological Outcome
Li et al., 2018[29]	Mice	Immunofluorescent staining, RT-qPCR, Western blot, and ELISA	LPS	Peripheral circulating exosomes contribute to neuroinflammation in conditions with systemic inflammation. Microglial cells and astrocytes in mouse brain are activated in a dose- and time- dependent manner by peripheral LPS.
Kelley et al., 2017[1]	Mice	Immunohistochemistry and electrophysiology	LPS	Pregnant mice exposed to LPS displayed altered hippocampal excitatory synaptic function. Synaptic transmission at CA3–CA1 synapses was increased due in part to an increase in the probability of glutamate release from the presynaptic CA3 axon terminals.
Lei et al., 2017[5]	Mice	RT-qPCR, immunohistochemistry, cytometry, MS FX PRO *in vivo* imager	LPS	The acute maternal immune response leads to placental infiltration of peripheral immune cells, suggesting that compromise of the maternal–fetal barrier occurs early in the course of an immune response.
Tomasoni et al., 2017[25]	KO mice	Immunofluorescent staining on free-floating sections and *in vitro* experiments	IL-1β, IL-6, IL-9, IL-10, IL-1R8, IFN-γ, CXCR4, TGF- β	During development, immune molecules such as CXCR4, IFN-γ, IL-1β, IL-6, IL-9, IL-10, and TGF-β affect neurogenesis, neuronal migration, axon guidance, synapse formation, activity-dependent refinement of circuits, and synaptic plasticity.

Zhao et al., 2015[30]	Mice	Immunofluorescent staining, RT-qPCR, and Western blot	IL1β, TNFα, CCL2, cinnamaldehyde, TLR4, IRAK, IL1 receptor associated kinase, MAPK, MPO, myeloperoxidase	Brain ischemic injury is undoubtedly associated with the expression of inflammatory mediators such as inflammatory cytokines, chemokines, and adhesion molecules. These mediators stimulate the expression of adhesion molecules on leukocytes and endothelial cells and cause the adherence and extravasation of leukocytes into brain parenchyma.
Nadeau-Vallée et al., 2013[34]	Mice	—	IL-1, IL-1β, IL-6, IL-8, IL-10, Pghs2, TNFα, and CRP	IL-1 is an important driver of neonatal morbidity in PTB. In a LPS-induced model of infection-associated PTB prevented PTB, neonatal mortality, and fetal brain inflammation.

IL = interleukin; IFN-γ = interferon; TNF-α = tumor necrosis factor-alpha; LPS = lipopolysaccharides; TGF- β = transforming growth factor β; CCL/CXCL = chemokines; CRP = C-reactive protein; ELISA = enzyme-linked immunosorbent assay; PCR = polymerase chain reaction; PTB = preterm birth.

Evidence from Human Studies—Abnormal CNS Development and Neuropsychiatric Disorders in Youth

The pathophysiology of neurodevelopmental and psychiatric disorders involves a complex cascade of events, which includes genetic and environmental factors. In parallel with animal models, evidence from population surveys strengthens the link between maternal immune activation or dysfunction and the later occurrence of altered neurodevelopment and neuropsychiatric disorders.[35–37]

Not only are infectious conditions typically involved in the immune system activation of pregnant women, consequently altering the neuroimmune cross-talk, but other non-infectious maternal conditions such as autoimmune diseases, stress-related disorders, substance use, and malnutrition have also been implicated in altered developmental processes.[35,36,38] This is also related to epigenetic mechanisms and may vary according to the type, intensity, and/or timing of exposure to risk factors (Figure 4.1).

Congenital infections have long been associated with CNS abnormalities with variable degrees of severity, including fetal death and major CNS malformation. In addition, later neurodevelopmental consequences may occur, such as impaired neurodevelopment and psychiatric disorders.[39] Classic TORCH (*toxoplasmosis, rubella, cytomegalovirus*, and herpes infections, as well as syphilis, *varicella-zoster*, and *parvovirus B19*) and, more recently, influenza and Zika viruses are infectious

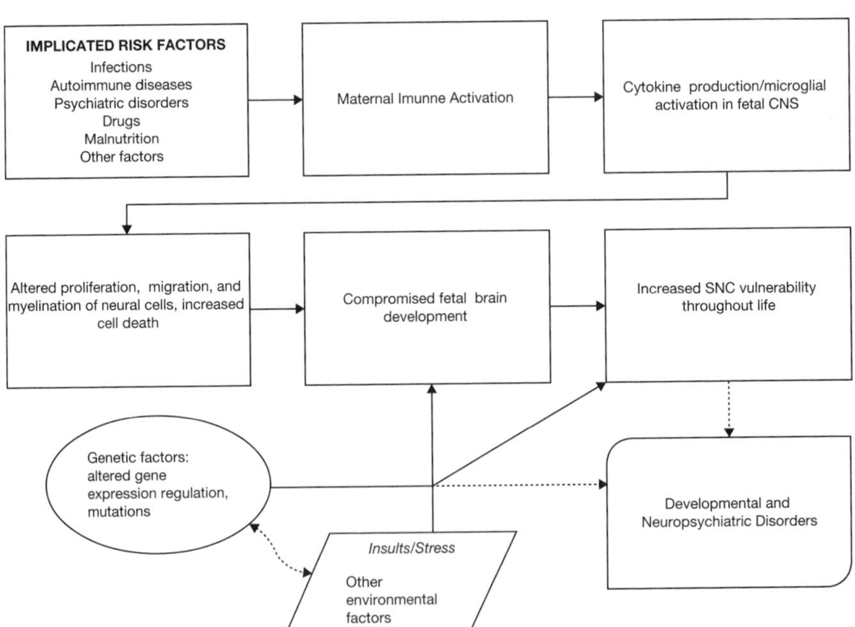

Figure 4.1 The pathophysiology of neurodevelopmental and psychiatric disorders.

agents typically associated with neurological complications.[36,38–40] Patients with major psychiatric disorders also frequently exhibit a history of prenatal and/or neonatal infections.[38] Additionally, early studies suggested that winter/spring births (marked by higher incidence of perinatal infections) are associated with an increase in schizophrenia risk, corroborating a role played by perinatal infection in its pathogenesis.[38] Maternal exposure to the influenza virus was reported to increase the risk of adult bipolar disorder,[41] and the risk, although modest, was later confirmed for bipolar disorder, but not for schizophrenia.[42]

Evidence from animal models and from epidemiological studies corroborates the hypothesis of congenital and/or maternal infections leading to maternal immune activation (MIA), disrupted fetal neurogenesis, and later development of neuropsychiatric morbidity. However, there is no evidence to support the role for a specific infectious agent in the pathogenesis of determined disorder. On the contrary, MIA and, subsequently, fetal CNS inflammation might be the final common pathway by which neural embryogenesis, networks, and ultimately functions are significantly or even definitively impaired by different factors, including infection.[35,43]

Among more recent studies, a 25-year retrospective epidemiological study, in Denmark, focusing on autism-related disorders (ARDs), showed an association between viral infections at the first trimester and maternal hospitalization during the second trimester with a higher likelihood of a diagnosis of ARDs in the offspring.[44] The second-trimester infections and hospitalizations were mostly related to bacteria; however, the site of infection did not influence the outcome. In a posterior analysis, maternal influenza infection, prolonged episodes of fever, and the use of various antibiotics during pregnancy were also associated with ARDs.[45] A large case–control study, part of the Childhood Autism Risk from Genetics and Environment (CHARGE) Study, including 538 children with ASD, 163 with developmental delays (DD), and 421 typically developing controls, found an association between maternal fever and risk for ASD and developmental delays, but this was not specific for influenza infection.[46]

Findings of inflammatory biomarkers later in life support this theory of the close relationship between MIA, CNS development and neuropsychiatric disorders in youth and across the lifespan. Autism is one of the most studied neuropsychiatric disorders from an immune/inflammatory perspective in pediatric patients. Studies share as common findings elevated IFN-g in plasma, supernatant of whole-blood cultures, and frozen brain tissue in youth with autism, compared to typically developing controls.[47–50] Those findings are not conclusive since other studies did not detect the same results. Findings for IL-12 and IFN-α were also contradictory.[51] IL-2 and IL-2R, IL-1b, and IL-13 do not seem to differentiate between subjects with autism and controls.[52–54] The evidence for TNF-a implication is also inconsistent. Many studies have examined levels of IL6, with contradictory findings. IL-6 has been shown to be both increased and decreased in the plasma, and increased in the CSF, lymphoblasts, and postmortem brain tissue.[54,55] IL-10 has been repeatedly found to be lower or similar in subjects with ASD compared to controls across all but one study among 10 pertinent studies, regardless of methodology and despite elevated pro-inflammatory mediators (PIMs).[48,49,52,53,56–62] Findings for other anti-inflammatory cytokines, such as IL-4 and IL5, were less consistent, with levels increased peripherally in the plasma, intracellularly.[52,63] IL-17, in two studies, showed increased levels peripherally in subjects with ASD.[64,65] IL-8 was elevated *in vitro*, as well as in the plasma, CSF, and postmortem brain tissue of patients with ASD, but not in lymphoblasts.[49,53,54,56,65] These diverse findings might echo the heterogeneity of ASDs, as diverse patients are grouped within the same diagnosis. In a meta-analysis (2015), authors concluded that patients with ARDs had significantly higher levels of IL-1β, IL-6, IL-8, INF-gama, eotaxin, and monocyte chemotactic protein-1 (MCP-1) in comparison to controls, while concentrations of transforming growth factor-β1 (TGF-β1) were lower.[66] Recent reviews from the same group and from other authors also support the central role of inflammation in autism,[67,68] including perspectives on treatment by using anti-inflammatory agents and cytokine inhibitors to modulate behavior symptoms, such as irritability and hyperactivity.[69]

In youth with mood disorders (major depressive disorder, dysthymia, bipolar disorder), the current available literature suggests elevation of inflammatory markers. Significantly elevated levels of IFN-g were detected in adolescents with major depressive disorder (MDD), and also in those with active suicidal ideation.[70] In another investigation, significantly elevated IL-1b, IL-2, and IL-10 were found in female youth with MDD and/or anxiety disorders. Additionally, IL-6 was elevated in the unmedicated group, when compared to controls.[71] Brambilla et al. reported elevated IL-1b in youth with dysthymia.[72] Findings regarding TNF-alpha levels suggest no differences between adolescents with MDD or controls,[71,72] but a decrease in TNF was observed in adolescents with dysthymia,[72] and also in those with MDD and suicidal intent.[70] In children and adolescents with bipolar disorder, one small study detected that high-sensitivity C reactive protein (CRP) (hsCRP), but not IL-6, was associated with manic symptom severity.[73] In a systematic review regarding depressive symptoms and inflammatory markers, the authors concluded that there was an altered pattern of circulating cytokines, especially IL1-β and TNF. However, the authors concluded that studies are heterogeneous and the relationship between psychiatric symptoms, environmental stressors, and cytokines should be further clarified.[74] Another systematic review on depression, suicide risk, and inflammation in pediatric patients obtained similar conclusions.[75] Recently, one study evaluated adolescents with bipolar disorder (BD) (n = 18), MDD (n = 13), or no psychiatric history ($n = 20$). Adolescents with BD had significantly higher spontaneous levels of NF-κB in peripheral blood mononuclear cells, and monocyte and lymphocyte populations, and higher plasma levels of IL-1β than healthy controls. Following stimulation with recombinant human TNF-α, participants with BD and MDD both had greater increases in NF-κB in monocytes than controls. Furthermore, greater stimulated increases of NF-κB in monocytes were associated with the current severity of depressive symptoms.[76]

In youth with obsessive-compulsive disorder (OCD) and/or Tourette's disorder (TD) or tic disorder, a common finding is the increase in inflammatory markers overall; two studies report increased plasma and serum IL-12, respectively, in comorbid TD and OCD. Only one of them additionally reported increased IL-12 in tic disorders alone. It is difficult to compare results between the available studies, since each group measured different markers.[77] Cheng et al.[78] reported that youth with tic disorders presented higher plasma IL-1, IL-6, and IL-17 than controls. Gabbay et al.[70] have replicated the IL-6 finding. Findings regarding TNF-a are inconsistent. A prospective, longitudinal study of patients with OCD and/or TD found no significant difference in hsCRP as compared to controls at baseline, during symptom exacerbation, or in follow-up.[70]

In a study of children and adolescents who were exposed to a motor vehicle accident, elevated IL-6 in the morning after the accident predicted development of post-traumatic stress disorder after six months.[79]

Regarding attention-deficit/hyperactivity disorder (ADHD), a recent review has been published on this topic.[80] Results on peripheral cytokine levels suggested a low-grade inflammation in patients with ADHD. Lower levels of IFN-γ and IL-13 were observed in medicated patients with ADHD when compared with medication-naïve patients, indicating the effect of treatment and/or clinical improvement on cytokine levels.[82]

In children and adolescents with schizophrenia and/or acute psychosis, a study reveals TNF was not elevated as compared to controls before or after treatment with clozapine; an imbalance of anti-inflammatory cytokines was detected in the CSF of children with schizophrenia, when compared to youth with ADHD.[82] In a study with 64 suicidal adolescents with acute psychosis and/or a mood disorder, elevated serum IL-1b and IL-8 were detected, in the comparison with controls.[83]

CONCLUSION

Maternal-fetal immune system interactions and bidirectional communication between immune system and CNS are important to brain development.[10–12] These interactions involve increased serum levels of pro-inflammatory cytokines and markers of endothelial cell activation.[84]

Evidence suggests that immune dysfunction may affect brain development and may play a role in neurodevelopmental/psychiatric disorders.[37,85] A single developmental insult can initiate a cascade of alterations that may not be detected structurally or functionally until later in life.[66] Thus, these effects may be manifested in the long term from the critical developmental window.[85] Up to now, the pathophysiological mechanisms are still not completely understood. Therefore, it is necessary to make joint efforts from several disciplines, including immunobiology, neurology, and psychiatry, to obtain a clear understanding of the role of immune mechanisms in CNS development.

REFERENCES

1. Kelley, M. H., Wu, W. W., Lei, J., et al. (2017) Functional changes of hippocampal synaptic signaling in offspring survivors of a mouse model of intrauterine inflammation. *J Neuroinflammation, 14*, 180.
2. Talati, A. N., Hackney, D. N., & Mesiano, S. (2017). Pathophysiology of preterm labor with intact membranes. *Semin Perinatol, 41*(7), 420–426.
3. Leviton, A., Allred, E. N., Yamamoto, H., et al. (2017). Antecedents and correlates of blood concentrations of neurotrophic growth factors in very preterm newborns. *Cytokine, 94*, 21–28.
4. Hodyl, N. A., Aboustate, N., Bianco-Miotto, T., et al. (2017). Child neurodevelopmental outcomes following preterm and term birth: what can the placenta tell us? *Placenta, 57*, 79–86.
5. Lei, J., Rosenzweig, J. M., Mishra, M. K., et al. (2017). Maternal dendrimer-based therapy for inflammation-induced preterm birth and perinatal brain injury. *Sci Rep,* 7(1), 6106.
6. Stewart, A., Tekes, A., Huisman, T. A., et al. (2013). Glial fibrillary acidic protein as a biomarker for periventricular white matter injury. *Am J Obstet Gynecol, 209*(1), 27.e1–e7.
7. Hielkema, T., & Hadders-Algra, M. (2016). Motor and cognitive outcome after specific early lesions of the brain—a systematic review. *Dev Med Child Neurol,* 58(4), 46–52.
8. Berger, I., Peleg, O., & Ofek-Shlomai, N. (2012). Inflammation and early brain injury in term and preterm infants. *Isr Med Assoc J, 14*(5), 318–323.
9. Marc, T. (2013). Brain development and the immune system: an introduction to inflammatory and infectious diseases of the child's brain. *Handb Clin Neurol, 112,* 1087–1089.
10. Dennis, M., Spiegler, B. J., Juranek, J. J., Bigler, E.d., Snead, O. C., & Fletcher, J. M. (2013). Age, plasticity, and homeostasis in childhood brain disorders. *Neurosci Biobehav Rev, 37*(10 p t2), 2760–7273.
11. Fung, T. C., Olson, C. A., & Hsiao, E. Y. (2017). Interactions between the microbiota, immune and nervous systems in health and disease. *Nat Neurosci, 20*(2), 145–155.

12. Banks, W. A. (2015). The blood-brain barrier in neuroimmunology: tales of separation and assimilation. *Brain Behav Immun, 44,* 1–8.
13. Burke, N. N., Fan, C. Y., & Trang, T. (2016). Microglial in health and pain: impact of noxious early life events. *Exp Physiol, 101*(8), 1003–1021.
14. Woodcock, T. M., Frugier, T., Nguyen, T. T., et al. (2017). The scavenging chemokine receptor ACKR2 has a significant impact on acute mortality rate and early lesion development after traumatic brain injury. *PLoS One, 12*(11), e0188305.
15. Magalhães, R. C., Pimenta, L. P., Barboza, I. G. et al. (2017). Inflammatory molecules and neurotrophic factors as biomarkers of neuropsychomotor development in preterm neonates: a systematic review. *Int J Dev Neurosci, 65,* 29–37.
16. Nadeau-Vallée, M., Chin, P. Y., Belarbi, L., et al. (2017). Antenatal suppression of IL-1 protects against inflammation-induced fetal injury and improves neonatal and developmental outcomes in mice. *J Immunol, 198*(5), 2047–2062.
17. Magalhães, R. C., Moreira, J. M., Vieira, É. L.M., Rocha, N. P., Miranda, D. M., & Simões E Silva, A. C. (2017). Urinary levels of IL-1β and GDNF in preterm neonates as potential biomarkers of motor development: a prospective study. *Mediat Inflamm, 2017,* 8201423.
18. Han, J., Harris, R. A., & Zhang, X. M. (2017). An update assessment of microglia depletion: current concepts and future directions. *Mol Brain, 10*(1), 25–32.
19. Wong, K., Noubade, R., Manzanillo, P., et al. (2017). Mice deficient in NRROS show abnormal microglial development and neurological disorders. *Nat Immunol, 18*(6), 633–641.
20. Hanamsagar, R., & Bilbo, S. D. (2017) Environment matters: microglia function and dysfunction in a changing world. *Curr Opin Neurobiol, 47,* 146–155.
21. Ulland, T. K., Wang, Y., & Colonna, M. (2015). Regulation of microglial survival and proliferation in health and diseases. *Semin Immunol, 27*(6), 410–415.
22. Liu, X., & Quan, N. (2018). Microglia and CNS interleukin-1: beyond immunological concepts. *Front Neurol, 9,* 8.
23. Mooser, C. A., Baptista, S., Arnoux, I., & Audinat, E. (2017). Microglia in CNS development: shaping the brain for the future. *Prog Neurobiol, 149–150,* 1–20.
24. Rzaska, M., Niewiadomski, S., & Karwacki, Z. (2017). Molecular mechanisms of bacterial infections of the central nervous system. *Anaesthesiol Intens Ther, 49*(5), 387–392.
25. Tomasoni, R., Morini, R., & Lopez-Atalaya, J. P., et al. (2017). Lack of IL-1R8 in neurons causes hyperactivation of IL-1 receptor pathway and induces MECP2-dependent synaptic defects. *Elife, 28,* 6 pii:e21735.
26. Leviton, A., Ryan, S., Allred, E. N., et al. (2017). Antecedents and early correlates of high and low concentrations of angiogenic proteins in extremely preterm newborns. *Clin Chim Acta, 471,* 1–5.
27. Plant, D. T., Pawlby, S., Sharp, D., Zunszain, P. A., & Pariante, C. M. (2016). Prenatal maternal depression is associated with offspring inflammation at 25 years: a prospective longitudinal cohort study. *Transl Psychiatry, 6*(11), e936.
28. Hagberg, H., & Mallard, C. (2005). Effect of inflammation on central nervous system development and vulnerability. *Curr Opin Neurol, 18*(2),117–123.

29. Li, J. J., Wang, B., Kodali, M. C., et al. (2018). In vivo evidence for the contribution of peripheral circulating inflammatory exosomes to neuroinflammation. *J Neuroinflammation, 15*, 8.
30. Zhao, J., Zhang, X., Dong, L., et al. (2015). Cinnamaldehyde inhibits inflammation and brain damage in a mouse model of permanent cerebral ischaemia. *Br J Pharmacol, 172*(20), 5009–5023.
31. Karve, I. P., Taylor, J. M., & Crack, P. J. (2016). The contribution of astrocytes and microglia to traumatic brain injury. *Br J Pharmacol, 173*(4), 692–702.
32. Lénárt, N., Brough, D., & Dénes, Á. (2016). Inflammasomes link vascular disease with neuroinflammation and brain disorders. *J Cereb Blood Flow Metab, 36*(10), 1668–1685.
33. Coronel-Restrepo, N., Posso-Osorio, I., Naranjo-Escobar, J., & Tobón, G. J. (2017). Autoimmune diseases and their relation with immunological, neurological and endocrinological axes. *Autoimmun Rev, 16*(7), 684–692.
34. Akimoto, N., Ifuku, M., Mori, Y., & Noda, M. (2013). Effects of chemokines (C-C motif) ligand 1 on microglial function. *Biochem Biophys Res Commun, 436*(3), 455–461.
35. Scola, G., & Duong, A. (2017). Prenatal maternal immune activation and brain development with relevance to psychiatric disorders. *Neuroscience, 346*, 403–408.
36. Hagberg, H., Gressens, P., & Mallard, C. (2012). Inflammation during fetal and neonatal life: implications for neurologic and neuropsychiatric disease in children and adults. *Ann Neurol, 71*(4), 444–457.
37. Marques, A. H., Bjørke-Monsen, A. L., Teixeira, A. L., & Silverman, M. N. (2015). Maternal stress, nutrition and physical activity: impact on immune function, CNS development and psychopathology. *Brain Res, 1617*, 28–46.
38. Brown, A. S. (2012). Epidemiologic studies of exposure to prenatal infection and risk of schizophrenia and autism. *Dev Neurobiol, 72*(10), 1272–1276.
39. Zhao, J., Chen, Y., Xu, Y., & Pi, G. (2013). Effect of intrauterine infection on brain development and injury. *Int J Dev Neurosci, 31*(7), 543–549.
40. Simões E Silva, A. C., Moreira, J. M., Romanelli, R. M., & Teixeira, A. L. (2016). Zika virus challenges for neuropsychiatry. *Neuropsychiatr Dis Treat, 12*, 1747–1760.
41. Parboosing, R., Bao, Y., Shen, L., Schaefer, C. A., & Brown, A. S. (2013). Gestational influenza and bipolar disorder in adult offspring. *JAMA Psychiatry, 70*, 677–685.
42. Selten, J. P., & Termorshuizen, F. (2017). The serological evidence for maternal influenza as risk factor for psychosis in offspring is insufficient: critical review and meta-analysis. *Schizophr Res, 183*, 2–9.
43. Prata, J., Santos, S. G., Almeida, M. I., Coelho, R., & Barbosa, M. A. (2017). Bridging autism spectrum disorders and schizophrenia through inflammation and biomarkers—pre-clinical and clinical investigations. *J Neuroinflammation, 14*(1), 179.
44. Atladottir, H. O., Thorsen, P., Østergaard, L., et al. (2010). Maternal infection requiring hospitalization during pregnancy and autism spectrum disorders. *J Autism Dev Disord, 40*, 1423–1430.
45. Atladottir, H., Henriksen, T. B., Schendel, D. E., & Parner, E. T. (2012) Autism after infection, febrile episodes, and antibiotic use during pregnancy: an exploratory study. *Pediatrics, 130*, e1447–e1454.

46. Zerbo, O., Iosif, A. M., Walker, C., Ozonoff, S., Hansen, R. L., & Hertz-Picciotto, I. (2013). Is maternal influenza or fever during pregnancy associated with autism or developmental delays? Results from the CHARGE (CHildhood Autism Risks from Genetics and Environment) study. *J Autism Dev Disord, 43,* 25–33.
47. El-Ansary A, & Al-Ayadhi L. (2012). Neuroinflammation in autism spectrum disorders. *J Neuroinflammation, 9,* 265.
48. Croonenberghs, J., Bosmans, E., Deboutte, D., Kenis, G., & Maes, M. (2002). Activation of the inflammatory response system in autism. *Neuropsychobiology, 45,* 1–6.
49. Vargas, D. L., Nascimbene, C., Krishnan, C., Zimmerman, A.W., & Pardo, C. A. (2005). Neuroglial activation and neuroinflammation in the brain of patients with autism. *Ann Neurol, 57,* 67–81.
50. Ashwood, P., Schauer, J., Pessah, I. N., & Van de Water, J. (2009). Preliminary evidence of the in vitro effects of BDE-47 on innate immune responses in children with autism spectrum disorders. *J Neuroimmunol, 208,* 130–135.
51. Singh, V. K. (1996). Plasma increase of interleukin-12 and interferon gamma. Pathological significance in autism. *J Neuroimmunol, 66,* 143–145.
52. Tostes, M. H., Teixeira, H. C., Gattaz, W. F., Brand~ao, M. A., & Raposo, N. R. (2012). Altered neurotrophin, neuropeptide, cytokines and nitric oxide levels in autism. *Pharmacopsychiatry, 45,* 241–243.
53. Ashwood, P., Krakowiak, P., Hertz-Picciotto, I., Hansen, R., Pessah, I. N., & Van de Water, J. (2011a). Altered T cell responses in children with autism. *Brain Behav Immun, 25,* 840–849.
54. Ashwood, P., Krakowiak, P., Hertz-Picciotto, I., Hansen, R., Pessah, I., & Van de Water, J. (2011b). Elevated plasma cytokines in autism spectrum disorders provide evidence of immune dysfunction and are associated with impaired behavioral outcome. *Brain Behav Immun, 25,* 40–45.
55. Manzardo, A. M., Henkhaus, R., Dhillon, S., & Butler, M. G. (2012). Plasma cytokine levels in children with autistic disorder and unrelated siblings. *Int J Dev Neurosci, 30,* 121–127.
56. Malik, M., Sheikh, A. M., Wen, G., Spivack, W., Brown, W. T., & Li, X. (2011). Expression of inflammatory cytokines, Bcl2 and cathepsin D are altered in lymphoblasts of autistic subjects. *Immunobiology, 216,* 80–85.
57. Wei, H., Zou, H., Sheikh, A. M., et al. (2011). IL-6 is increased in the cerebellum of autistic brain and alters neural cell adhesion, migration and synaptic formation. *J Neuroinflammation, 8,* 52.
58. Li, X., Chauhan, A., Sheikh, A. M., et al. (2009). Elevated immune response in the brain of autistic patients. *J Neuroimmunol, 207,* 111–116.
59. Jyonouchi, H., Geng, L., Cushing-Ruby, A., & Quraishi, H. (2008). Impact of innate immunity in a subset of children with autism spectrum disorders: a case control study. *J Neuroinflammation, 5,* 52.
60. Molloy, C. A., Morrow, A. L., Meinzen-Derr, J., et al. (2006). Elevated cytokine levels in children with autism spectrumdisorder. *J Neuroimmunol, 172,* 198–205.
61. Zimmerman, A. W., Jyonouchi, H., Comi, A. M., et al. (2005). Cerebrospinal fluid and serum markers of inflammation in autism. *Pediatr Neurol, 33,* 195–201.
62. Jyonouchi, H., Sun, S., & Le, H. (2001). Proinflammatory and regulatory cytokine production associated with innate and adaptive immune responses in children with

autism spectrum disorders and developmental regression. *J Neuroimmunol, 120,* 170–179.

63. Misener, V. L., Schachar, R., Ickowicz, A., et al. (2004). Replication test for association of the IL-1 receptor antagonist gene, IL1RN, with attention-deficit/hyperactivity disorder. *Neuropsychobiology, 50,* 231–234.
64. Al-Ayadhi, L. Y., & Mostafa, G. A. (2012). Elevated serum levels of interleukin-17A in children with autism. *J Neuroinflammation, 9,* 158.
65. Suzuki, K., Matsuzaki, H., Iwata, K., et al. (2011). Plasma cytokine profiles in subjects with high-functioning autism spectrum disorders. *PLoS One, 6,* e20470.
66. Masi, A., Quintana, D. S., Glozier, N., Lloyd, A. R., Hickie, I. B., & Guastella, A. J. (2015). Cytokine aberrations in autism spectrum disorder: a systematic review and meta-analysis. *Mol Psychiatry, 20*(4), 440–446.
67. Masi, A., Breen, E. J., Alvares, G. A., et al. (2017). Cytokine levels and associations with symptom severity in male and female children with autism spectrum disorder. *Mol Autism, 8,* 63.
68. Masi, A., Glozier, N., Dale, R., & Guastella, A. J. (2017). The immune system, cytokines, and biomarkers in autism spectrum disorder. *Neurosci Bull, 33*(2), 194–204.
69. Gładysz, D., Krzywdzińska, A., & Hozyasz, K. K. (2018). Immune abnormalities in autism spectrum disorder—could they hold promise for causative treatment? *Mol Neurobiol.*
70. Gabbay, V., Klein, R. G., Guttman, L. E., et al. (2009). A preliminary study of cytokines in suicidal and nonsuicidal adolescents with major depression. *J Child Adolesc Psychopharmacol, 19,* 423–430.
71. Henje Blom, E., Lekander, M., Ingvar, M., Asberg, M., Mobarrez, F., & Serlachius, E. (2012). Proinflammatory cytokines are elevated in adolescent females with emotional disorders not treated with SSRIs. *J Affect Disord, 136,* 716–723.
72. Brambilla, F., Monteleone, P., & Maj, M. (2004). Interleukin-1beta and tumor necrosis factor-alpha in children with major depressive disorder or dysthymia. *J Affect Disord, 78,* 273–277.
73. Goldstein, B. I., Collinger, K. A., Lotrich, F., et al. (2011). Preliminary findings regarding proinflammatory markers and brain-derived neurotrophic factor among adolescents with bipolar spectrum disorders. *J Child Adolesc Psychopharmacol, 21,* 479–484.
74. Mills, N. T., Scott, J. G., Wray, N. R., Cohen-Woods, S., & Baune, B. T. (2013). Research review: the role of cytokines in depression in adolescents: a systematic review. *J Child Psychol Psychiatry, 54*(8), 816–835.
75. Kim, J. W., Szigethy, E. M., Melhem, N. M., Saghafi, E. M., & Brent, D. A. (2014). Inflammatory markers and the pathogenesis of pediatric depression and suicide: a systematic review of the literature. *J Clin Psychiatry, 75*(11), 1242–1253.
76. Miklowitz, D. J., Portnoff, L. C., Armstrong, C. C., et al. (2016). Inflammatory cytokines and nuclear factor-kappa B activation in adolescents with bipolar and major depressive disorders. *Psychiatry Res, 241,* 315–322.
77. Leckman, J. F., Katsovich, L., Kawikova, I., et al. (2005). Increased serum levels of interleukin-12 and tumor necrosis factor-alpha in Tourette's syndrome. *Biol Psychiatry, 57,* 667–673.

78. Cheng, Y. H., Zheng, Y., He, F., et al. (2012). Detection of autoantibodies and increased concentrations of interleukins in plasma from patients with Tourette's syndrome. *J Mol Neurosci, 48*, 219–224.
79. Pervanidou, P., Kolaitis, G., Charitaki, S., et al. (2007). Elevated morning serum interleukin (IL)-6 or evening salivary cortisol concentrations predict posttraumatic stress disorder in children and adolescents six months after a motor vehicle accident. *Psychoneuroendocrinology, 32*(8–10), 991–999.
80. Anand, D., Colpo, G. D., Zeni, G., Zeni, C. P., & Teixeira, A. L. (2017). Attention-deficit/hyperactivity disorder and inflammation: what does current knowledge tell us? A systematic review. *Front Psychiatry, 8*, 228.
81. Oades, R. D., Dauvermann, M. R., Schimmelmann, B. G., Schwarz, M. J., & Myint, A. M. (2010). Attention-defcit hyperactivity disorder (ADHD) and glial integrity: S100B, cytokines and kynurenine metabolism—effects of medication. *Behav Brain Funct, 6*, 29.
82. Mittleman, B. B., Castellanos, F. X., Jacobsen, L. K., Rapoport, J. L., Swedo, S. E., & Shearer, G. M. (1997). Cerebrospinal fluid cytokines in pediatric neuropsychiatric disease. *J Immunol, 159*, 2994–2999.
83. Falcone, T., Fazio, V., Lee, C., et al. (2010). Serum S100B: a potential biomarker for suicidality in adolescents? *PLoS One, 5*, e11089.
84. Khandaker, G. M., & Dantzer, R. (2016). Is there a role for immune-to-brain communication in schizophrenia? *Psychopharmacology (Berl), 233*(9), 1559–1573.
85. Miranda, A., Roque, S., Serre-Miranda, C., Pêgo, J. M., & Correia-Pinto, J. (2017). Inflammatory response and long-term behavioral assessment after neonatal CO2-pneumothorax study in a rodent model. *J Pediatr Surg*, pii:S0022.3468(17)30507.9.

5

The Immune System as a Sensor Able to Affect Other Homeostatic Systems

ADRIANA DEL REY AND HUGO BESEDOVSKY ■

INTRODUCTION

It is today well established that the functioning of the immune system can be influenced by the brain and brain-associated neuroendocrine mechanisms. However, regulatory mechanisms are based on an active communication between regulatory centers and the system to be regulated. Although the immune system is permanently in operation, its basal conditions are frequently disturbed by events from inside and outside the organism that need homeostatic regulatory adjustments. Certain oscillatory physiological conditions known to affect immunity, such as circadian rhythms and the female sexual cycle, are programmed. However, during unpredictable daily common situations such as body injury, tissue damage, and infections, non–pre-established adjustments need to be rapidly attained. For example, a redistribution of energy provision towards the immune system would be necessary following an overt immune activation. Indeed, immune responses involve costly energetic processes,[1] such as, for example, immune cell proliferation, replacement, and clonal expansion, recirculation and homing, and production of growth factors and regulatory and effector molecules. Thus, under unpredictable life events that involve the immune system, it is necessary that mechanisms that control immunity interact with those that control metabolic, cardio-respiratory, and other physiological processes in order to reach functional balances that allow adaptation and, finally, survival. It is expected that an inappropriate adjustment of such complex interactions could contribute to brain pathology. Indeed, immune dysfunctions are associated with psychiatric disorders. At present, a most exciting aspect of the research is to elucidate whether or not this association is causal, and whether immune mechanisms contribute to the etiology of psychiatric diseases, not only by inducing neuroinflammation and neurodegeneration, but also by disrupting physiological mechanisms that allow adaptation to daily life.

In the following discussion, we cover aspects concerning the capacity of the immune system to provide information to the brain about its operation, since this is an essential step in the chain of events that lead to neuroendocrine immunoregulation. We also discuss our view on how immune information is processed at central levels, and how it could be integrated with other information that the brain permanently receives. Considering the need to adjust the distribution of energy between the immune and nervous systems during health and disease, the role of brain-borne cytokines, such as IL-1, in this process will be used as an example.

THE SENSING CAPACITY OF THE IMMUNE SYSTEM

The phylogenetically ancient mechanisms of natural immunity in mammals are based on cells that express germ line-encoded pattern-recognition receptors (PRRs). References on mechanisms linked to immune recognition can be found in several reviews, for example.[2,3] The view of natural immunity as a non-specific system has changed radically with the identification of PRRs, such as Toll-like receptors (TLR), which detect certain components expressed by pathogens (pathogen-associated molecular patterns; PAMPs). Other PPRs, not necessarily sensing pathogens but also other foreign- or self-dangerous products, such as certain cytosolic PRRs, retinoic acid inducible gene I (RIG-I)-like receptors, and nucleotide-binding oligomerization domain-(NOD)-like receptors, were later discovered. Natural immune reactivity was followed in evolution by the more sophisticated adaptive immunity system, based on a huge repertoire that allows immune recognition and responses that are not inherited but mostly generated by random gene rearrangements. This was a "clever" way to increase the probability of successfully copping with newly emerging, dangerous infectious agents that are in continuous variation. Thus, both natural and specific immune recognition in an interlinked fashion allow the immune system to sense modifications of self-components and those caused by the intrusion of external challenges.

THE IMMUNE SYSTEM CAN AFFECT THE FUNCTIONING OF OTHER HOMEOSTATIC SYSTEMS

The aforementioned sophisticated sensing capacity of the immune system explains its large degree of autonomy to process the information received and to generate efferent responses to different types of insults. However, from a physiological point of view, it seems essential that the brain receives information about ongoing processes devoted to eliminating stimuli that threaten the stability of the organism, so as to coordinate the necessary homeostatic adjustments. This is particularly relevant for the immune system, due to its adaptive function in changeable environments, the high energetic and metabolic cost of immunity, and the fact that neuroendocrine agents produced under brain control can affect the activity of immune cells. Long ago, we proposed that if the immune system is under neuroendocrine regulation, its state of activity should be perceived at central levels, and, when a given intensity is reached, it should elicit neuroendocrine responses.[4,5]

Several neuroendocrine changes observed during infectious and other diseases that involve the participation of the immune system have been known for a long time. The initial view was that these effects were the consequence of the stress of being and feeling sick, and/or of the disease itself. However, this conclusion has been revised, since there is now clear evidence that products released by immune cells mediate most of the effects detected. Probably the first indication in this context was that fever during infections is mediated by endogenous pyrogens (for review[6]). Later, it was shown that immune mediators can affect mechanisms that exert immunoregulatory effects, independently of their pyrogenic actions, as is the case, for example, of the stimulation of the hypothalamus-pituitary-adrenal (HPA) axis.[7,8] We have shown, for example, that endogenous interleukin-1 (IL-1) mediates the stimulation of this axis following inoculation of New Castle Disease virus, which is innocuous for mice and humans, but lethal for chickens.[8,9] A comparable immune-mediated activation of the HPA axis was later detected during several other viral infections, including murine cytomegalovirus (MCMV), lymphocytic choriomeningitis virus (LCMV), influenza, *Herpes simplex* virus type 1 (HSV-1), and human immunodeficiency virus (HIV) for review.[10] It was also shown that the response of the HPA axis to low amounts of lipopolysaccharide (LPS) from gram-negative bacteria is mediated by IL-1,[11] and that, when this endotoxin is given at subpyrogenic doses, this response is mediated by activated macrophages,[12] an effect that today can be interpreted as dependent on TLR4.

To study the possibility that the adaptive branch of the immune system can elicit neuroendocrine responses, it was also necessary to deal with the confounding factor that, under natural conditions, immune responses are frequently associated with body damage and altered organ functions, and that the illness per se can elicit neuroendocrine responses linked to the stress of being sick. The approach to circumventing this problem was to immunize animals with innocuous antigens that can elicit a strong immune response without causing any disease. One of the antigens used was sheep red blood cells (SRBC), a model of immunization that permited us to use animals that received the same number of syngeneic red blood cells as control. Following this approach, an increase in glucocorticoid blood levels[13]; in the rate of firing of neurons, predominantly in the ventromedial hypothalamic nucleus[14]; and changes in the concentration of hypothalamic noradrenergic neurotransmitters were observed during the immune response.[4] We have also shown that products released during a human allogeneic immune response can mediate the activation of this axis.[15] Later, we showed that the immune response elicited by inoculation of allogeneic cells evokes changes in neuronal activity in brain regions different from those elicited by other types of T cell–dependent immune responses.[16] The type of cells and mediators involved in immune responses to different antigenic stimuli, as well as their kinetics, makes it difficult to perform a strict evaluation of the immunoregulatory relevance of the neuroendocrine responses that occur at different times. However, there are some models of immune response in which neuroendocrine responses have been studied at critical steps, such as induction, expansion, and generation of effector molecules and cells, and extinction.

The response of the HPA axis to T-dependent antigens is a relatively late phenomenon, but it can be preceded by stimulation of this axis during inflammatory events that often occur in parallel to antigenic presentation and that favor specific clonal expansion. There is little doubt at present that increased endogenous levels of glucocorticoids are anti-inflammatory, inhibit the production of pro-inflammatory cytokines, and affect the number and activity of immune cells,[17] an effect that also occurs following administration of high pharmacological doses of the hormone.[18] However, there is evidence indicating that this hormone plays a more complex role during immune processes. For example, low concentration of glucocorticoids can favor the initiation of the immune response, as shown *in vitro*,[19] and pretreatment with this hormone can enhance T cell effector functions.[20,21] Glucocorticoids can also stimulate TLR2 transcription and, to a lesser extent, also that of TLR4, in different cell types, including inflammatory cells (for review[22]), and regulate the expression of the NLRP3 (nucleotide-binding domain and leucine-rich repeat-containing family, pyrin domain containing 3) inflammasome under conditions of increased local

ATP concentrations.[23] A possible interpretation of these effects is that an early glucocorticoid-mediated increase in the sensitivity of inflammatory cells favors the recognition of invading microorganisms and the initiation of the process tending to their neutralization[24] by facilitating antigen presentation and the initiation of specific clonal expansion.[21] Following an early increase in glucocorticoid levels, most likely mediated by the release of pro-inflammatory cytokines by antigen-presenting cells, there is a normalization in the release of the hormone, since increased corticosteroid levels are again noticed only days after immunization, indicating that there is a time gap during which endogenous glucocorticoids do not interfere with IL-2 production[25] and its effect on T-cell clonal expansion. There is evidence that immune-derived products can also mediate the late phase of the stimulation of the HPA axis during the immune response. For example, a several-fold increase in adrenocorticotropin (ACTH) and glucocorticoid blood levels is triggered when supernatants from highly purified cultures of stimulated lymphocytes derived from less than one milliliter of human blood are injected into mice.[15] Furthermore, it has been demonstrated that Th2 lymphocytes can produce IL-1, which on one hand, can favor the production of IL-2, and on the other, stimulate the HPA axis.[8] In our view, the late increase in glucocorticoid levels just before the immune response is fully expressed would serve to shape the response. The late surge of this hormone would reduce the probability of bystander effects, such as, for example, the recruitment of unrelated lymphocyte clones and the control of the cytokine-mediated activation of the NLR3 inflammasome during the specific clonal expansion of immune cells. More direct evidence for the effect of glucocorticoids in shaping the specificity of an immune response derives from studies on sequential antigenic competition. In these studies, we showed that the inhibition of the immune response to a second, unrelated, antigen is largely dependent on the increase in glucocorticoid levels caused by the immune response to the first antigen.[19] Furthermore, endogenous glucocorticoids are likely to exert such control without interfering with the ongoing immune response, since during antigen-driven clonal expansion, activated T cells, for example, become resistant to the apoptotic effect of this hormone and the sensitivity to glucocorticoids declines during late phases of the specific immune response.[26,27] The immunoregulatory effect of endogenous glucocorticoids would be also manifested at a late phase, since this hormone can contribute to its termination by supporting the production of regulatory T cells.[28–30] It should be mentioned, however, that not all types of immune response trigger the stimulation of the HPA axis. For example, transplantation immunity results in inhibition of the activity in this axis, an effect that is expected to favor graft rejection.[31,32]

The immune response triggers not only endocrine responses, but also sympathetic autonomic responses. "Hard-wired" direct effects of autonomic nerves that innervate immune organs have been detected, and it is well known that immune cells are in close contact with nerve fibers that innervate immune organs, such as the spleen and lymph nodes (for references[33]). The fact that sympathetic nerves respond with changes in the release of noradrenaline (NA), its main neurotransmitter, during immune processes indicates that the effects of the sympathetic nervous system (SNS) are of central relevance for immunoregulation. For example, acute inflammation elicits a quick increase in NA release,[34] while the content and concentration of this neurotransmitter are decreased in lymphoid organs, such as the spleen, during the adaptive immune response to innocuous antigens and during several infectious and autoimmune/inflammatory diseases.[35–39] A loss of sympathetic fibers in immune organs during these pathologies may indicate that prolonged immune processes could lead to the atrophy of these fibers. While the anti-inflammatory effects of NA, predominantly mediated by β-adrenergic receptors, are widely recognized,[40,41] the role of the neurotransmitter in adaptive immunity is more complex. Based on studies on the effects of sympathetic denervation and of noradrenergic agonists during antigen-specific immune responses (for review[40,41]), we have proposed that the decrease in noradrenergic activity during the immune response is a way of releasing immune cells from the inhibitory effects of NA, thus favoring the takeoff of the adaptive response (for references[35,36]). Furthermore, we have observed that cytokines such as IL-1 can inhibit the local release of NA in the spleen.[42,43] This effect results in a diminished arteriolar vasoconstrictor tonus, which, by increasing splenic blood flow, can favor antigen trapping, priming of immune cells, and recirculation of activated lymphoid cells.[44] However, it is expected that the effect of sympathetic nerves on immune cells is differentially exerted during different types of immune responses. For example, we have observed that the immune response to enterotoxin B, a superantigen derived from gram-positive bacteria, is paralleled by biphasic changes in the activity of the SNS.[45] Although also in this model splenic NA concentration is markedly decreased 2–10 days after inoculation of the enterotoxin, this change is preceded by a significant increase in splenic NA concentration a few hours after enterotoxin administration. Furthermore, it should be added that the expression of noradrenergic receptors differs between different types of immune cells.[41]

Thus, the evidence discussed shows that the immune system, which is itself a homeostatic system, can affect the functioning of the endocrine and the peripheral nervous systems, the two other major homeostatic systems of the body.

THE SENSORY CAPACITY OF THE IMMUNE SYSTEM ALLOWS PROVIDING INFORMATION TO THE BRAIN

The evidence that immune responses elicited at peripheral levels can send messages to the brain brought us to propose long ago that the immune system is another sensory system.[4] Indeed, it fulfills the requirement to be regarded as such, considering that, like the "classical" senses, the immune system can receive, process, and send information to the brain about external and internal stimuli. The information sent to the brain by the classical senses is in most cases cognitive when several filtering mechanisms and thresholds are surpassed. In the case of the immune system, the information that it sends to the brain is non-cognitive per se, but it is perceived by the brain anyhow, as revealed by the neuroendocrine responses it elicits. It is, however, difficult to conceive that the immune system could send to the brain specific information about the myriad of immunogenic stimuli received. Rather, it seems more likely that it sends information about the type of responses that are put in motion to neutralize them. Indeed, different types of stimuli elicit different types of innate and adaptive immune responses, which can reflect the type of the ligands that immune cell receptors recognize. Immune information needs to be processed into appropriate signals before it is sent to defined brain centers via humoral or neural routes. The abundant literature showing that immune cell products such as interleukins, interferons, and chemokines can affect the brain and associated neuroendocrine mechanisms will not be reviewed here. This information can be found in several reviews (for example[35,46–49]). The need to process peripheral information before sending it to the brain also occurs in other sensorial systems; for example, the visual and auditory systems. The brain does not "see" light or "hear" sounds unless the information is transformed into electrochemical signals that can be centrally detected. Furthermore, there are physiological responses that originate in the eye and the internal ear that are oriented to refine the focus and the orientation of the stimuli that are received, such as ocular and acoustic reflexes. The fact that immune cells are not concentrated in one organ is also not an exception, considering the diffuse distribution of tactile and pain-sensing receptors. Furthermore, information derived from immune cells present in mucosal and epithelial tissues and other innervated tissue may be fused with the information supplied by stimulation of tactile and pain receptors, providing an anatomical representation of this combination of signals in the brain.

In summary, in our view, there is enough evidence to consider the immune system a neuro-sensorial receptor system since it detects and reacts to components of the external and internal world and informs the brain about the type of the ongoing immune processes.

PROCESSING AND INTEGRATION OF IMMUNE-DERIVED INFORMATION AT BRAIN LEVELS

In contrast to the present knowledge about the classical sensorial systems, it is still necessary to understand how immune information is processed at central levels, how it is integrated with other sensorial and intrinsic inputs, and which brain areas are involved. We have today only some clues to attempt answering these questions.

Humoral and neural pathways have been proposed as a way to convey immune information to the brain.[50] The production of immune mediators, such as cytokines, in the brain could be part of this communication system. Cytokine production in the brain can be triggered not only following peripheral immune stimulation, but also by neural cells that have been stimulated by neuro-sensorial signals. For example, peripheral administration of LPS induces the expression and production of several cytokines in the central nervous system (CNS).[51] Also, peripheral administration of IL-1β induces the expression of its own gene in the brain,[52] and cytokines such as IL-1 and IL-6 are also produced in the hypothalamus during peripheral specific immune responses.[37] Furthermore, increased expression of cytokines such as IL-1β, IL-6, IL-1, IL-18, and IL-1ra is observed during increased neuronal activity, such as during long-term potentiation (LTP) in vivo and in hippocampal slices.[51,53] The cytokines induced in this way are relevant for synaptic plasticity, since IL-1 is necessary to support LTP maintenance,[54,55] while IL-6 affects this process in an opposite way.[56] In line with these results are the findings that IL-1 also supports, while IL-6 inhibits, learning and memory consolidation.[51,53,56–59] Furthermore, we have recently shown that the expression of these cytokines is increased following learning a hippocampal-dependent task,[53] an effect that is dissociable from the stress of the learning paradigm. These results indicate that the production of immunoregulatory cytokines in the brain is a physiological process crucial for brain functions based on synaptic plasticity, such as learning and memory.

It has also been reported that IL-4 and interferon (IFN)-γ produced by T cells located in the meninges are necessary for the maturation of brain functions, and these cytokines have been linked to learning.[60] However, there is no evidence for direct connections and feedback interactions between neural cells involved in learning acquisition and meningeal T cells. Also, the effect could be the result of immune cell redistribution caused, for example, by the stress of learning a task. In any case, even if there were not such direct effects, IL-4 produced by T cells in the meningeal space or in the periphery could reach the brain parenchyma via the cerebrospinal fluid. In this way, these mediators can influence the production or action of cytokines such as IL-1, IL-1ra, IL-6, and IL-18, which, as mentioned, are produced by neural cells during learning a task and during other processes linked to synaptic plasticity such as LTP.

In our view, the physiological effects of brain-borne cytokines produced following peripheral immune or central neuronal signals on brain functions should be considered in the context of the at-present widely accepted concept of the tripartite synapse, which includes astrocytes[61] as the third party. For a long time, it has been considered that astrocytes, a large (probably the largest) population among the different neural cell types in the brain, exert only a supportive role for neuronal activity. Today it is known that, due to their extended distribution and close contact with neurons, these cells are the main components of the neuronal environment and the micro-architecture of the brain parenchyma; store and provide energetic substrates; and control neural cell development, synaptogenesis, and synaptic activity. Importantly, due to their immune functions, astrocytes are also part of the intrinsic defense system of the brain.[62,63] Thus, because of their dual neural and immune functions, astrocytes can be categorized as "neuro-immune" cells. These functions include the production of a variety of transmitters with immune and neural effects, such as IL-1, IL-6, and TNFα.[64–66]

We have proposed that the tripartite synapse plays a central role in processing immune signals in the brain and in their integration with neuro-sensorial signals.[51] The fact that the stimulation of glucose transport by IL-1 produced either by astrocytes or neurons can be transferred from one cell type to the other,[66] suggests that this cytokine plays a main role in the tripartite synapse by mediating a reciprocal control of energy supply between neural cells. The final effect of cytokines produced during the activation of the tripartite synapse on brain mechanisms and on neuroendocrine immunoregulation will depend on how, when, and where in the brain such stimulation occurs, and on the type of synapsis affected. As discussed, when the increased production of IL-1 and other cytokines would be neuronally triggered, their effects on the tripartite synapse would be to modulate physiological brain functions such as learning and memory. When their production in the brain was immunologically triggered, their effect would predominate in brain areas in which these mediators can reset homeostasis and exert immunoregulatory actions by eliciting neuroendocrine responses (Figure 5.1).

Since the brain is constantly exposed to neuro-sensorial stimuli, the adaptive or maladaptive outcome following reception of immune-derived signals should be based on the integration of this information. In our view, the tripartite synapse, particularly in brain areas affected by immune products, such as the hypothalamus and the hippocampus, offers the ideal molecular and cellular bases for the integration of immune and neuro-sensorial signals.

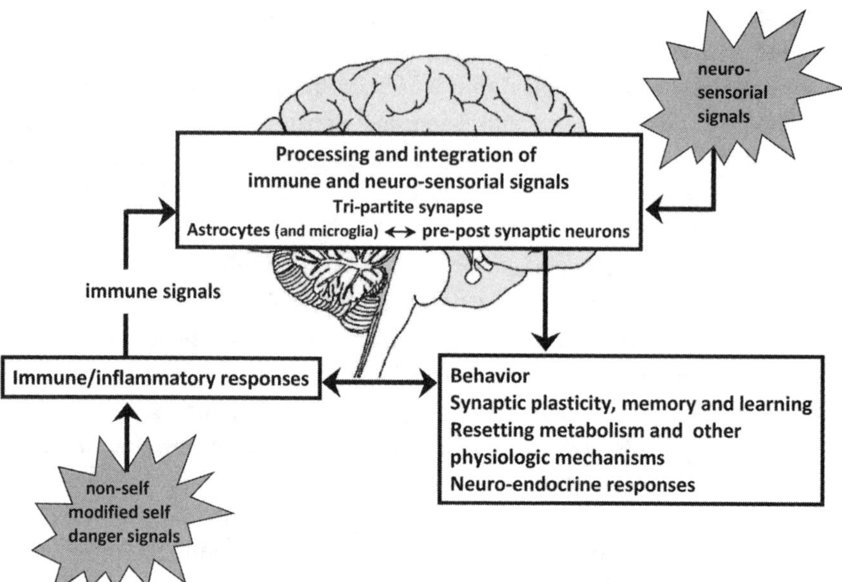

Figure 5.1 Processing and integration of immune and neuro-sensorial signals at brain levels. The brain receives immune and neural signals generated either at peripheral or central levels. The evidence discussed in the text supports the view that the integration and processing of these types of signals occur at the level of the tripartite synapse, with astrocytes as third party. Astrocytes are neuro-immune cells because they can directly affect synaptic transmission or indirectly affect brain functions via the production of cytokines and other mediators. Tripartite synapses are found in the whole brain. If tripartite synapses are activated by immune and/or neural signals at the level of the hypothalamic neuro-endocrine centres, it is expected that an immunoregulatory and metabolic effects would predominate. If tripartite synapses are activated by these signals in the hippocampus for example, the outcome is expected to be predominantly expressed in processes such as learning and memory.

THE CAPACITY OF THE IMMUNE SYSTEM TO RESET HOMEOSTASIS

Immune responses are frequently prolonged and require energy support for their maintenance.[1] It is therefore conceivable that many homeostatic systems need to be adjusted during the course of immune responses. Under these circumstances, physiological systems need to be remodeled, not only to provide energy to support inflammatory-cell turnover and clonal expansion of lymphocytes, but also for the mobilization of immune cells, so that they can reach the tissues and organs where they can eliminate, if they are successful, infectious agents (for example). It has been estimated that activated immune cells use more than 20% of the glucose-derived energy available in the body (for references[67]).

The concept of allostasis, which denominates a process addressed to achieve stability through physiological or behavioral changes, derives from adjustments necessary in a highly variable environment.[68] When this process is maintained for a long period, it has a functional cost (allostatic load) and can favor the expression of chronic diseases.[69] It should be considered, however, that immune and inflammatory responses are also prolonged responses tending to maintain the constancy of molecular and cellular body constituents. Allostatic adjustments during immune responses would be very costly if, every time they occurred, it would be necessary to violate homeostatic rules that tend to maintain the *status quo*. Indeed, as mentioned, there are neuroendocrine responses coupled with immune responses that serve to mediate metabolic adjustments and provide immunoregulatory signals. These multiple adjustments could be done progressively, but such a process would imply that a large proportion of the effort should be concentrated on counteracting the regulatory homeostatic mechanisms that tend to keep the system at the pre-set level of regulation. Thus linear, step-by-step, efforts would have a high cost, particularly during processes based on an acute increase in the activity of the immune system. Based on experimental data, we propose an alternative to the concept of allostasis when it is linked to immunoregulation that would serve to minimize the allostatic load during immune responses. Under these conditions, allostatic changes, rather than being based on linear adjustments, would lead to a preprogrammed switch towards a more adaptive set-point. This way of mediating adaptation during the immune process implies that physiological counter-regulatory mechanisms must be adjusted to this new set-point, a process that should be reversible when the previous conditions are reestablished. Still, when such reversibility is interfered with during prolonged situations, the resettled set-points would contribute to pathologies by overloading allostasis. Although the difference would seem subtle, we view allostasis, at least when referring to immune responses, as a mechanism of adaptation based on changes in regulatory set-points.

Glucoregulation is the mechanism that inspired Claude Bernard and Walter Cannon to define homeostasis. It is now known that, by decreasing glucose blood levels, insulin elicits brain-integrated counter-regulatory responses mediated by hormones and neurotransmitters that tend to return glycemia to the pre-set level. When insulin release is abnormally prolonged, such as during fasting, a persistent hypoglycemia is the result of a balance between the effects of the hormone and the counter-regulatory mechanisms, with the consequence that many physiological functions are altered, eventually leading to hypoglycemic shock. This seems not to be the case of IL-1, which is produced following immune stimulation and during increased synaptic activity. This cytokine induces a profound and long-lasting insulin-independent hypoglycemia in mice that is not paralleled by overt neurological symptoms. This is a unique property of an immune-derived product that is elicited by endogenous levels of IL-1; for example, following activation of TLR4. Furthermore, IL-1, a cytokine that plays a central process in the inflammasome and in the metabolic syndrome, induces the production of other cytokines such as tumor-necrosis factor alpha (TNFα) and IL-6, which also exert multiple systemic effects. We have shown that IL-1 changes the set-point of glucoregulation to a lower level. Indeed, IL-1, either administered or endogenously produced, induces a reduction in glucose levels of about 50–60%, and there is not a quick return to euglycemia, even following a glucose load, but to the previous hypoglycemic values. Furthermore, the decrease in glucose blood levels does not go beyond a limit, even when the minimal dose that induces hypoglycemia is increased more than 20 times and no hypoglycemic shock is observed. This effect, clearly different from that of insulin, shows that IL-1 can induce the resetting of a tightly controlled mechanism such as glucose homeostasis. It is possible that this resetting is directed at redistributing resources needed to supply fuel to highly demanding immune cells. In mice, this process leads to hypoglycemia, and it is paralleled neither by a prolonged counter-regulation mediated by hormones, such as glucocorticoids, glucagon, and catecholamines, nor by a compensatory increase in food intake. In other species, such as the rat, IL-1 mediates a resetting of glucose homeostasis by another mechanism. In this species, IL-1 induces only a 10–15% decrease in the concentration of blood glucose, but this is paralleled by a decrease in insulin levels and somewhat compensated for by adrenal hormones, since administration of the cytokine to adrenalectomized rats leads to a profound hypoglycemia (for references[67]). Also, an insulin-independent hypoglycemia caused by products from immune cells is observed in rats during sepsis induced by cecal ligation. Interestingly, IL-1 and TNFα cause insulin resistance in humans with sepsis (for references[67]).

There is evidence of other resetting effects of IL-1, for example, the baroreceptor reflex, which controls cardiovascular functions,[70] and during fever, which is defined as a change in the set-point of thermoregulation.[6] Thus, immune responses contribute to reset homeostasis by releasing cytokines and other products that, besides their intrinsic immune functions, can also induce neuroendocrine responses, which, in turn, can support host defenses. Since the immune system operates at the interface between health and disease, the immune response can be protective and adaptive, particularly during acute pathologies. However, when the immune system is hyperactive over a long time, it can trigger non-adaptive neuroendocrine-induced metabolic disruptions.

OVERVIEW AND PERSPECTIVES

We have summarized selected aspects concerning immune–brain communication under the view of a network of interactions. Furthermore, we proposed that, instead of transient neuroendocrine immunoregulatory effects, there is a resetting of regulatory adaptive systems when immune responses are prolonged. Such a resetting may indicate that during intense immune responses, the immune system can take command of homeostasis.

Within the context of this book, it should probably be emphasized again that the examples chosen in this chapter relate the three major homeostatic systems: the immune, the endocrine, and the nervous systems.

The activity of the HPA axis, one of the best-studied neuroendocrine mechanisms affected by products of activated immune cells, is also altered during psychiatric pathologies. The stimulation of the HPA axis is also part of the adaptive response to stress, but an excessive and sustained increased activity can lead to disturbances in brain mechanisms that include behavior, cognition, mood, and personality. Thus, it is not surprising that a deregulation of this axis is noticed in a broad spectrum of neuropsychiatric disorders, among which are major depression, anxiety disorders, acute psychosis, and delirium. There is a large amount of literature addressing this aspect (for references and review, see, for example,[71]). As previously discussed, glucocorticoids play an important role in immunoregulation, and many cytokines, in a surprisingly redundant way (for review[72,10]) stimulate the HPA axis. It is therefore conceivable that cytokines released during immune processes can contribute to the deregulation of the HPA axis during psychiatric pathologies.

Although maybe not so abundant, there is also evidence of an association between alterations in the activity of the autonomic nervous system and psychiatric disorders (for review[73]). As discussed, a reduction in noradrenergic activity in immune organs is observed during specific immune responses and during the development of chronic, lymphoproliferative, and infectious diseases. In other models, such as stimulation with endotoxins and enterotoxins, there is a quick increase in splenic sympathetic activity. Thus, it is rather tempting to speculate that immune signals could be involved in alterations in autonomic activity during certain psychiatric pathologies.

Finally, the association between metabolic disruptions and psychiatric illnesses, such as schizophrenia, bipolar disorder, depressive disorder, post-traumatic stress disorder, and Alzheimer's disease, is remarkable.[74–77] We have given here as an example the capacity of IL-1 to reset glucose homeostasis.[52] However, cytokines such as IL-1, IL-6, and TNFα mediate metabolic pathologies associated with cognitive and behavioral deficits.[77,78] The myeloid differentiation factor 88 (MyD88)–)-mediated pathway is involved in IL-1 signal transduction, and MyD88 knockout mice provide a good example of the concomitant expression of immune, metabolic, and neurological disorders. Parallel to immune deficiencies,[79,80] MyD88 knockout mice have neurological and behavioral alterations.[81] A reduction in MyD88 expression also results in an accelerated progression of pathology in a mouse model of Alzheimer's disease.[82] In addition,

MyD88-deficient mice show a tendency to develop type 2 diabetes, as frequently observed in patients with Alzheimer's disease.[83]

It is still not known how immune-derived and neuro-sensorial signals that converge in the brain are integrated and stored. We have proposed that the tripartite synapse that includes astrocytes, which have both neural and immune properties, acts as a relay system that performs these functions, and that cytokines in the brain such as IL-1 and IL-6 play a central role in orchestrating the reception, interpretation, and transmission of messages. However, we are still far from understanding how such a relay system operates when the brain receives different types of signals. In fact, the final neuroendocrine and behavioral responses will depend on the nature and weight of these signals, which could also be conditioned by previous experiences. An even more formidable task would be to clarify whether the perception of the outcome of the combination of immune and neuro-sensorial signals is altered in patients with psychiatric diseases.

ACKNOWLEDGMENTS

This work was supported by a grant from the Deutsche Forschungsgemeinschaft (DFG RE 1451) to Adriana del Rey.

REFERENCES

1. Pearce, E. L., Poffenberger, M. C., Chang, C. H., & Jones, R. G. (2013). Fueling immunity: insights into metabolism and lymphocyte function. *Science, 342,* 1242454. doi:10.1126/science.1242454
2. Kawai, T., & Akira, S. (2010). The role of pattern-recognition receptors in innate immunity: update on Toll-like receptors. *Nat Immunol, 11,* 373–384.
3. Thaiss, C. A., Levy, M., Itav, S., & Elinav, E. (2016). Integration of innate immune signaling. *Trends Immunol, 37,* 84–101.
4. Besedovsky, H., del Rey, A., Sorkin, E., Da Prada, M., Burri, R., & Honegger, C. (1983). The immune response evokes changes in brain noradrenergic neurons. *Science, 221,* 564–566.
5. Del Rey, A., & Besedovsky, H. O. (2017). Immune-neuro-endocrine reflexes, circuits, and networks: physiologic and evolutionary implications. *Front Horm Res, 48,* 1–18.
6. Roth, J., & Blatteis, C. M. (2014). Mechanisms of fever production and lysis: lessons from experimental LPS fever. *Compr Physiol, 4,* 1563–1604.
7. Berkenbosch, F., van Oers, J., del Rey, A., Tilders, F., & Besedovsky, H. (1987). Corticotropin-releasing factor-producing neurons in the rat activated by interleukin-1. *Science, 238,* 524–526.
8. Besedovsky, H., del Rey, A., Sorkin, E., & Dinarello, C. A. (1986). Immunoregulatory feedback between interleukin-1 and glucocorticoid hormones. *Science, 233,* 652–654.

9. Besedovsky, H. O., & del Rey, A. (1989). Mechanism of virus-induced stimulation of the hypothalamus-pituitary-adrenal axis. *J Steroid Biochem, 34*, 235–239.
10. Silverman, M. N., Pearce, B. D., Biron, C. A., & Miller, A. H. (2005). Immune modulation of the hypothalamic-pituitary-adrenal (HPA) axis during viral infection. *Viral Immunol, 18*, 41–78.
11. Rivier, C., Chizzonite, R., & Vale, W. (1989). In the mouse, the activation of the hypothalamic-pituitary-adrenal axis by a lipopolysaccharide (endotoxin) is mediated through interleukin-1. *Endocrinology, 125*, 2800–2805.
12. Derijk, R., Van Rooijen, N., Tilders, F. J., Besedovsky, H. O., del Rey, A., & Berkenbosch, F. (1991). Selective depletion of macrophages prevents pituitary-adrenal activation in response to subpyrogenic, but not to pyrogenic, doses of bacterial endotoxin in rats. *Endocrinology, 129*, 330–338.
13. Besedovsky, H., Sorkin, E., Keller, M., & Muller, J. (1975). Changes in blood hormone levels during the immune response. *Proc Soc Exp Biol Med, 150*, 466–470.
14. Besedovsky, H., Sorkin, E., Felix, D., & Haas, H. (1977). Hypothalamic changes during the immune response. *Eur J Immunol, 7*, 323–325.
15. Besedovsky, H. O., del Rey, A., Sorkin, E., Lotz, W., & Schwulera, U. (1985). Lymphoid cells produce an immunoregulatory glucocorticoid increasing factor (GIF) acting through the pituitary gland. *Clin Exp Immunol, 59*, 622–628.
16. Furukawa, H., Yamashita, A., del Rey, A., & Besedovsky, H. (2004). c-Fos expression in the rat cerebral cortex during systemic GvH reaction. *Neuroimmunomodulation, 11*, 425–433.
17. del Rey, A., Besedovsky, H., & Sorkin, E. (1984). Endogenous blood levels of corticosterone control the immunologic cell mass and B cell activity in mice. *J Immunol, 133*, 572–575.
18. Coutinho, A. E., & Chapman, K. E. (2011). The anti-inflammatory and immunosuppressive effects of glucocorticoids: recent developments and mechanistic insights. *Mol Cell Endocrinol, 335*, 2–13.
19. Besedovsky, H. O., del Rey, A., & Sorkin, E. (1979). Antigenic competition between horse and sheep red blood cells as a hormone-dependent phenomenon. *Clin Exp Immunol, 37*, 106–113.
20. Almawi, W. Y., Hess, D. A., Assi, J. W., Chudzik, D. M., & Rieder, M. J. (1999). Pretreatment with glucocorticoids enhances T cell effector function: possible implication for immune rebound accompanying glucocorticoid withdrawal. *Cell Transplant, 8*, 637–647.
21. Wiegers, G. J., & Reul, J. M. (1998). Induction of cytokine receptors by glucocorticoids: functional and pathological significance. *Trends Pharmacol Sci, 19*, 317–321.
22. Chinenov, Y., & Rogatsky, I. (2007). Glucocorticoids and the innate immune system: crosstalk with the Toll-like receptor signaling network. *Mol Cell Endocrinol, 275*, 30–42.
23. Busillo, J. M., Azzam, K. M., & Cidlowski, J. A. (2011). Glucocorticoids sensitize the innate immune system through regulation of the NLRP3 inflammasome. *J Biol Chem, 286*, 38703–38713.
24. Busillo, J. M., & Cidlowski, J. A. (2013). The five Rs of glucocorticoid action during inflammation: ready, reinforce, repress, resolve, and restore. *Trends Endocrinol Met, 24*, 109–119.

25. Huber, M., Beuscher, H. U., Rohwer, P., Kurrle, R., Rollinghoff, M., & Lohoff, M. (1998). Costimulation via TCR and IL-1 receptor reveals a novel IL-1alpha-mediated autocrine pathway of Th2 cell proliferation. *J Immunol, 160,* 4242–4247.
26. Jamieson, C. A., & Yamamoto, K. R. (2000). Crosstalk pathway for inhibition of glucocorticoid-induced apoptosis by T cell receptor signaling. *Proc Natl Acad Sci U S A, 97,* 7319–7324.
27. Strauss, G., Osen, W., & Debatin, K. M. (2002). Induction of apoptosis and modulation of activation and effector function in T cells by immunosuppressive drugs. *Clin Exp Immunol, 128,* 255–266.
28. Chen, X., Murakami, T., Oppenheim, J. J., & Howard, O. M. (2004). Differential response of murine CD4+CD25+ and CD4+CD25– T cells to dexamethasone-induced cell death. *Eur J Immunol, 34,* 859–869.
29. Chen, X., Oppenheim, J. J., Winkler-Pickett, R. T., Ortaldo, J. R., & Howard, O. M. (2006). Glucocorticoid amplifies IL-2-dependent expansion of functional FoxP3(+)CD4(+)CD25(+) T regulatory cells in vivo and enhances their capacity to suppress EAE. *Eur J Immunol, 36,* 2139–2149.
30. Karagiannidis, C., Akdis, M., Holopainen, P., et al. (2004). Glucocorticoids upregulate FOXP3 expression and regulatory T cells in asthma. *J Allergy Clin Immunol, 114,* 1425–1433.
31. Besedovsky, H. O., Sorkin, E., & Keller, M. (1978). Changes in the concentration of corticosterone in the blood during skin-graft rejection in the rat. *J Endocrinol, 76,* 175–176.
32. Zakarian, S., Eleazar, M. S., & Silvers, W. K. (1989). Regulation of pro-opiomelanocortin biosynthesis and processing by transplantation immunity. *Nature, 339,* 553–556.
33. Janig, W. (2014). Sympathetic nervous system and inflammation: a conceptual view. *Auton Neurosci, 182,* 4–14.
34. MacNeil, B. J., Jansen, A. H., Greenberg, A. H., & Nance, D. M. (1996). Activation and selectivity of splenic sympathetic nerve electrical activity response to bacterial endotoxin. *Am J Physiol, 270,* R264–R270.
35. Besedovsky, H. O., & del Rey, A. (2007). Physiology of psychoneuroimmunology: a personal view. *Brain Behav Immun, 21,* 34–44.
36. Del Rey, A., & Besedovsky, H. O. (2008). Sympathetic nervous system–immune interactions in autoimmune lymphoproliferative diseases. *Neuroimmunomodulation, 15,* 29–36.
37. Del Rey, A., Wolff, C., Wildmann, J., et al. (2008). Disrupted brain-immune system-joint communication during experimental arthritis. *Arthritis Rheum, 58,* 3090–3099.
38. Roggero, E., Perez, A. R., Pollachini, N., et al. (2016). The sympathetic nervous system affects the susceptibility and course of *Trypanosoma cruzi* infection. *Brain Behav Immun, 58,* 228–236.
39. Wolff, C., Straub, R. H., Hahnel, A., et al. (2015). Mimicking disruption of brain-immune system-joint communication results in collagen type II-induced arthritis in non-susceptible PVG rats. *Mol Cell Endocrinol, 415,* 56–63.
40. Elenkov, I. J., Wilder, R. L., Chrousos, G. P., & Vizi, E. S. (2000). The sympathetic nerve—an integrative interface between two supersystems: the brain and the immune system. *Pharmacol Rev, 52,* 595–638.

41. Sanders, V. M. (2012). The beta2-adrenergic receptor on T and B lymphocytes: do we understand it yet? *Brain Behav Immun, 26,* 195–200.
42. Rogausch, H., del Rey, A., Kabiersch, A., & Besedovsky, H. O. (1995). Interleukin-1 increases splenic blood flow by affecting the sympathetic vasoconstrictor tonus. *Am J Physiol, 268,* R902–R908.
43. Rogausch, H., del Rey, A., Kabiersch, A., Reschke, W., Ortel, J., & Besedovsky, H. (1997). Endotoxin impedes vasoconstriction in the spleen: role of endogenous interleukin-1 and sympathetic innervation. *Am J Physiol, 272,* R2048–R2054.
44. Rogausch, H., del Rey, A., Oertel, J., & Besedovsky, H. O. (1999). Norepinephrine stimulates lymphoid cell mobilization from the perfused rat spleen via beta-adrenergic receptors. *Am J Physiol, 276,* R724–R730.
45. del Rey, A., Kabiersch, A., Petzoldt, S., & Besedovsky, H. O. (2002). Involvement of noradrenergic nerves in the activation and clonal deletion of T cells stimulated by superantigen in vivo. *J Neuroimmunol, 127,* 44–53.
46. Deverman, B. E., & Patterson, P. H. (2009). Cytokines and CNS development. *Neuron, 64,* 61–78.
47. Erta, M., Quintana, A., & Hidalgo, J. (2012). Interleukin-6, a major cytokine in the central nervous system. *Int J Biol Sci, 8,* 1254–1266.
48. Reyes-Vazquez, C., Prieto-Gomez, B., & Dafny, N. (2012). Interferon modulates central nervous system function. *Brain Res, 1442,* 76–89.
49. Rostene, W., Kitabgi, P., & Parsadaniantz, S. M. (2007). Chemokines: a new class of neuromodulator? *Nature Rev Neurosci, 8,* 895–903.
50. Dantzer, R., Konsman, J. P., Bluthe, R. M., & Kelley, K. W. (2000). Neural and humoral pathways of communication from the immune system to the brain: parallel or convergent? *Auton Neurosci, 85,* 60–65.
51. Besedovsky, H. O., & del Rey, A. (2011). Central and peripheral cytokines mediate immune–brain connectivity. *Neurochem Res, 36,* 1–6.
52. Del Rey, A., Roggero, E., Randolf, A., et al. (2006). IL-1 resets glucose homeostasis at central levels. *Proc Natl Acad Sci USA, 103,* 16039–16044.
53. Del Rey, A., Balschun, D., Wetzel, W., Randolf, A., & Besedovsky, H. O. (2013). A cytokine network involving brain-borne IL-1beta, IL-1ra, IL-18, IL-6, and TNFalpha operates during long-term potentiation and learning. *Brain Behav Immun, 33,* 15–23.
54. Schneider, H., Pitossi, F., Balschun, D., Wagner, A., del Rey, A., & Besedovsky, H. O. (1998). A neuromodulatory role of interleukin-1beta in the hippocampus. *Proc Natl Acad Sci USA, 95,* 7778–7783.
55. Ross, F. M., Allan, S. M., Rothwell, N. J., & Verkhratsky, A. (2003). A dual role for interleukin-1 in LTP in mouse hippocampal slices. *J Neuroimmunol, 144,* 61–67.
56. Balschun, D., Wetzel, W., del Rey, A., et al. (2004). Interleukin-6: a cytokine to forget. *FASEB J, 18,* 1788–1790.
57. Donzis, E. J., & Tronson, N. C. (2014). Modulation of learning and memory by cytokines: signaling mechanisms and long term consequences. *Neurobiol Learn Mem, 115,* 68–77.
58. McAfoose, J., & Baune, B. T. (2009). Evidence for a cytokine model of cognitive function. *Neurosci Biobehav Rev, 33,* 355–366.
59. Yirmiya, R., & Goshen, I. (2011). Immune modulation of learning, memory, neural plasticity and neurogenesis. *Brain Behav Immun, 25,* 181–213.

60. Derecki, N. C., Cardani, A. N., Yang, C. H., et al. (2010). Regulation of learning and memory by meningeal immunity: a key role for IL-4. *J Exp Med, 207,* 1067–1080.
61. Perea, G., Navarrete, M., & Araque, A. (2009). Tripartite synapses: astrocytes process and control synaptic information. *Trends Neurosci, 32,* 421–431.
62. Gimsa, U., Mitchison, N. A., & Brunner-Weinzierl, M. C. (2013). Immune privilege as an intrinsic CNS property: astrocytes protect the CNS against T cell-mediated neuroinflammation. *Mediators Inflammat,* 2013, 320519. doi:10.1155/2013/320519
63. Verkhratsky, A., Nedergaard, M., & Hertz, L. (2015). Why are astrocytes important? *Neurochem Res, 40,* 389–401.
64. Lau, L. T., & Yu, A. C. (2001). Astrocytes produce and release interleukin-1, interleukin-6, tumor necrosis factor alpha and interferon-gamma following traumatic and metabolic injury. *J Neurotrauma, 18,* 351–359.
65. Rothwell, N. J., & Luheshi, G. N. (2000). Interleukin 1 in the brain: biology, pathology and therapeutic target. *Trends Neurosci, 23,* 618–625.
66. del Rey, A., Verdenhalven, M., Lorwald, A. C., et al. (2016). Brain-borne IL-1 adjusts glucoregulation and provides fuel support to astrocytes and neurons in an autocrine/paracrine manner. *Mol Psychiatry, 21,* 1309–1320.
67. Besedovsky, H. O., & del Rey, A. (2014). Physiologic versus diabetogenic effects of interleukin-1: a question of weight. *Curr Pharm Des, 20,* 4733–4740.
68. Sterling, P., & Eyer, J. (1988). Allostasis: a new paradigm to explain arousal pathology. In S. Fisher & J. Reason (Eds.), *Handbook of life stress, cognition and health* (pp. 629–649). New York: John Wiley & Sons.
69. McEwen, B. S. (2007). Physiology and neurobiology of stress and adaptation: central role of the brain. *Physiol Rev, 87,* 873–904.
70. Rogausch, H., Vo, N. T., del Rey, A., & Besedovsky, H. O. (2000). Increased sensitivity of the baroreceptor reflex after bacterial endotoxin. *Ann NY Acad Sci, 917,* 165–168.
71. Pereira, A. M., & Meijer, O. C. (2017). Glucocorticoid regulation of neurocognitive and neuropsychiatric function. In E. B. Geer (Ed.), *The hypothalamic-pituitary-adrenal axis in health and disease* (pp. 27–41). Cham, Switzerland: Springer International Publishing.
72. Besedovsky, H. O., & del Rey, A. (1996). Immune-neuro-endocrine interactions: facts and hypotheses. *Endocr Rev, 17,* 64–102.
73. Alvares, G. A., Quintana, D. S., Hickie, I. B., & Guastella, A. J. (2016). Autonomic nervous system dysfunction in psychiatric disorders and the impact of psychotropic medications: a systematic review and meta-analysis. *J Psychiatry Neurosci, 41,* 89–104.
74. Akter, K., Lanza, E. A., Martin, S. A., Myronyuk, N., Rua, M., & Raffa, R. B. (2011). Diabetes mellitus and Alzheimer's disease: shared pathology and treatment? *Br J Clin Pharmacol, 71,* 365–376.
75. Chen, Z., & Zhong, C. (2013). Decoding Alzheimer's disease from perturbed cerebral glucose metabolism: implications for diagnostic and therapeutic strategies. *Prog Neurobiol, 108,* 21–43.

76. Harris, L. W., Guest, P. C., Wayland, M. T., et al. (2013). Schizophrenia: metabolic aspects of aetiology, diagnosis and future treatment strategies. *Psychoneuroendocrinology, 38,* 752–766.
77. Ho, C. S. H., Zhang, M. W. B., Mak, A., & Ho, R. C. M. (2014). Metabolic syndrome in psychiatry: advances in understanding and management. *B J Psych Advances, 20,* 101–112.
78. Steiner, J., Bernstein, H. G., Schiltz, K., et al. (2014). Immune system and glucose metabolism interaction in schizophrenia: a chicken–egg dilemma. *Prog Neuropsychopharmacol Biol Psychiatry, 48,* 287–294.
79. Adachi, O., Kawai, T., Takeda, K., et al. (1998). Targeted disruption of the MyD88 gene results in loss of IL-1- and IL-18-mediated function. *Immunity, 9,* 143–150.
80. Kawai, T., Adachi, O., Ogawa, T., Takeda, K., & Akira, S. (1999). Unresponsiveness of MyD88-deficient mice to endotoxin. *Immunity, 11,* 115–122.
81. Drouin-Ouellet, J., LeBel, M., Filali, M., & Cicchetti, F. (2012). MyD88 deficiency results in both cognitive and motor impairments in mice. *Brain Behav Immun, 26,* 880–885.
82. Michaud, J. P., Richard, K. L., & Rivest, S. (2011). MyD88-adaptor protein acts as a preventive mechanism for memory deficits in a mouse model of Alzheimer's disease. *Mol Neurodegener, 6,* 5. doi.org/10.1186/1750-1326-6-5
83. Hosoi, T., Yokoyama, S., Matsuo, S., Akira, S., & Ozawa, K. (2010). Myeloid differentiation factor 88 (MyD88)-deficiency increases risk of diabetes in mice. *PLoS One, 5,* e12537.

6

Microbiota-Gut-Brain Axis and Neuropsychiatric Disorders

GILLIARD LACH, TIMOTHY G. DINAN, AND JOHN F. CRYAN

INTRODUCTION

Early in the twentieth century, the term *dysbiosis* was first defined by Elie Metchnikoff to denote an imbalance in the intestinal microbial community. Together with the pioneering studies of Robert Koch and Louis Pasteur, among others, the work of Metchnikoff was fundamental to establishing the concept that the gut microbiota and the immune system live in constant opposition against each other. Thus, modifying the intestinal microbiota could improve human health.[1] However, it was not until the last two decades—largely due to advances in sequencing technology—that a full appreciation of the importance of the gut microbiota in health and disease, and in programming the body's main systems, including the brain, occurred.

THE GUT MICROBIOTA

The adult mammalian gut is home to a well-populated and diverse world of microbes living in harmony with each other. This beneficial cohabitation is the fruit of a vigorous process of maturation and development that starts from birth and is shaped by events throughout the lifespan. In addition, several factors can modulate the gut microbiota diversity and its interaction with the immune system. Some factors arising in early life, such as mode of birth, breastfeeding, weaning age, and maternal lifestyle in infants can interfere in the maturation of the microbiota and immune system,[2] while other factors dependent on the use of antibiotics, dietary habits, exercise, illness, and aging can produce dysbiosis, facilitating or inhibiting the proliferation of commensal bacteria.[3–5] The interactive dynamics of the various factors shape the microbial landscape to create a unique signature, which may have an impact on health and disease (Figure 6.1).

The diversity of a normal adult microbiota is reflected in thousands of bacterial species divided into 55 phyla.[6] However, the vast majority of the microbes in the gut are assigned to only eight bacterial phyla, mainly the phyla Firmicutes (e.g., *Lactobacillus, Clostridium, Enterococcus* genus), Bacteroidetes (*Bacteroides* genus) and *Actinobacteria* (i.e., *Bifidobacterium* genus) representing over 95% of the intestinal community.[7,8] The overrepresentation of few phyla shows a rich and highly specialized niche that mediates a symbiotic relationship between host and microbe to provide an efficient protective role by restraining potential invasion (i.e., of pathogenic bacteria) and stimulating the host's immune responses.[5,9] Moreover, bacteria genera such as *Lactobacillus* and *Bifidobacterium* are among the most common commensal bacteria used to promote beneficial effects in brain function and behavior[9] (Table 6.1). These microbe–host interactions happen through different signaling pathways, which include the modulation of the immune system,[10,11] enteroendocrine cells of the gut,[12] the hypothalamic-pituitary-adrenal (HPA) axis,[13] production of bacterial metabolites such as short-chain fatty acids (SCFAs),[14] transformation of bile acids,[15] tryptophan metabolism,[16] and via vagus nerve signalling.[17] Additionally, it has been shown recently that epigenetic factors and bacterial peptidoglycan derived from the commensal bacteria may also play a role in gut–brain communication.[18,19]

Neural Pathways

The vagus nerve is perhaps the best characterized neural pathway of bidirectional communication between the gut and the brain. Information from the gut is delivered tonically to the nucleus of the solitary tract via sensory fibers of the vagus nerve,[20] while rupture of the vagal signaling abolishes the gut-to-brain communication.[17,21] For example, *Lactobacillus reuteri,* which was shown to increase oxytocin release in the hypothalamus, had its effect blunted after vagotomy.[22] Vagotomy also blocked anxiolytic-like behavior induced by *Bifidobacterium longum* and *Lactobacillus rhamnosus* in mice,[17,23] while transient inactivation of the dorsal vagal complex attenuated social isolation deficit induced by peripheral administration of lipopolysaccharides (LPS), which are found in the outer membrane of gram-negative bacteria.[22,24]

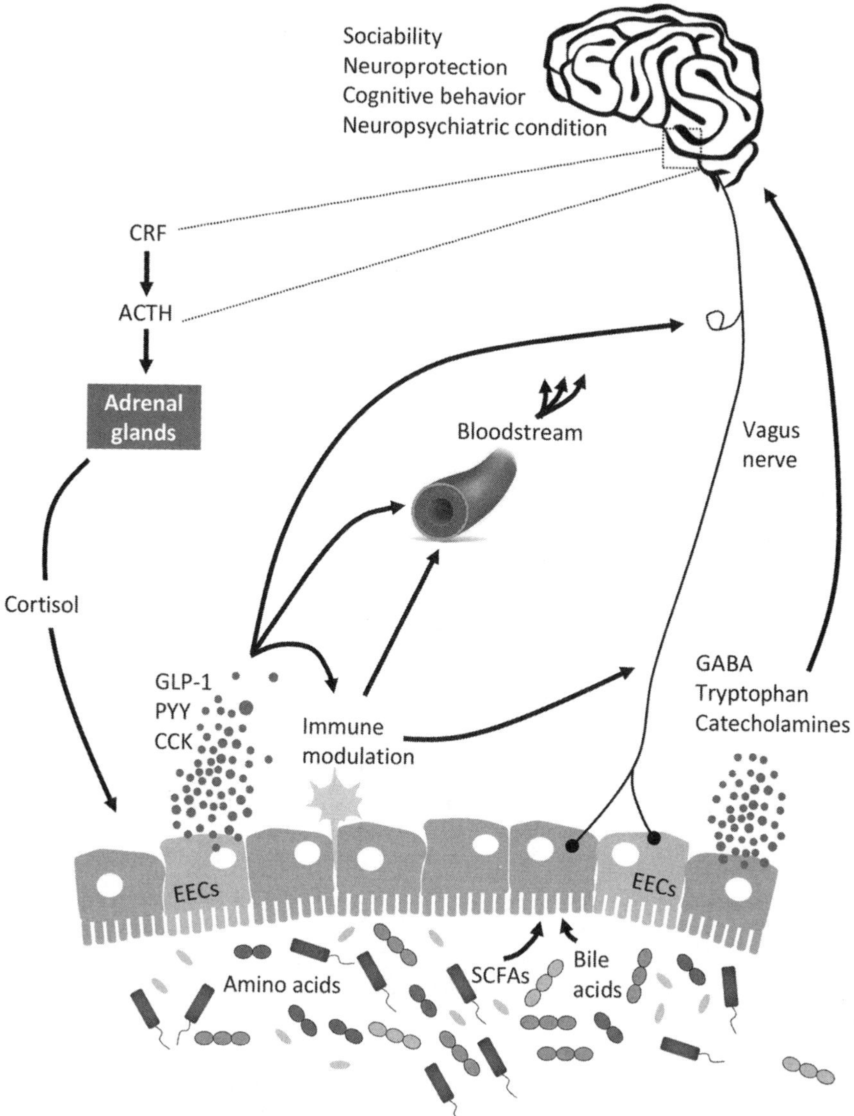

Figure 6.1 *Cross-talk between diet-derived nutrients, the microbiota and its metabolites, and the brain.* The fermentation of carbohydrates can result in the synthesis of SCFAs, which can directly activate FFAR on EECs. Bile acids derived from fatty acid metabolism can also interact with specific receptors in the intestinal wall. Both SCFA and bile acids can thus stimulate gut peptides' secretion as well as immunomodulatory responses. In addition, the gut microbiota can modulate the production of a myriad of neurotransmitters and neuroactive substances that can regulate host peripheral and central functions via direct and indirect mechanisms.

Abbreviations: EECs, enteroendocrine cells; FFAR, free fatty acid receptor; SCFAs, short-chain fatty acids.

Table 6.1 Preclinical Evidences Targeting the Gut Microbiota for the Treatment of Neuropsychiatric Disorders

Microbial Manipulation	Outcomes	Reference
3'Sialyllactose and 6'Sialyllactose	Anxiolytic effect in mice	35
B. longum or *L. fermentum* or *L. helveticus*	Anxiolytic effect in mice	23, 36, 37
FOS+GOS	Antidepressant and anxiolytic in mice	38
B. infantis or *L. plantarum* or *L. rhamnosus*	Antidepressant-like behavior in rats and mice	17, 39, 40
Butyric acid	Antidepressant-like behavior in rats	15
L. helveticus and *B. longum*	Attenuated HPA-axis activation in mice	42
Kefir	Protective effects against anxiety and cognitive impairment in rats	43
L. helveticus or *L. rhamnosus* or *C. butyricum*	Prevented memory deficit in memory impaired-mice	36, 44, 45, 46, 47
VSL#3	Attenuated age-related deficits in aged mice	48
L. fermentum	Reduced Aβ fibril formation in mice	46, 49
Bacillus sp.	Increase production of L-DOPA in vitro	50
Bacteroides fragilis/ thetaiotaomicron	Normalize ASD-like behavior in mice	51
L. reuteri	Reversed antisocial behaviors in mice	52

B = Bifidobacterium; *L* = Lactobacillus; *C* = Clostridium; FOS = fructo-oligosaccharides; GOS = galacto-oligosaccharides; *HPA* = hypothalamic-pituitary-adrenal; *Aβ* = β-amyloid; *L-DOPA* = levodopa; *ASD* = autism-spectrum disorder; VSL#3 = is a mixture of 8 different strains of bacteria.

Enteroendocrine Signaling

Enteroendocrine cells (EECs) are specialized cells located throughout the gut, which can produce different peptides and other signaling molecules. Such molecules include serotonin, cholecystokinin (CCK), glucagon-like peptide (GLP)-1, and peptide YY (PYY) (Table 6.1). EECs represent approximately 1% of epithelial cells in the gut lumen and express specific chemosensory receptors that modulate afferent nerve fibers, enabling them to detect and transduce information from the gut to the central nervous system (CNS).[12,25] Following secretion, peptides released by EECs diffuse throughout the lamina propria and ultimately reach the bloodstream to stimulate sensory neurons to potentially act through the vagus nerve, thereby creating an intersection for gut peptidergic signaling.[26,27] Gut peptide signaling can be modulated by the microbiota.[28] Germ-free mice have reduced numbers of EECs in the ileum and colon; thereby, these mice exhibit attenuated expression of PYY, GLP-1, and CCK.[29] Breton and colleagues also found that a protein produced by *Escherichia coli* may facilitate the release of GLP-1 and PYY.[30] The prebiotics oligofructose and inulin-type fructan increase the levels of GLP-1 in animals and humans.[3,31–33] Moreover, SCFAs have been shown to increase secretion of CCK, PYY, and GLP-1 through binding to SCFAs' cognate receptors expressed upon EECs.[14,34] Taken together, these data suggest that increase the production of SCFAs may modulate peptide release through EECs.[17,23,35–52]

Microbiota and Immune Pathways

The intestinal lamina propria is densely infiltrated by innate lymphoid cells that can produce an array of cytokines.[53] These cells communicate with the epithelium, intestinal microbiota, and enteric nervous system through the expression of a variety of receptors for neurotransmitters, including dopamine, serotonin, neuropeptide Y, substance P, and vasoactive intestinal peptide.[53–55] This apparatus confers upon the immune system a major role in maintaining the microbiota safely within their gut luminal area. On the other hand, signals from the microbiota govern development and functional integrity of the immune system.[55]

The importance of the gut microbiota on the immune system can be appreciated in germ-free studies. In these animals, the lymphoid tissue in the gut remains rudimentary, with no lymphoid cells projected to the lumen, impairing the initiation of the immune system.[56] In agreement, germ-free mice present grossly enlarged caeca, failure of secondary lymphoid development, impaired antibody responses, diminished number of T and B cells, and defective production of cytokines.[55,57] Indeed, early life is a critical window of development that allows the full-scale establishment of an adequate host-microbial homeostasis, and has long-term health implications. Neonates have several immune immaturities, including decreased expression of costimulatory molecules, diminished dendritic cell differentiation, and impaired phagocytosis.[58,59] A few days after colonization, the gut microbiota "teach" the host immune system through specific strains called *pathobionts* of potential inflammatory allies.[60] Pathobionts activate mucosal immune priming for the bacterial sampling process to minimize their exposure to the systemic immune system by transiently breaching the epithelial barrier, and stimulate production of antimicrobial proteins and a variety of pro-inflammatory and regulatory immune components.[61,62] When the gut microbiota community stabilizes, the immune system is continuously stimulated by several structural components of the microbial cells and responds by producing lymphocytes and cytokines.[63]

Interestingly, selective constituents of the gut microbiota shape specific aspects of adaptive and innate immunity throughout the lifespan. For example, only germ-free mice colonized with *Citrobacter rodentium, Escherichia coli,* or other commensal bacteria that can adhere to intestinal epithelium promote effective T_H17 responses,[64,65] suggesting a mechanism of induction by distinct bacteria to foster tissue-specific inflammation. On the other hand, members of the *Clostridia* class and the strain *Bacteroides fragilis* promote immunosuppressive cytokines and inhibit natural killer T-cell proliferation through bacterial production of SCFAs and polysaccharide A, respectively.[66–68] Low-level exposure of LPS to the immune cells is essential in the establishment and maintenance of mucosal homeostasis.[69] Moreover, components of LPS-sensing receptors complex are expressed by peripheral sensory neurons, including nodose ganglion of the vagus nerve, and dorsal root ganglion of the spinal cord.[70]

It is crucial for the CNS to communicate with the immune system in order to maintain homeostasis.[11,55] The immune system has at least three ways to conduct a dialogue with the brain. First, cytokines could be transported in the blood to act on the brain or LPS could trigger Toll-like receptors (TLRs) expressed by microglial cells in the CNS. Second, products of gut immune cells could directly communicate with afferent sensory fibers and/or enteric nerves to trigger signals that are ultimately relayed to the brain via the vagal pathway. Third, circulating immune cells have the potential to traffic to the brain where they can liberate cytokines and other inflammatory mediators.

In an animal model of brain inflammation, researchers observed that antibiotic treatment mitigated CNS inflammation through decreasing Th17 cells in the gut via modulation of invariant T-cell populations.[71] Moreover, the mucosal immune system has also a key role in ischemic stroke by reducing pro-inflammatory intestinal γδ T cells. These cells travel to the brain following ischemic injury, where they propagate pro-inflammatory cytokines that exacerbate tissue damage. However, in mice with reduced bacterial diversity, impaired differentiation of IL-17-producing-γδ T cells in the gut was found, with fewer of these cells migrating to the meninges or cervical lymph nodes.[72] Dysbiosis also worsened recovery after spinal cord injury and enhanced pro-inflammatory response to the lesion epicenter, while probiotic supplementation enhanced anti-inflammatory recruitment.[73] Depletion of the microbiota with antibiotics also reduced brain infiltration of $Ly6C^{hi}$ monocytes,[74] which has been associated with deficits in neurogenesis in adult mice.[74]

Despite the peripheral immune influence in the CNS, the brain has its own immune system, with microglial cells in particular playing a key role in the central immune response as the resident phagocytic cells. Once activated, microglia are involved in the synaptic remodeling to improve neuronal network signaling. In addition, microglia have recently been shown to be regulated by the microbiota.[10] In germ-free or antibiotic-treated mice, microglia cells have an immature phenotype and also fail to display an activated phenotype following systemic

LPS administration.[10] However, administration of the SCFA butyric reduces microglial activation following LPS.[75] Moreover, transplantation of gut microbiota from high-fat-diet–fed mice to normal host mice results in microglial inflammation and neuroinflammation,[76] suggesting the importance of an appropriate diet, i.e., low in fat and rich in non-digestible carbohydrates.

Gut Microbiota and Neuropsychiatric Disorders

Accumulating evidence suggests that the dysregulation of the microbiota-gut-brain axis contributes to the development of neuropsychiatric disorders.[28,77–83] Moreover, the burden of psychiatric disorders in patients with gastrointestinal (GI) diseases, and vice versa, is substantial. Both irritable bowel syndrome (IBS) and inflammatory bowel disease (IBD) are highly comorbid with depression and anxiety.[84] Children diagnosed with autism spectrum disorder (ASD) have been found to possess an altered gut microbial composition,[85–87] and many schizophrenic patients also present related GI complaints. These examples of gut–brain disturbances raise questions about whether the gut could be involved in the pathophysiology of neuropsychiatric disorders, which are generally attributed to aberrations of neurodevelopment and synaptic plasticity.

Gut Microbiota in Anxiety and Depression

Besides the comorbidity between IBD and depression and anxiety, these psychiatric conditions are often accompanied by changes in colonic motility, which, in turn, alters the composition and stability of the gut microbiota, as well as the colonic physiology and morphology.[88–90] Moreover, anxiety and depression can alter the intestinal barrier, which could allow a microbiota-driven pro-inflammatory response through the translocation of certain bacterial products from the gut.[28,91–93] For example, Maes and colleagues found in two different studies that translocation of LPS plays a role in the pathophysiology of depression in vulnerable individuals by progressively amplifying immune pathways.[92,93]

Maybe the most convincing data regarding the influence of the gut microbiota in anxiety and depression come from fecal microbiota transplantation studies. When fecal microbiota of depressive patients were transplanted into rats or mice, the depressive as well as the anxiety-related phenotypes were also transferred into the animals.[94–96] These animals also displayed an elevated kynurenine/tryptophan ratio, indicating that perhaps the depressed microbiota are facilitating the conversion of tryptophan into the deleterious metabolite kynurenine.[94] Such metabolites are part of the metabolism of key neurotransmitters such as serotonin, which has many implications for the enteric nervous system (ENS) and CNS function, and the physiology of anxiety and depression.[16,28]

Other bacterial communities are changed in the mood disorders. Yu and colleagues found that depressive-like rats had the relative abundances of the phyla Bacteroidetes significantly increased, whereas Firmicutes was markedly decreased.[97] Bailey and Coe found that maternal separation decreased fecal *Lactobacillus* in rhesus monkeys a few days post-separation.[88] Early-life stress is also able to disrupt the gut microbiota in adults rats.[98,99] In support of targeting the microbiome for psychobiotic potential[100] (Table 6.2), it has recently been shown that women who received a *Lactobacillus rhamnosus* during pregnancy and lactation had significantly lower depression and anxiety scores in the post-partum period.[101]

Interestingly, antidepressants have been shown to have antimicrobial effects *in vitro*, possibly modulating the pathophysiology of the anxiety and depression by reshaping not only brain biochemistry, but also the gut microbiota.[102] Conversely, certain antibiotics, such as β-lactams and tetracyclines, have potential antidepressant properties in rodents and humans,[103,104] while other antibiotics such as fluoroquinolones have been suggested to be associated with the development of depression and anxiety.[105,106]

Although altered gut microbiota and its effects on host metabolic phenotype during psychiatric illness remain unknown, these observations highlight the potential association of the gut microbiota with the pathophysiology of depression

Table 6.2 Clinical Evidences Targeting the Gut Microbiota for the Treatment of Neuropsychiatric Disorders

Microbial Manipulation	**Outcomes**	**Reference**
L. helveticus and *B. longum*	Reduced psychological distress	107
B-GOS	Lowered cortisol reactivity to positive versus negative stimuli	108
Fermented milk product*	Reduced brain response to an emotional faces-attention task	109
L. casei Shirota	Reduced anxiety scores and improved mood	110, 111
L. casei Shirota	Reduce constipation in PD patients	112
Mixture of *L. acidophilus, L. rhamnosus,* and *B. longum*	Improvement in the severity of autism, and GI symptoms	113

L = Lactobacillus; *B* = Bifidobacterium; *B-GOS* = Bimuno-galacto-oligosaccharides; *PD* = Parkinson's disease; *GI* = gastrointestinal

*Containing *B. animalis, Streptococcus thermophiles, L. bulgaricus,* and *L. lactis.*

and anxiety. Thus, the manipulation of the gut microbiota may be a useful tool to decode the physiological role of the gut–brain axis in psychiatric disorders. The use of germ-free rodents, animals with pathogenic bacterial infections, animals exposed to probiotic agents or to antibiotics, prebiotics, and fecal microbiota transferred from another animal or human may lead to our understanding the gut microbiota's role in these neuropsychiatric disorders.[107–113]

Gut Microbiota in Parkinson's Disease

Parkinson's disease (PD) is characterized by dopaminergic loss of the *pars compacta* of the substantia nigra and widespread neuroinflammation.[114] PD patients exhibit a range of motor-related symptoms, such as bradykinesia and tremor, with consequent difficulties in walking and gait, and impairments in posture.[115] Non-motor-related symptoms, which are commonly observed years before the motor hallmarks, include depression, cognitive impairment, sleep disturbances, and constipation, suggestive of GI dysfunction.[81,115] Approximately 80% of PD patients suffer from constipation.[116] In PD, constipation is associated with α-synuclein accumulation and neurodegeneration of nerve fibers and ganglia of the ENS, increased local inflammation, oxidative stress, and neurodegeneration.[117–119]

Preclinical evidence suggests that signals from gut microbes are required for α-synuclein-dependent motor deficits in an animal model of PD.[119] Moreover, the role of gut microbes in the aggregation of α-synuclein dependent on intact vagus nerve pathway.[120] Dopamine, a prime neurotransmitter in PD physiology, is synthetized in the brain by dopamine-producing enzymes, whose synthesis/inhibition is controlled by gut microbiota.[121] Interestingly, colonization of α-synuclein-overexpressing mice with microbiota from PD patients enhanced physical impairments, compared to microbiota transplanted from healthy human donors.[119] This suggests that alterations in the microbiome may represent a risk factor for PD, probably due to failure in the biosynthesis of dopamine.

One of the most well-known microbial associations of PD is that of *Helicobacter pylori*. PD patients have increased probability of *Helicobacter pylori* infection, which in turn, delays the absorption of levodopa, a primary source of the dopamine metabolism, and a first-line pharmacological treatment.[122] Deficit in the absorption of levodopa will result in reduced availability of dopamine, causing motor impairment.[123] Moreover, *Prevotella* bacteria are also changed in PD patients.[124] *Prevotella* synthetizes mucin and SCFAs, and reduction in its abundance has been linked with increased gut permeability.[124] The fecal microbiota of PD patients were also found to contain higher levels of pro-inflammatory–associated bacteria from the genus *Faecalibacterium* and lower levels of anti-inflammatory–associated bacteria from the genera *Blautia* and *Roseburia* when compared with healthy controls.[125] On the other hand, a recent study demonstrated elevated concentration of the SCFA butyrate, suggesting a role for SCFA in PD.[126] Accordingly, administration of SCFAs exacerbated motor deficits in germ-free mice that overexpress α-synuclein, and also displayed less activated microglia relative to conventional colonized mice of the same genotype.[119] However, treatment with sodium butyrate also reduced the neurotoxicity induced by α-synuclein in dopaminergic neurons.[127] Overall, these studies suggest that reductions in the gut microbiota diversity play a role in inducing a pro-inflammatory intestinal state, initiating a cascade that may culminate with the development and/or deterioration of PD.

Gut Microbiota in Alzheimer's Disease

Alzheimer's disease (AD) is characterized by the accumulation of amyloid beta (Aβ) peptide plaques and toxic tau fibrils.[128] Aβ deposits contribute to neuronal dysfunction and are considered the source of the local inflammatory response that affects many areas in the brain, including the hippocampus; the frontal, temporal and parietal cortices; and the cholinergic basal forebrain.[129] AD symptoms frequently begin with the loss of ability to form new memories, and culminate in severe long-term memory loss.[128] AD patients also have non-cognitive symptoms, such as physical disability.[128]

The inflammatory profile of AD suggests a causal link between dysregulation of the microbiota and systemic inflammation, which may initiate or exacerbate the neurodegeneration occurring in the brain of AD patients. The gut microbiota may also influence AD through molecular mimicry. For instance, the bacterial extracellular amyloid protein curli is sensed by the A11 Aβ oligomer antibody, which interestingly suggests that there are configurations of amyloid proteins that have the same structure as microbial proteins.[130] Cattaneo and colleagues identified an increase of pro-inflammatory *Escherichia/Shigella* bacterial genera in fecal samples from AD patients with prominent Aβ deposits.[131] The bacterial profile and Aβ deposits were correlated positively with increased expression of the pro-inflammatory cytokines IL-1β and CXCL2 in the bloodstream.[131] Moreover, levels of the *Escherichia coli* metabolite were elevated in the brain parenchyma and blood vessels of AD patients.[132] The authors also found that LPS co-localizes with $A\beta_{1-40}$ in amyloid plaques, suggesting that bacterial components are translocated from the gut to the brain in AD,[132] corroborating data of increased gut permeability observed in elderly people.

Preclinical data reinforce the view that the microbiota–gut–brain axis may play a role in AD. Germ-free and antibiotic-treated mice have deficits in memory tasks, and reduced expression of brain-derived neurotrophic factor (BDNF) in the hippocampus, as well as microglial immaturity.[37,44,82,83,133] BDNF is crucial for synaptic plasticity and cognitive function, and it has been found to be decreased in AD patients, correlating with increased Aβ deposits.[134] Moreover, in a recent study with aged germ-free mice overexpressing amyloid precursor protein and presenilin 1 (APP/PS1), reduced levels of $A\beta_{42}$ were found when compared with conventional mice, along with a reduction in microglial activation.[135] Germ-free APP/PS1 mice displayed increased levels of Aβ-degrading enzymes and other degrading enzymes, suggesting that, in the absence of microbiota, the murine brain is better equipped to target the amyloid pathology.[135] When germ-free APP/PS1 mice were colonized with the microbiota from aged, conventionally raised APP/PS1, increased cerebral Aβ levels were found.[135] Reduced microglial and astrocyte accumulation surrounding amyloid plaques in the hippocampus was also found in APP/PS1 transgenic mice under antibiotic treatment.[136] Interestingly, antibiotic treatment was found to decrease insoluble Aβ plaques while leading to a corresponding increase in the levels of soluble Aβ plaque deposition and circulating cytokines, suggesting that the gut microbiota have some bearing on amyloid load in the brain.[136]

In the 5xFAD transgenic mouse model of AD (which rapidly develops amyloid plaques), levels of Aβ precursor protein are elevated in the brain and also in the GI tract.[137] Moreover, 5xFAD mice displayed increased levels of the pro-inflammatory *Clostridium leptum* species.[137] Taken together, the depletion of the gut microbiota leads to microglial immaturity through impaired innate immune responses and can contribute to the pathogenesis of AD.

Gut Microbiota, Sociability, and Neurodevelopmental Disorders

There is growing evidence to support a role for the microbiota in regulating social behavior in mammals, including that relevant to ASD.[15,80] One reason is the higher prevalence of GI abnormalities in ASD, reported in up to 90% of the patients.[51] This suggests an association between ASD and dysregulation of the gut microbiota. Indeed, ASD children show greater gut microbiota diversity, with altered levels of Bacteroidetes and Firmicutes.[138,139] However, large-scale studies are needed with appropriate consideration given to dietary differences in ASD patients.[4,87] Indeed, different studies found that ASD subjects displayed higher levels of *Clostridium*.[140,141] Interestingly, ASD children treated with vancomycin had improved sociability and decreased GI symptoms.[142] On the other hand, epidemiological studies have demonstrated that the use of antibiotics in one's early life can be a trigger for the development of ASD,[143] indicating a probable deficit in the immune system maturation. Finally, a small pilot study demonstrated that transfer of a standardized human gut microbiota mixture could improve GI and behavioral symptoms in autistic children.[144]

In preclinical studies, autistic symptoms can be modelled in rodents, and alterations in the gut microbiome composition are associated with ASD-like phenotypes. Germ-free and antibiotic-treated mice present social deficits in social behavior tests, indicating an autism-like phenotype.[78,133,52] These deficits can be reversed following recolonization with microbiota, confirming the gut microbiota's role in social behaviors,[78] although a study reported opposing behavioral results, suggesting differences in animal breeding/husbandry techniques and the behavioral protocol used.[79]

In the *in utero* valproate animal model of autism, the pups display an autistic-like phenotype along with intestinal inflammation and increases in the Firmicutes/Bacteroidetes ratio and butyrate concentrations, suggesting alterations in microbial metabolism.[80,145] However, another study using BTBR mice, which display autistic-like features, found that those mice had a reduced Firmicutes/Bacteroidetes ratio, along with an increased abundance of species such as *Akkermansia mucinphilia* and reduced *Bifidobacterium*, suggesting that specific microbiota dysregulation is behind the autistic phenotype as observed in ASD children.[146] In fact, we have shown a reduction in the relative abundance of very particular bacterial taxa; namely, bile-metabolizing *Bifidobacterium* and *Blautia species* are associated with GI dysfunction and impaired social interactions in BTBR mice.[15] Another animal model of autism, the maternal immune activation (MIA), has also been reported to change gut microbiota.[51,147] Interestingly, treating MIA offspring with *Bacteroides fragilis* transplanted from humans corrected gut permeability, microbial composition, and autistic-like behaviors.[51]

It can be difficult to distinguish whether these modifications are the cause of the condition, a direct result of the condition, or whether they can be attributed to the atypical eating behavior exhibited by many autistic children. However, these results emphasize the potentially important changes of behavior that could be dependent on GI factors.

CONCLUSION

We are at the very early stages of understanding the complexities of communication along the microbiota-gut-brain axis. However, there is already strong evidence to support the influence of the enteric microbiome on brain function in health and disease, suggesting that the gut microbiome plays a crucial role in normal brain development, as well as the modulation of host physiological systems important in stress-related disorders. Understanding how the gut microbes might influence the upstream cascade to the brain should be a central objective of future research in this field. Rising interest in this area of research will no doubt lead to greater insights into the mechanism(s) underlying microbiome-gut-brain communication, and provide a novel understanding of the potential for microbe-based therapeutic strategies that may aid in the treatment of neuropsychiatric disorders.

ACKNOWLEDGMENTS

This work was supported by Science Foundation Ireland in the form of a center grant (APC Microbiome Institute; Grant No. SFI/12/RC/2273). The advice of Simon Spichak is gratefully acknowledged.

REFERENCES

1. Metchnikoff E. *The prolongation of life: optimistic studies*. New York: G.P. Putnam's Sons; 1910.
2. Nicholson JK, Holmes E, Kinross J, et al. Host-gut microbiota metabolic interactions. *Science*. 2012;336(6086):1262–1267. doi:10.1126/science.1223813
3. Sandhu KV, Sherwin E, Schellekens H, Stanton C, Dinan TG, Cryan JF. Feeding the microbiota-gut-brain axis: diet, microbiome, and neuropsychiatry. *Transl Res*. 2017;179:223–244. doi:10.1016/j.trsl.2016.10.002
4. Sherwin E, Dinan TG, Cryan JF. Recent developments in understanding the role of the gut microbiota in brain health and disease. *Ann N Y Acad Sci*. 2018;1420(1):5–25. doi:10.1111/nyas.13416
5. Dinan TG, Cryan JF. Gut instincts: microbiota as a key regulator of brain development, ageing and neurodegeneration. *J Physiol*. 2017 Jan 15;595(2):489–503. doi:10.1113/JP273106
6. Turnbaugh PJ, Ley RE, Hamady M, Fraser-Liggett CM, Knight R, Gordon JI. The Human Microbiome Project. *Nature*. 2007;449(7164):804–810. doi:10.1038/nature06244
7. Qin J, Li R, Raes J, et al. A human gut microbial gene catalogue established by metagenomic sequencing. *Nature*. 2010;464(7285):59–65. doi:10.1038/nature08821
8. Eckburg PB, Bik EM, Bernstein CN, et al. Diversity of the human intestinal microbial flora. *Science*. 2005;308(5728):1635–1638. doi:10.1126/science.1110591
9. Scott KP, Antoine J-M, Midtvedt T, van Hemert S. Manipulating the gut microbiota to maintain health and treat disease. *Microb Ecol Health Dis*. 2015;26:25877. doi:10.3402/mehd.v26.25877
10. Erny D, Hrabě de Angelis AL, Jaitin D, et al. Host microbiota constantly control maturation and function of microglia in the CNS. *Nat Neurosci*. 2015;18(7):965–977. doi:10.1038/nn.4030

11. El Aidy S, Dinan TG, Cryan JF. Gut microbiota: the conductor in the orchestra of immune–neuroendocrine communication. *Clin Ther.* 2015;37(5):954–967. doi:10.1016/j.clinthera.2015.03.002
12. Latorre R, Sternini C, De Giorgio R, Greenwood-Van Meerveld B. Enteroendocrine cells: a review of their role in brain-gut communication. *Neurogastroenterol Motil.* 2016;28(5):620–630. doi:10.1111/nmo.12754
13. Sudo N, Chida Y, Aiba Y, et al. Postnatal microbial colonization programs the hypothalamic-pituitary-adrenal system for stress response in mice. *J Physiol.* 2004;558(1):263–275. doi:10.1113/jphysiol.2004.063388
14. Stilling RM, van de Wouw M, Clarke G, Stanton C, Dinan TG, Cryan JF. The neuropharmacology of butyrate: the bread and butter of the microbiota-gut-brain axis? *Neurochem Int.* 2016;99:110–132. doi:10.1016/j.neuint.2016.06.011
15. Golubeva AV., Joyce SA, Moloney G, et al. Microbiota-related changes in bile acid & tryptophan metabolism are associated with gastrointestinal dysfunction in a mouse model of autism. *E Bio Medicine.* 2017;24:166–178. doi:10.1016/j.ebiom.2017.09.020
16. O'Mahony SM, Clarke G, Borre YE, Dinan TG, Cryan JF. Serotonin, tryptophan metabolism and the brain-gut-microbiome axis. *Behav Brain Res.* 2015;277:32–48. doi:10.1016/j.bbr.2014.07.027
17. Bravo JA, Forsythe P, Chew MV, et al. Ingestion of *Lactobacillus* strain regulates emotional behavior and central GABA receptor expression in a mouse via the vagus nerve. *Proc Natl Acad Sci U S A.* 2011;108(38):16050–16055. doi:10.1073/pnas.1102999108
18. Stilling RM, Dinan TG, Cryan JF. Microbial genes, brain & behaviour—epigenetic regulation of the gut-brain axis. *Genes Brain Behav.* 2014;13(1):69–86. doi:10.1111/gbb.12109
19. Arentsen T, Qian Y, Gkotzis S, et al. The bacterial peptidoglycan-sensing molecule Pglyrp2 modulates brain development and behavior. *Mol Psychiatry.* 2017;22(2):257–266. doi:10.1038/mp.2016.182
20. Goehler LE, Gaykema RPAA, Opitz N, Reddaway R, Badr N, Lyte M. Activation in vagal afferents and central autonomic pathways: early responses to intestinal infection with *Campylobacter jejuni. Brain Behav Immun.* 2005;19(4):334–344. doi:10.1016/j.bbi.2004.09.002
21. Bercik P, Denou E, Collins J, et al. The intestinal microbiota affect central levels of brain-derived neurotropic factor and behavior in mice. *Gastroenterology.* 2011;141(2):599–609. doi:10.1053/j.gastro.2011.04.052
22. Poutahidis T, Kearney SM, Levkovich T, et al. Microbial symbionts accelerate wound healing via the neuropeptide hormone oxytocin. *PLoS One.* 2013;8(10):e78898. doi:10.1371/journal.pone.0078898
23. Bercik P, Park AJ, Sinclair D, et al. The anxiolytic effect of *Bifidobacterium longum* NCC3001 involves vagal pathways for gut–brain communication. *Neurogastroenterol Motil.* 2011;23(12):1132–1139. doi:10.1111/j.1365-2982.2011.01796.x
24. Marvel FA, Chen CC, Badr N, Gaykema RPA, Goehler LE. Reversible inactivation of the dorsal vagal complex blocks lipopolysaccharide-induced social withdrawal and c-Fos expression in central autonomic nuclei. *Brain Behav Immun.* 2004;18(2):123–134. doi:10.1016/j.bbi.2003.09.004

25. Bellono NW, Bayrer JR, Leitch DB, et al. Enterochromaffin cells are gut chemosensors that couple to sensory neural pathways. *Cell.* 2017;170(1):185–198. e16. doi:10.1016/j.cell.2017.05.034
26. Okano-Matsumoto S, McRoberts JA, Taché Y, Adelson DW. Electrophysiological evidence for distinct vagal pathways mediating CCK-evoked motor effects in the proximal versus distal stomach. *J Physiol.* 2011;589(Pt 2):371–393. doi:10.1113/jphysiol.2010.196832
27. Dockray GJ. Enteroendocrine cell signalling via the vagus nerve. *Curr Opin Pharmacol.* 2013;13(6):954–958. doi:10.1016/j.coph.2013.09.007
28. Lach G, Schellekens H, Dinan TG, Cryan JF. Anxiety, depression, and the microbiome: a role for gut peptides. *Neurotherapeutics.* 2018 Jan;15(1):36–59. doi:10.1007/s13311-017-0585-0
29. Duca FA, Swartz TD, Sakar Y, Covasa M. Increased oral detection, but decreased intestinal signaling for fats in mice lacking gut microbiota. *PLoS One.* 2012;7(6):e39748. doi:10.1371/journal.pone.0039748
30. Breton J, Tennoune N, Lucas N, et al. Gut commensal *E. coli* proteins activate host satiety pathways following nutrient-induced bacterial growth. *Cell Metab.* 2016;23(2):324–334. doi:10.1016/j.cmet.2015.10.017
31. Cani PD, Dewever C, Delzenne NM. Inulin-type fructans modulate gastrointestinal peptides involved in appetite regulation (glucagon-like peptide-1 and ghrelin) in rats. *Br J Nutr.* 2004;92(3):521–526. doi:10.1079/BJN20041225
32. Cani PD, Neyrinck AM, Maton N, Delzenne NM. Oligofructose promotes satiety in rats fed a high-fat diet: involvement of glucagon-like peptide-1. *Obes Res.* 2005;13(6):1000–1007. doi:10.1038/oby.2005.117
33. Cani PD, Joly E, Horsmans Y, Delzenne NM. Oligofructose promotes satiety in healthy human: a pilot study. *Eur J Clin Nutr.* 2006;60(5):567–572. doi:10.1038/sj.ejcn.1602350
34. Nøhr MK, Pedersen MH, Gille A, et al. GPR41/FFAR3 and GPR43/FFAR2 as cosensors for short-chain fatty acids in enteroendocrine cells vs FFAR3 in enteric neurons and FFAR2 in enteric leukocytes. *Endocrinology.* 2013;154(10):3552–3564. doi:10.1210/en.2013-1142
35. Tarr AJ, Galley JD, Fisher SE, Chichlowski M, Berg BM, Bailey MT. The prebiotics 3'Sialyllactose and 6'Sialyllactose diminish stressor-induced anxiety-like behavior and colonic microbiota alterations: evidence for effects on the gut–brain axis. *Brain Behav Immun.* 2015;50:166–177. doi:10.1016/j.bbi.2015.06.025
36. Ohland CL, Kish L, Bell H, et al. Effects of *Lactobacillus helveticus* on murine behavior are dependent on diet and genotype and correlate with alterations in the gut microbiome. *Psychoneuroendocrinology.* 2013;38(9):1738–1747. doi:10.1016/j.psyneuen.2013.02.008
37. Wang T, Hu X, Liang S, et al. *Lactobacillus fermentum* NS9 restores the antibiotic induced physiological and psychological abnormalities in rats. *Benef Microbes.* 2015;6(5):707–717. doi:10.3920/BM2014.0177
38. Burokas A, Arboleya S, Moloney RD, et al. Targeting the microbiota-gut-brain axis: prebiotics have anxiolytic and antidepressant-like effects and reverse the impact of chronic stress in mice. *Biol Psychiatry.* 2017;39(0):763–781. doi:10.1016/j.biopsych.2016.12.031

39. Desbonnet L, Garrett L, Clarke G, Kiely B, Cryan JF, Dinan TG. Effects of the probiotic *Bifidobacterium infantis* in the maternal separation model of depression. *Neuroscience.* 2010;170(4):1179–1188. doi:10.1016/j.neuroscience.2010.08.005
40. Liu Y-W, Liu W-H, Wu C-C, et al. Psychotropic effects of *Lactobacillus plantarum* PS128 in early life-stressed and naïve adult mice. *Brain Res.* 2016;1631:1–12. doi:10.1016/j.brainres.2015.11.018
41. Wei Y Bin, Melas PA, Wegener G, Mathe AA, Lavebratt C. Antidepressant-like effect of sodium butyrate is associated with an increase in tet1 and in 5-hydroxymethylation levels in the BDNF gene. *Int J Neuropsychopharmacol.* 2015;18(2):1–10. doi:10.1093/ijnp/pyu032
42. Ait-Belgnaoui A, Colom A, Braniste V, et al. Probiotic gut effect prevents the chronic psychological stress-induced brain activity abnormality in mice. *Neurogastroenterol Motil.* 2014;26(4):510–520. doi:10.1111/nmo.12295
43. Noori N, Bangash MY, Motaghinejad M, Hosseini P, Noudoost B. Kefir protective effects against nicotine cessation-induced anxiety and cognition impairments in rats. *Adv Biomed Res.* 2014;3(1):251. doi:10.4103/2277-9175.146377
44. Gareau MG, Wine E, Rodrigues DM, et al. Bacterial infection causes stress-induced memory dysfunction in mice. *Gut.* 2011;60(3):307–317. doi:10.1136/gut.2009.202515
45. Davari S, Talaei SA, Alaei H, Salami M. Probiotics treatment improves diabetes-induced impairment of synaptic activity and cognitive function: behavioral and electrophysiological proofs for microbiome-gut-brain axis. *Neuroscience.* 2013;240:287–296. doi:10.1016/j.neuroscience.2013.02.055
46. Yana MH, Wang X, Zhu X. Mitochondrial defects and oxidative stress in Alzheimer disease and Parkinson disease. *Free Radic Biol Med.* 2013;62:90–101. doi:10.1016/j.freeradbiomed.2012.11.014
47. Liu J, Sun J, Wang F, et al. Neuroprotective effects of *Clostridium butyricum* against vascular dementia in mice via metabolic butyrate. *Biomed Res Int.* 2015;2015:412946. doi:10.1155/2015/412946
48. Distrutti E, O'Reilly JA, McDonald C, et al. Modulation of intestinal microbiota by the probiotic VSL#3 resets brain gene expression and ameliorates the age-related deficit in LTP. *PLoS One.* 2014;9(9):e106503. doi:10.1371/journal.pone.0106503
49. Mori T, Koyama N, Guillot-Sestier MV, Tan J, Town T. Ferulic acid is a nutraceutical β-secretase modulator that improves behavioral impairment and Alzheimer-like pathology in transgenic mice. Ikezu T, ed. *PLoS One.* 2013;8(2):e55774. doi:10.1371/journal.pone.0055774
50. Surwase SN, Jadhav JP. Bioconversion of l-tyrosine to l-DOPA by a novel bacterium Bacillus sp. JPJ. *Amino Acids.* 2011;41(2):495–506. doi:10.1007/s00726-010-0768-z
51. Hsiao EY, McBride SW, Hsien S, et al. Microbiota modulate behavioral and physiological abnormalities associated with neurodevelopmental disorders. *Cell.* 2013;155(7):1451–1463. doi:10.1016/j.cell.2013.11.024
52. Buffington SA, Di Prisco GV, Auchtung TA, Ajami NJ, Petrosino JF, Costa-Mattioli M. Microbial reconstitution reverses maternal diet-induced social and synaptic deficits in offspring. *Cell.* 2016;165(7):1762–1775. doi:10.1016/j.cell.2016.06.001

53. Spits H, Cupedo T. Innate lymphoid cells: emerging insights in development, lineage relationships, and function. *Annu Rev Immunol.* 2012;30(1):647–675. doi:10.1146/annurev-immunol-020711-075053
54. Geuking MB, Köller Y, Rupp S, McCoy KD. The interplay between the gut microbiota and the immune system. *Gut Microbes.* 2014;5(3):411–418. doi:10.4161/gmic.29330
55. Lee YK, Mazmanian SK. Has the microbiota played a critical role in the evolution of the adaptive immune system? *Science.* 2010;330(6012):1768–1773. doi:10.1126/science.1195568
56. Macpherson AJ, Harris NL. Opinion: interactions between commensal intestinal bacteria and the immune system. *Nat Rev Immunol.* 2004;4(6):478–485. doi:10.1038/nri1373
57. Ikeda M, Hamada K, Sumitomo N, Okamoto H, Sakakibara B. Serum amyloid A, cytokines, and corticosterone responses in germfree and conventional mice after lipopolysaccharide injection. *Biosci Biotechnol Biochem.* 1999;63(6):1006–1010. doi:10.1271/bbb.63.1006
58. Velilla PA, Rugeles MT, Chougnet CA. Defective antigen-presenting cell function in human neonates. *Clin Immunol.* 2006;121(3):251–259. doi:10.1016/j.clim.2006.08.010
59. Kotiranta-Ainamo A, Rautonen J, Rautonen N. Imbalanced cytokine secretion in newborns. *Biol Neonate.* 2004;85(1):55–60. doi:10.1159/000074959
60. El Aidy S, van Baarlen P, Derrien M, et al. Temporal and spatial interplay of microbiota and intestinal mucosa drive establishment of immune homeostasis in conventionalized mice. *Mucosal Immunol.* 2012;5(5):567–579. doi:10.1038/mi.2012.32
61. El Aidy S, Derrien M, Aardema R, et al. Transient inflammatory-like state and microbial dysbiosis are pivotal in establishment of mucosal homeostasis during colonisation of germ-free mice. *Benef Microbes.* 2014;5(1):67–77. doi:10.3920/BM2013.0018
62. Mazmanian SK, Cui HL, Tzianabos AO, Kasper DL. An immunomodulatory molecule of symbiotic bacteria directs maturation of the host immune system. *Cell.* 2005;122(1):107–118. doi:10.1016/j.cell.2005.05.007
63. Heumann D, Barras C, Severin A, Glauser MP, Tomasz A. Gram-positive cell walls stimulate synthesis of tumor necrosis factor alpha and interleukin-6 by human monocytes. *Infect Immun.* 1994;62(7):2715–2721.
64. Atarashi K, Tanoue T, Ando M, et al. Th17 cell induction by adhesion of microbes to intestinal epithelial cells. *Cell.* 2015;163(2):367–380. doi:10.1016/j.cell.2015.08.058
65. Ivanov II, Atarashi K, Manel N, et al. Induction of intestinal Th17 cells by segmented filamentous bacteria. *Cell.* 2009;139(3):485–498. doi:10.1016/j.cell.2009.09.033
66. Furusawa Y, Obata Y, Fukuda S, et al. Commensal microbe-derived butyrate induces the differentiation of colonic regulatory T cells. *Nature.* 2013;504(7480):446–450. doi:10.1038/nature12721
67. Smith PM, Howitt MR, Panikov N, et al. The microbial metabolites, short-chain fatty acids, regulate colonic Treg cell homeostasis. *Science.* 2013;341(6145):569–573. doi:10.1126/science.1241165

68. Round JL, Mazmanian SK. Inducible Foxp3+ regulatory T cell development by a commensal bacterium of the intestinal microbiota. *Proc Natl Acad Sci U S A.* 2010;107(27):12204–12209. doi:10.1073/pnas.0909122107
69. Bested AC, Logan AC, Selhub EM, et al. Intestinal microbiota, probiotics and mental health: from Metchnikoff to modern advances: Part II—contemporary contextual research. *Gut Pathog.* 2013;5(1):3. doi:10.1186/1757-4749-5-3
70. Yang NJ, Chiu IM. Bacterial signaling to the nervous system via toxins and metabolites. *J Mol Biol.* 2017;429(5):587–605. doi:10.1016/j.jmb.2016.12.023
71. Yokote H, Miyake S, Croxford JL, Oki S, Mizusawa H, Yamamura T. NKT cell-dependent amelioration of a mouse model of multiple sclerosis by altering gut flora. *Am J Pathol.* 2008;173(6):1714–1723. doi:10.2353/ajpath.2008.080622
72. Houlden A, Goldrick M, Brough D, et al. Brain injury induces specific changes in the caecal microbiota of mice via altered autonomic activity and mucoprotein production. *Brain Behav Immun.* 2016;57:10–20. doi:10.1016/j.bbi.2016.04.003
73. Kigerl KA, Hall JCE, Wang L, Mo X, Yu Z, Popovich PG. Gut dysbiosis impairs recovery after spinal cord injury. *J Exp Med.* 2016;213(12):2603–2620. doi:10.1084/jem.20151345
74. Möhle L, Mattei D, Heimesaat MM, et al. Ly6Chi monocytes provide a link between antibiotic-induced changes in gut microbiota and adult hippocampal neurogenesis. *Cell Rep.* 2016;15(9):1945–1956. doi:10.1016/j.celrep.2016.04.074
75. Kim H-S, Whang S-Y, Woo M-S, Park J-S, Kim W-K, Han I-O. Sodium butyrate suppresses interferon-gamma-, but not lipopolysaccharide-mediated induction of nitric oxide and tumor necrosis factor-alpha in microglia. *J Neuroimmunol.* 2004;151(1–2):85–93. doi:10.1016/j.jneuroim.2004.02.006
76. Bruce-Keller AJ, Salbaum JM, Luo M, et al. Obese-type gut microbiota induce neurobehavioral changes in the absence of obesity. *Biol Psychiatry.* 2015;77(7):607–615. doi:10.1016/j.biopsych.2014.07.012
77. Lach G, Morais LH, Costa APR, Hoeller AA. Envolvimento da flora intestinal na modulação de doenças psiquiátricas. *Vittalle—Rev Ciên Saúde.* 2017;29(1):64–82.
78. Desbonnet L, Clarke G, Shanahan F, Dinan TG, Cryan JF. Microbiota is essential for social development in the mouse. *Mol Psychiatry.* 2014;19(2):146–148. doi:10.1038/mp.2013.65
79. Arentsen T, Raith H, Qian Y, Forssberg H, Heijtz RD. Host microbiota modulates development of social preference in mice. *Microb Ecol Heal Dis.* 2015;26:29719. doi:10.3402/mehd.v26.29719
80. de Theije CGM, Wopereis H, Ramadan M, et al. Altered gut microbiota and activity in a murine model of autism spectrum disorders. *Brain Behav Immun.* 2014;37:197–206. doi:10.1016/j.bbi.2013.12.005
81. Felice VD, Quigley EM, Sullivan AM, O'Keeffe GW, O'Mahony SM. Microbiota-gut-brain signalling in Parkinson's disease: implications for non-motor symptoms. *Parkinsonism Relat Disord.* 2016;27:1–8. doi:10.1016/j.parkreldis.2016.03.012
82. Fröhlich EE, Farzi A, Mayerhofer R, et al. Cognitive impairment by antibiotic-induced gut dysbiosis: analysis of gut microbiota–brain communication. *Brain Behav Immun.* 2016;56:140–155. doi:10.1016/j.bbi.2016.02.020
83. Hoban AE, Moloney RD, Golubeva AV, et al. Behavioural and neurochemical consequences of chronic gut microbiota depletion during adulthood in the rat. *Neuroscience.* 2016;339:463–477. doi:10.1016/j.neuroscience.2016.10.003

84. Garakani A, Win T, Virk S, Gupta S, Kaplan D, Masand PS. Comorbidity of irritable bowel syndrome in psychiatric patients: a review. *Am J Ther.* 2003;10(1):61–67. doi:10.1097/00045391-200301000-00014
85. de Magistris L, Familiari V, Pascotto A, et al. Alterations of the intestinal barrier in patients with autism spectrum disorders and in their first-degree relatives. *J Pediatr Gastroenterol Nutr.* 2010;51(4):418–424. doi:10.1097/MPG.0b013e3181dcc4a5
86. Mangiola F, Ianiro G, Franceschi F, Fagiuoli S, Gasbarrini G, Gasbarrini A. Gut microbiota in autism and mood disorders. *World J Gastroenterol.* 2016;22(1):361–368. doi:10.3748/wjg.v22.i1.361
87. Mayer EA, Padua D, Tillisch K. Altered brain-gut axis in autism: comorbidity or causative mechanisms? *BioEssays.* 2014;36(10):933–939. doi:10.1002/bies.201400075
88. Bailey MT, Coe CL. Maternal separation disrupts the integrity of the intestinal microflora in infant rhesus monkeys. *Dev Psychobiol.* 1999;35(2):146–155. doi:10.1002/(SICI)1098-2302(199909)35:2<146::AID-DEV7>3.0.CO;2-G
89. Park AJ, Collins J, Blennerhassett PA, et al. Altered colonic function and microbiota profile in a mouse model of chronic depression. *Neurogastroenterol Motil.* 2013;25(9):733–e575. doi:10.1111/nmo.12153
90. O'Malley D, Julio-Pieper M, Gibney SM, Dinan TG, Cryan JF. Distinct alterations in colonic morphology and physiology in two rat models of enhanced stress-induced anxiety and depression-like behaviour. *Stress.* 2010;13(2):114–122. doi:10.3109/10253890903067418
91. Kelly JR, Kennedy PJ, Cryan JF, Dinan TG, Clarke G, Hyland NP. Breaking down the barriers: the gut microbiome, intestinal permeability and stress-related psychiatric disorders. *Front Cell Neurosci.* 2015;9:392. doi:10.3389/fncel.2015.00392
92. Maes M, Kubera M, Leunis J-C, Berk M. Increased IgA and IgM responses against gut commensals in chronic depression: further evidence for increased bacterial translocation or leaky gut. *J Affect Disord.* 2012;141(1):55–62. doi:10.1016/j.jad.2012.02.023
93. Maes M, Kubera M, Leunis J-C. The gut-brain barrier in major depression: intestinal mucosal dysfunction with an increased translocation of LPS from gram negative enterobacteria (leaky gut) plays a role in the inflammatory pathophysiology of depression. *Neuro Endocrinol Lett.* 2008;29(1):117–124.
94. Kelly JR, Borre Y, O' Brien C, et al. Transferring the blues: depression-associated gut microbiota induces neurobehavioural changes in the rat. *J Psychiatr Res.* 2016;82:109–118. doi:10.1016/j.jpsychires.2016.07.019
95. Zheng P, Zeng B, Zhou C, et al. Gut microbiome remodeling induces depressive-like behaviors through a pathway mediated by the host's metabolism. *Mol Psychiatry.* 2016;21(6):786–796. doi:10.1038/mp.2016.44
96. De Palma G, Lynch MDJ, Lu J, et al. Transplantation of fecal microbiota from patients with irritable bowel syndrome alters gut function and behavior in recipient mice. *Sci Transl Med.* 2017;9(379):1–15. doi:10.1126/scitranslmed.aaf6397
97. Yu M, Jia H, Zhou C, et al. Variations in gut microbiota and fecal metabolic phenotype associated with depression by 16S rRNA gene sequencing and LC/MS-based metabolomics. *J Pharm Biomed Anal.* 2017;138:231–239. doi:10.1016/j.jpba.2017.02.008
98. O'Mahony SM, Marchesi JR, Scully P, et al. Early life stress alters behavior, immunity, and microbiota in rats: implications for irritable bowel syndrome and psychiatric illnesses. *Biol Psychiatry.* 2009;65(3):263–267. doi:10.1016/j.biopsych.2008.06.026

99. De Palma G, Blennerhassett P, Lu J, et al. Microbiota and host determinants of behavioural phenotype in maternally separated mice. *Nat Commun.* 2015;6:7735. doi:10.1038/ncomms8735
100. Dinan TG, Stanton C, Cryan JF. Psychobiotics: a novel class of psychotropic. *Biol Psychiatry.* 2013;74(10):720–726. doi:10.1016/j.biopsych.2013.05.001
101. Slykerman RF, Hood F, Wickens K, et al. Effect of *Lactobacillus rhamnosus* HN001 in pregnancy on postpartum symptoms of depression and anxiety: a randomised double-blind placebo-controlled trial. *EBioMedicine.* 2017;24:159–165. doi:10.1016/j.ebiom.2017.09.013
102. Munoz-Bellido JL, Munoz-Criado S, Garcìa-Rodrìguez JA. Antimicrobial activity of psychotropic drugs. Selective serotonin reuptake inhibitors. *Int J Antimicrob Agents.* 2000;14(3):177–180. doi:10.1016/S0924-8579(99)00154-5
103. Ferreira Mello BS, Monte AS, McIntyre RS, et al. Effects of doxycycline on depressive-like behavior in mice after lipopolysaccharide (LPS) administration. *J Psychiatr Res.* 2013;47(10):1521–1529. doi:10.1016/j.jpsychires.2013.06.008
104. Miyaoka T, Wake R, Furuya M, et al. Minocycline as adjunctive therapy for patients with unipolar psychotic depression: an open-label study. *Prog Neuro-Psychopharmacology Biol Psychiatry.* 2012;37(2):222–226. doi:10.1016/j.pnpbp.2012.02.002
105. Ahmed AIA, van der Heijden FMMA, van den Berkmortel H, Kramers K. A man who wanted to commit suicide by hanging himself: an adverse effect of ciprofloxacin. *Gen Hosp Psychiatry.* 2011;33(1):82.e5–e7. doi:10.1016/j.genhosppsych.2010.07.002
106. Kaur K, Fayad R, Saxena A, et al. Fluoroquinolone-related neuropsychiatric and mitochondrial toxicity: a collaborative investigation by scientists and members of a social network. *J Community Support Oncol.* 2016;14(2):54–65. doi:10.12788/jcso.0167
107. Messaoudi M, Violle N, Bisson J-F, Desor D, Javelot H, Rougeot C. Beneficial psychological effects of a probiotic formulation (*Lactobacillus helveticus* R0052 and *Bifidobacterium longum* R0175) in healthy human volunteers. *Gut Microbes.* 2011;2(4):256–261. doi:10.4161/gmic.2.4.16108
108. Schmidt K, Cowen PJ, Harmer CJ, Tzortzis G, Errington S, Burnet PWJ. Prebiotic intake reduces the waking cortisol response and alters emotional bias in healthy volunteers. *Psychopharmacology (Berl).* 2015;232(10):1793–1801. doi:10.1007/s00213-014-3810-0
109. Tillisch K, Labus J, Kilpatrick L, et al. Consumption of fermented milk product with probiotic modulates brain activity. *Gastroenterology.* 2013;144(7):1394–1401.e4. doi:10.1053/j.gastro.2013.02.043
110. Benton D, Williams C, Brown A. Impact of consuming a milk drink containing a probiotic on mood and cognition. *Eur J Clin Nutr.* 2007;61(3):355–361. doi:10.1038/sj.ejcn.1602546
111. Rao AV, Bested AC, Beaulne TM, et al. A randomized, double-blind, placebo-controlled pilot study of a probiotic in emotional symptoms of chronic fatigue syndrome. *Gut Pathog.* 2009;1(1):6. doi:10.1186/1757-4749-1-6
112. Cassani E, Privitera G, Pezzoli G, et al. Use of probiotics for the treatment of constipation in Parkinson's disease patients. *Minerva Gastroenterol Dietol.* 2011;57(2):117–121.

113. Shaaban SY, El Gendy YG, Mehanna NS, et al. The role of probiotics in children with autism spectrum disorder: a prospective, open-label study. *Nutr Neurosci.* 2017:1–6. doi:10.1080/1028415X.2017.1347746
114. Rousseaux MWC, Shulman JM, Jankovic J. Progress toward an integrated understanding of Parkinson's disease. *F1000Research.* 2017;6:1121. doi:10.12688/f1000research.11820.1
115. Ba F, Obaid M, Wieler M, Camicioli R, Martin WRW. Parkinson disease: the relationship between non-motor symptoms and motor phenotype. *Can J Neurol Sci/J Can Sci Neurol.* 2016;43(2):261–267. doi:10.1017/cjn.2015.328
116. Ueki A, Otsuka M. Life style risks of Parkinson's disease: association between decreased water intake and constipation. *J Neurol.* 2004;251(S7):vII18–23. doi:10.1007/s00415-004-1706-3
117. Cersosimo MG, Benarroch EE. Pathological correlates of gastrointestinal dysfunction in Parkinson's disease. *Neurobiol Dis.* 2012;46(3):559–564. doi:10.1016/j.nbd.2011.10.014
118. Forsyth CB, Shannon KM, Kordower JH, et al. Increased intestinal permeability correlates with sigmoid mucosa alpha-synuclein staining and endotoxin exposure markers in early Parkinson's disease. Oreja-Guevara C, ed. *PLoS One.* 2011;6(12):e28032. doi:10.1371/journal.pone.0028032
119. Sampson TR, Debelius JW, Thron T, et al. Gut microbiota regulate motor deficits and neuroinflammation in a model of Parkinson's disease. *Cell.* 2016;167(6):1469–1480.e12. doi:10.1016/j.cell.2016.11.018
120. Holmqvist S, Chutna O, Bousset L, et al. Direct evidence of Parkinson pathology spread from the gastrointestinal tract to the brain in rats. *Acta Neuropathol.* 2014;128(6):805–820. doi:10.1007/s00401-014-1343-6
121. Eisenhofer G, Aneman Å, Friberg P, et al. Substantial production of dopamine in the human gastrointestinal tract. *J Clin Endocrinol Metab.* 1997;82(11):3864–3871. doi:10.1210/jcem.82.11.4339
122. Dobbs RJ, Dobbs SM, Weller C, et al. Helicobacter hypothesis for idiopathic Parkinsonism: before and beyond. *Helicobacter.* 2008;13(5):309–322. doi:10.1111/j.1523-5378.2008.00622.x
123. Hashim H, Azmin S, Razlan H, et al. Eradication of *Helicobacter pylori* infection improves levodopa action, clinical symptoms and quality of life in patients with Parkinson's disease. *PLoS One.* 2014;9(11):e112330. doi:10.1371/journal.pone.0112330
124. Scheperjans F, Aho V, Pereira PAB, et al. Gut microbiota are related to Parkinson's disease and clinical phenotype. *Mov Disord.* 2015;30(3):350–358. doi:10.1002/mds.26069
125. Keshavarzian A, Green SJ, Engen PA, et al. Colonic bacterial composition in Parkinson's disease. *Mov Disord.* 2015;30(10):1351–1360. doi:10.1002/mds.26307
126. Unger MM, Spiegel J, Dillmann K-U, et al. Short chain fatty acids and gut microbiota differ between patients with Parkinson's disease and age-matched controls. *Parkinsonism Relat Disord.* 2016;32:66–72. doi:10.1016/j.parkreldis.2016.08.019
127. Paiva I, Pinho R, Pavlou MA, et al. Sodium butyrate rescues dopaminergic cells from alpha-synuclein-induced transcriptional deregulation and DNA damage. *Hum Mol Genet.* 2017;26(12):2231–2246. doi:10.1093/hmg/ddx114

128. Abraham CR, Selkoe DJ, Potter H. Immunochemical identification of the serine protease inhibitor α1-antichymotrypsin in the brain amyloid deposits of Alzheimer's disease. *Cell.* 1988;52(4):487–501. doi:10.1016/0092-8674(88)90462-X
129. Cameron B, Landreth GE. Inflammation, microglia, and Alzheimer's disease. *Neurobiol Dis.* 2010;37(3):503–509. doi:10.1016/j.nbd.2009.10.006
130. Friedland RP. Mechanisms of molecular mimicry involving the microbiota in neurodegeneration. *J Alzheimer's Dis.* 2015;45(2):349–362. doi:10.3233/JAD-142841
131. Cattaneo A, Cattane N, Galluzzi S, et al. Association of brain amyloidosis with pro-inflammatory gut bacterial taxa and peripheral inflammation markers in cognitively impaired elderly. *Neurobiol Aging.* 2017;49:60–68. doi:10.1016/j.neurobiolaging.2016.08.019
132. Zhan X, Stamova B, Jin L-W, DeCarli C, Phinney B, Sharp FR. Gram-negative bacterial molecules associate with Alzheimer disease pathology. *Neurology.* 2016;87(22):2324–2332. doi:10.1212/WNL.0000000000003391
133. Desbonnet L, Clarke G, Traplin A, et al. Gut microbiota depletion from early adolescence in mice: implications for brain and behaviour. *Brain Behav Immun.* 2015;48:165–173. doi:10.1016/j.bbi.2015.04.004
134. Michalski B, Corrada MM, Kawas CH, Fahnestock M. Brain-derived neurotrophic factor and TrkB expression in the "oldest-old," the 90+ Study: correlation with cognitive status and levels of soluble amyloid-beta. *Neurobiol Aging.* 2015;36(12):3130–3139. doi:10.1016/j.neurobiolaging.2015.08.022
135. Harach T, Marungruang N, Duthilleul N, et al. Reduction of Abeta amyloid pathology in APPPS1 transgenic mice in the absence of gut microbiota. *Sci Rep.* 2017;7:41802. doi:10.1038/srep41802
136. Minter MR, Zhang C, Leone V, et al. Antibiotic-induced perturbations in gut microbial diversity influence neuro-inflammation and amyloidosis in a murine model of Alzheimer's disease. *Sci Rep.* 2016;6(1):30028. doi:10.1038/srep30028
137. Brandscheid C, Schuck F, Reinhardt S, et al. Altered gut microbiome composition and tryptic activity of the 5xFAD Alzheimer's mouse model. *J Alzheimer's Dis.* 2017;56(2):775–788. doi:10.3233/JAD-160926
138. Finegold SM, Dowd SE, Gontcharova V, et al. Pyrosequencing study of fecal microflora of autistic and control children. *Anaerobe.* 2010;16(4):444–453. doi:10.1016/j.anaerobe.2010.06.008
139. Berding K, Donovan SM. Microbiome and nutrition in autism spectrum disorder: current knowledge and research needs. *Nutr Rev.* 2016;74(12):723–736. doi:10.1093/nutrit/nuw048
140. Parracho HMRT, Bingham MO, Gibson GR, McCartney AL. Differences between the gut microflora of children with autistic spectrum disorders and that of healthy children. *J Med Microbiol.* 2005;54(10):987–991. doi:10.1099/jmm.0.46101-0
141. Tomova A, Husarova V, Lakatosova S, et al. Gastrointestinal microbiota in children with autism in Slovakia. *Physiol Behav.* 2015;138:179–187. doi:10.1016/j.physbeh.2014.10.033

142. Sandler RH, Finegold SM, Bolte ER, et al. Short-term benefit from oral vancomycin treatment of regressive-onset autism. *J Child Neurol.* 2000;15(7):429–435. doi:10.1177/088307380001500701
143. Vuong HE, Hsiao EY. Emerging roles for the gut microbiome in autism spectrum disorder. *Biol Psychiatry.* 2016;81(5):1–13. doi:10.1016/j.biopsych.2016.08.024
144. Kang D-W, Adams JB, Gregory AC, et al. Microbiota transfer therapy alters gut ecosystem and improves gastrointestinal and autism symptoms: an open-label study. *Microbiome.* 2017;5(1):10. doi:10.1186/s40168-016-0225-7
145. De Theije CGM, Koelink PJ, Korte-Bouws GAH, et al. Intestinal inflammation in a murine model of autism spectrum disorders. *Brain Behav Immun.* 2014;37:240–247. doi:10.1016/j.bbi.2013.12.004
146. Newell C, Bomhof MR, Reimer RA, Hittel DS, Rho JM, Shearer J. Ketogenic diet modifies the gut microbiota in a murine model of autism spectrum disorder. *Mol Autism.* 2016;7(1):37. doi:10.1186/s13229-016-0099-3
147. Hsiao EY, McBride SW, Chow J, Mazmanian SK, Patterson PH. Modeling an autism risk factor in mice leads to permanent immune dysregulation. *Proc Natl Acad Sci U S A.* 2012;109(31):12776–12781. doi:10.1073/pnas.1202556109

7

Investigating Immune Changes in the Psychiatric Patients

ANTONIO L. TEIXEIRA, ISABELLE BAUER, AKIF CAMKURT, AND SUDHAKAR SELVARAJ

INTRODUCTION

Despite all the recent advances in neuroscience, the diagnosis of psychiatric disorders still relies on subjective statements of patients and/or observation of their behaviors. This reflects in part the limited understanding of the pathogenesis of these disorders and/or the way they are traditionally conceptualized. In this context, biomarkers, i.e., a measurable characteristic (e.g., molecules) that indicates physiological or pathogenic processes for psychiatric disorders, remain elusive.[1]

There is a relative consensus about the need for biomarkers in psychiatry to advance the field. However, the classic strategy of investigating the association of biological findings with psychiatric disorders has been of very limited value. Even major psychiatric disorders, such as schizophrenia and bipolar disorder, are not single disorders/diseases but syndromes comprising clinically similar but biologically distinct conditions.[2] To address this challenge, the U.S. National Institute of Mental Health (NIMH) proposed the Research Domain Criteria (RDoC), a new framework to categorize psychiatric disorders based on observable behaviors and biological measures, theoretically leading to an etiology-based classification. Instead of the classic strategy, the RDoC proposes investigating the association of biological measures with well-defined signs and symptoms across the psychiatric disorders.[3] The RDoC is an ongoing project and has its own problems (e.g., clinical or translational meaning), but it started a fundamental debate in psychiatry.

Beyond the concept of a laboratory test for a specific psychiatric disorder, there are other reasons to detect biomarkers in psychiatry, including: identifying biological subtypes or subgroups of patients, staging and stratifying patients, and predicting treatment response or prognosis, ultimately leading to a more personalized approach. Among potential biomarkers, immune markers have been regarded as very promising due to the role played by the immune system in the physiopathology of psychiatric disorders and the relatively easy access to immune biomarkers.[4] This chapter will discuss the emerging translational and clinical implications of immune markers relevant to primary psychiatric illnesses.

PERIPHERAL INFLAMMATORY MARKERS: FOCUS ON CRP AND CYTOKINES

There is now a substantial body of evidence showing altered circulating levels of immune cells and molecules in patients with psychiatric disorders, usually indicating a low-grade systemic inflammation. The most-studied immune markers are C-reactive protein (CRP), the cytokines tumor necrosis factor-alpha (TNF-α), interleukin 1 beta (IL-1β), IL-6, IL-8, IL-10, and the neutrophil:lymphocyte ratio (NLR). Besides NLR, other inflammation-related indexes, such as the monocyte:lymphocyte ratio and platelet:lymphocyte ratio, can be easily calculated from blood cell count assay that is routinely performed in clinical practice.

CRP is another molecule widely available as a routine laboratory test. CRP or hs-CRP (high sensitivity CRP) is an acute-phase protein synthesized by the liver in response to inflammatory mediators such as IL-6. CRP is traditionally measured down to concentrations around 3–5 mg/L, whereas hs-CRP measures down to levels of 0.3 mg/L, being able to detect very low-grade inflammation. High CRP levels are seen in different clinical conditions characterized by systemic inflammation. The American Heart Association and U.S. Centers for Disease Control and Prevention (CDC) defined risk groups for cardiovascular diseases based on hs-CRP levels: Low risk: less than 1.0 mg/L; Average risk: 1.0–3.0 mg/L; and High risk: above 3.0 mg/L.

Several systematic reviews and meta-analyses evaluating inflammatory/immune mediators in different psychiatric disorders show increased circulating levels of pro-inflammatory molecules in line with a low-grade systemic inflammation.[5–7] More importantly, there is a significant overlap of the observed changes across different conditions. For instance, a meta-analysis comparing blood cytokines in acutely and chronically ill patients with schizophrenia, bipolar disorder, and major depression disorder (MDD) found elevated levels of IL-6 and TNF-α in acutely ill patients with these three major psychiatric disorders.[8] In chronically ill patients, only the levels of IL-6 were significantly increased in these conditions. Of note, IL-6 levels decreased after treatment in schizophrenia and in MDD. Increased CRP levels were also reported in MDD, bipolar disorder, and schizophrenia.[9] Regarding cellular parameters, patients with MDD and bipolar disorder exhibit increased NLR as compared to controls, reflecting systemic inflammation.[10]

There are studies investigating immune markers in the cerebrospinal fluid (CSF). However, due to concerns with CSF sampling, most studies have a limited sample size. In line with blood assessment, increased CSF levels of IL-6 and IL-8 have been reported in patients with schizophrenia and MDD, while CSF levels of IL-1β were significantly increased in patients with schizophrenia and bipolar disorder (BD) compared to controls.[11] Altogether, these studies highlight the lack of specificity of immune markers, undermining their potential role as diagnostic markers. A similar observation is valid for other systemic biomarkers related to processes influenced by immune mechanisms such as markers of oxidative stress.[12,13]

Despite not contributing to specific psychiatric diagnoses, immune markers have been evaluated as risk markers for detecting the development of future conditions and complications, such as suicide. A reliable risk marker can lead to early intervention and follow-up, possibly influencing outcomes. Prospective large-scale population studies have shown that elevated serum levels of IL-6 are associated with increased future risk of heart disease[14] and diabetes mellitus type 2.[15] Similarly, a longitudinal population birth cohort study found that children with higher serum levels of IL-6 at age nine had elevated risk of depression and psychotic disorders at age 18, after controlling for demographic factors.[16] Accordingly, low-grade systemic inflammation as measured by IL-6 levels might be a risk marker for the future psychiatric disorders, also indicating a common mechanism underlying the high comorbidity between somatic and psychiatric illnesses of adult life.[16] Other longitudinal studies confirmed the association between inflammatory markers, mainly CRP, and subsequent depressive symptoms in the community.[17] Also related to risk assessment, altered levels of IL-6 and IL-1β cytokines in blood, CSF, and postmortem brain samples were associated with suicidality.[18]

Resistance to treatment is a major issue in psychiatry. Early detection of treatment-resistant patients could optimize treatment strategies, potentially fostering a faster and safer recovery. Two recent clinical trials suggested that CRP levels can be a useful marker for antidepressant selection. Uher et al.[19] reported that CRP levels separated responders from non-responders to antidepressant treatment. Patients with CRP levels less than 1 mg/L responded well to the selective serotonin reuptake inhibitor (SSRI) escitalopram with greater reduction in depressive symptoms, while those with CRP levels above 1 mg/L had a better response to the tricyclic antidepressant (TCA) nortriptyline.

Similarly, Jha et al.[20] found that depressed patients with CRP >1 mg/L at baseline responded better to bupropion plus escitalopram combination compared to those on escitalopram alone. Another study that investigated the efficacy of the TNF antagonist infliximab reported no significant difference in efficacy when compared with placebo in treatment-resistant depression.[21] After a secondary post-hoc analysis based on baseline hs-CRP levels, infliximab-treated patients with hs-CRP above 5 mg/L had a greater reduction in depressive symptoms compared to placebo. Although preliminary, these studies provide evidence supporting a role for inflammatory markers, mainly CRP, as predictors of antidepressant response.

Overlapping inflammatory markers across psychiatric disorders point to a likely common and/or shared immune dysfunction. The fact that these markers might predict future development of major psychiatric syndromes corroborates this assumption. Due to the multifactorial nature of psychiatric disorders, it is not surprising that no immune marker has been found to be useful as a diagnostic biomarker. It remains to be established whether there are inflammation-related subtypes of major psychiatric disorders with unique therapeutics and prognoses. It is worth highlighting that there are several confounding factors involved in the peripheral measurement of inflammatory markers, including changes due to diurnal variations, diet, stress, menstrual cycle, body weight, glycemic state, and sleep patterns, among others, that are difficult to control in research and clinical settings. For the assessment of parameters not available in routine labs, such as cytokines, issues related to the standardization of blood sampling, storage, and analytical techniques can also be a problem.

AUTOANTIBODIES

The investigation of the association of immune antibodies with psychiatric disorders is not new. In the 1960s, Fessel studied the role of autoimmunity in schizophrenia and reported "anti-brain" factors in patients' sera.[22] Later, autoantibodies targeting brain dopamine were proposed as an etiological mechanism for psychotic illnesses.[23]

In the 1990s, Susan Swedo's group at the NIMH proposed a subtype of pediatric obsessive-compulsive disorder (OCD) and tic disorders with acute onset thought to be triggered by antibodies produced in response to Group A beta-hemolytic streptococci targeting the basal ganglia.[24] This condition was later named *pediatric autoimmune neuropsychiatric disorders associated with streptococcal infections* (PANDAS), and it responded to intravenous immunoglobulin and plasmapheresis.[25] The role of anti–basal ganglia antibodies as biomarkers of a subgroup of OCD has been highly discussed.[26]

More recently, Joseph Dalmau's group described autoantibodies directed against the specific subunit GluN1 of glutamate anti-N-methyl-D-aspartate (NMDA) receptors in patients presenting with a rapid-onset neuropsychiatric syndrome characterized by psychosis and/or mood changes followed by motor signs (dyskinesias, catatonia), autonomic dysfunction, seizures, and delirium. In some patients, this clinical syndrome progresses to frank neurological deterioration requiring intensive care support and mechanical ventilation.[27,28] Women are more commonly affected (up to 80%), and the median age of onset is 21 years, with a peak between 18 and 23 years.[27]

The exact prevalence of anti-NMDA receptor antibody encephalitis is not clear. It has been proposed that an estimate of 8% of patients with a primary diagnosis of psychosis had anti-NMDA receptor antibodies.[29] Some maintain that prevalence might be underestimated, as patients with initial psychiatric presentation without neurological signs are often missed. Patients with anti-NMDA-receptor antibody encephalitis respond to immunotherapy, including steroids, plasmapheresis, and intravenous immunoglobulins, but a substantial number of patients continues to have cognitive deficits after recovery and have a risk of relapse.[27] The possibility that some of the severely ill psychiatric patients may have an identifiable pathology that can be tested with serum biomarkers and treated with successful immunotherapies brought significant attention, and independent groups are currently investigating this hypothesis.

Further research is required to define the prevalence of this condition and to investigate if routine screening for anti-NMDA antibodies of all patients with the first presentation of psychosis is clinically feasible. Such anti-NMDA-antibody screening may yield better outcome if applied to a specific subset of patients presenting with psychosis and, for instance, catatonic symptoms. Furthermore, laboratory techniques for the measurement of the antibody need to be standardized. Another open question refers to the best matrix for anti-NMDA-antibody assessment, as serum/blood seems to exhibit lower levels of these antibodies than CSF. While the first source is more accessible, the latter seems to be more specific and sensitive. Finally, prospective longitudinal studies are clearly required to characterize whether immune-based treatments significantly change the clinical outcome of psychosis.[30]

NEUROIMAGING MARKERS OF NEUROINFLAMMATION

Magnetic resonance imaging (MRI) and positron emission tomography (PET) techniques have been applied to study neuroinflammation that is characterized, among other processes, by microglial activation in the brain, and that has been implicated in the pathogenesis of several psychiatric disorders. The peripheral benzodiazepine receptor (PBR), also called the 18 kDa translocator protein (TSPO), is a mitochondrial membrane protein expressed at low levels in glial cells of healthy brains.[31] PBR/TSPO expression has been shown to increase after activation of glial cells following brain injury and other neuroinflammatory conditions, and has therefore been proposed as an *in vivo* biomarker of cellular activation during neuroinflammation.[31] Advances in brain-imaging techniques such as PET allow the study of PBR/TSPO using selective radiotracers.

The first investigation of TSPO in schizophrenia employed the first-generation TSPO radioligand (R)-[11C]PK11195, and found elevated (R)-[11C]PK11195 binding in total brain gray matter in a cohort of 10 patients within five years of schizophrenia diagnosis.[32] Doorduin et al. found similar increases in (R)-[11C]PK11195 binding in schizophrenia.[33]

The second-generation TSPO ligand 11C-PBR28 is an imidazopyridine derivative shown to have higher brain uptake and improved specific binding than (R)-[11C]PK11195.[34,35] However, the results with the newer second-generation TSPO radiotracers [11C]DAA1106, [11C]PBR28, [11C]PBR111, and [18F]-FEPPA have been so far mixed. While some studies showed an increased TSPO binding in cortical gray matter in patients with schizophrenia compared to healthy controls,[36,37] others reported decreased[38] or no change.[39–42] Differences in PET ligands, quantification methodology, and clinical confounders may underlie differences among studies.[43] Also, the role of TSPO seems to be far more complex than previously anticipated.[39]

In patients with MDD of at least moderate severity, two studies[44,45] found elevated TSPO binding in various cortico-limbic regions, but one study in patients with mild depression did not find any difference when compared with healthy control subjects.[46] In bipolar disorder, using [11C]-(R)-PK11195, Haarman et al. found a significantly increased binding potential in the right hippocampus of euthymic bipolar patients.[47] Similarly, TSPO increase has been reported in obsessive-compulsive disorder.[48]

Together, these findings support glial/microglial activation and altered neuroimmune-inflammatory processes in major psychiatric disorders. There are several unanswered questions, including:

1. Whether there is any correlation between central and peripheral inflammatory changes;
2. the timing of microglial changes; and, critically,
3. whether the microglial activation is primary, leading to symptoms and cognitive impairment, or secondary to the onset of clinical symptoms; and
4. whether these changes are protective or damaging.

Further prospective studies with more specific targets will be required before translation into the clinical practice.

MRI has been widely used in the diagnosis, monitoring, and prognosis of inflammatory neurological diseases such as multiple sclerosis, a disease characterized by multifocal white matter lesions. Advanced MRI techniques such as contrast MRI, diffusion imaging, magnetization transfer, and MR spectroscopy have been used to study surrogate markers of neuroinflammation. For instance, gadolinium-enhanced MRI has been used to study blood–brain barrier (BBB) permeability, but an old study did not show any BBB damage in schizophrenia patients.[49] A recent study suggested combining MRI and PET to gather both morphological and physiological information that may be of utility when studying neuroinflammation.[50] The implementation of this multimodal approach is still in its infancy, but it holds potential for future studies.

MR spectroscopy (MRS) has been used to identify changes in glial cell content with increase in myoinositol (glial marker) and choline (marker for cell membrane metabolism and cellular turnover), and reduction N-acetyl aspartate (marker for neuronal viability) as markers of the activation or migration of glial cells in inflamed areas. This MRS profile is consistently found in populations with bipolar disorder[51,52] and schizophrenia.[53] This supports the hypothesis that neuroinflammation may underlie the pathophysiology of these conditions. However, MRS metabolite changes are not always specific to neuroinflammation (e.g., inter-individual differences, equipment, normal fluctuations over time), thus clinical interpretation is problematic.[54]

Diffusion tensor imaging (DTI) is widely used to assess white matter integrity in neuroinflammatory diseases such as multiple sclerosis. DTI provides information on the anisotropy of water diffusion and helps to identify fiber orientation by detecting the preferred direction of diffusing protons. DTI measures can also be used to estimate the loss of axons and demyelination.[55] In multiple sclerosis, findings point toward reduced fractional anisotropy in temporo-frontal regions.[56] Intriguingly, reduced fractional anisotropy in frontal regions, cingulum, corpus callosum, corticospinal tracts, and superior longitudinal fascicles is consistently reported in mood disorders[57,58] and was found to accurately classify pediatric BD individuals from healthy volunteers using machine learning methods.[59]

A recent review on neuroimaging and neuroinflammation discussed the benefits of combining DTI with optical coherence tomography (OCT). OCT is a reliable and minimally invasive technique that uses infrared light to quantify retinal layer thickness and assess neuro-axonal degeneration. Alongside DTI measures of white matter integrity and connectivity, these OCT measures could help identify the presence of activated microglia activity and, potentially, a strong inflammatory response.[60]

CHALLENGES AND PERSPECTIVES

Besides the traditional indications, such as ruling out metabolic or infectious encephalopathy, investigating causes of delirium, following up treatment (e.g., lithium level), the role of laboratory or auxiliary tests in psychiatry is still very limited. Although considerable progress have been made in the neurobiological understanding of psychiatric disorders, we are still lacking a valid biomarker to be incorporated in the psychiatric practice. Some major challenges and obstacles for immune biomarkers studies include the following:

1. A particular marker has been associated with different psychiatric disorders and, therefore, lacks discriminative property;
2. Low reproducibility of the results, reflecting in part methodological issues, but also the intrinsic heterogeneity of the psychiatric disorders; and
3. Multiple confounding factors can influence measures, such as medical comorbidities, treatment, drug use, and diet, among others.

The heterogeneity of psychiatric disorders/syndromes means that biomarkers may help unravel their intricate neurobiological basis. More importantly, the studies have shown that dysregulation in immune pathways is somehow associated with several psychiatric disorders. The successful development of companion biomarkers for psychiatric disorders will require the integration of multiple platforms (e.g., protein-based, cellular-based, transcriptome, metabolomics) aiming at different immune mechanisms. The study of immune mechanisms/markers in psychiatry may help determine subtypes or endophenotype candidates for psychiatric disorders, fostering so-called personalized medicine. If an elusive "immune-subtype" is recognized, personalized approaches can take place for these patients.

REFERENCES

1. Teixeira, A. L., Salem, H., Frey, B. N., Barbosa, I. G., & Machado-Vieira, R. (2016). Update on bipolar disorder biomarker candidates. *Expert Rev Mol Diagn, 16*(11), 1209–1220.
2. Nasrallah, H. A. (2013). Lab tests for psychiatric disorders: few clinicians are aware of them. *Curr Psychiatry Rep, 12*, 5–7.
3. Cuthbert, B. N., & Insel, T. R. (2013). Toward the future of psychiatric diagnosis: the seven pillars of RDoC. *BMC Med, 11*, 126.
4. Ivison, S., Des Rosiers, C., Lesage, S., Rioux, J. D., & Levings, M. K. (2017). Biomarker-guided stratification of autoimmune patients for biologic therapy. *Curr Opin Immunol, 49*, 56–63.

5. Sayana, P., Colpo, G. D., Simoes, L. R., Giridharan, V. V., Teixeira, A. L., Quevedo, J., et al. (2017). A systematic review of evidence for the role of inflammatory biomarkers in bipolar patients. *J Psychiatr Res, 92*, 160–182.
6. Rodrigues-Amorim, D., Rivera-Baltanas, T., Spuch, C., Caruncho, H. J., Gonzalez-Fernandez, A., Olivares, J. M., et al. (2018). Cytokines dysregulation in schizophrenia: a systematic review of psychoneuroimmune relationship. *Schizophr Res, 197*, 19–33.
7. Misiak, B., Stanczykiewicz, B., Kotowicz, K., Rybakowski, J. K., Samochowiec, J., & Frydecka, D. (2018). Cytokines and C-reactive protein alterations with respect to cognitive impairment in schizophrenia and bipolar disorder: A systematic review. *Schizophr Res, 192*, 16–29.
8. Goldsmith, D. R., Rapaport, M. H., & Miller, B. J. (2016). A meta-analysis of blood cytokine network alterations in psychiatric patients: comparisons between schizophrenia, bipolar disorder and depression. *Mol Psychiatry, 21*(12), 1696–1709.
9. Fernandes, B. S., Steiner, J., Molendijk, M. L., Dodd, S., Nardin, P., Goncalves, C. A., et al. (2016). C-reactive protein concentrations across the mood spectrum in bipolar disorder: a systematic review and meta-analysis. *Lancet Psychiatry, 3*(12), 1147–1156.
10. Mazza, M. G., Lucchi, S., Tringali, A. G. M., Rossetti, A., Botti, E. R., & Clerici, M. (2018). Neutrophil/lymphocyte ratio and platelet/lymphocyte ratio in mood disorders: a meta-analysis. *Prog Neuropsychopharmacol Biol Psychiatry, 84*(Pt A), 229–236.
11. Wang, A. K., & Miller, B. J. (2018). Meta-analysis of cerebrospinal fluid cytokine and tryptophan catabolite alterations in psychiatric patients: comparisons between schizophrenia, bipolar disorder, and depression. *Schizophr Bull, 44*(1), 75–83.
12. Brown, N. C., Andreazza, A. C., & Young, L. T. (2014). An updated meta-analysis of oxidative stress markers in bipolar disorder. *Psychiatry Res, 218*(1–2), 61–68.
13. Camkurt, M. A., Findikli, E., Izci, F., Kurutas, E. B., & Tuman, T. C. (2016). Evaluation of malondialdehyde, superoxide dismutase and catalase activity and their diagnostic value in drug naïve, first episode, non-smoker major depression patients and healthy controls. *Psychiatry Res, 238*, 81–85.
14. Danesh, J., Kaptoge, S., Mann, A. G., Sarwar, N., Wood, A., Angleman, S. B., et al. (2008). Long-term interleukin-6 levels and subsequent risk of coronary heart disease: two new prospective studies and a systematic review. *PLoS Med, 5*(4), e78.
15. Pradhan, A. D., Manson, J. E., Rifai, N., Buring, J. E., & Ridker, P. M. (2001). C-reactive protein, interleukin 6, and risk of developing type 2 diabetes mellitus. *JAMA, 286*(3), 327–334.
16. Khandaker, G. M., Pearson, R. M., Zammit, S., Lewis, G., & Jones, P. B. (2014). Association of serum interleukin 6 and C-reactive protein in childhood with depression and psychosis in young adult life: a population-based longitudinal study. *JAMA Psychiatry, 71*(10), 1121–1128.
17. Valkanova, V., Ebmeier, K. P., & Allan, C. L. (2013). CRP, IL-6 and depression: a systematic review and meta-analysis of longitudinal studies. *J Affect Disord, 150*(3), 736–744.
18. Black, C., & Miller, B. J. (2015). Meta-analysis of cytokines and chemokines in suicidality: distinguishing suicidal versus nonsuicidal patients. *Biol Psychiatry, 78*(1), 28–37.

19. Uher, R., Tansey, K. E., Dew, T., Maier, W., Mors, O., Hauser, J., et al. (2014). An inflammatory biomarker as a differential predictor of outcome of depression treatment with escitalopram and nortriptyline. *Am J Psychiatry, 171*(12), 1278–1286.
20. Jha, M. K., Minhajuddin, A., Gadad, B. S., Greer, T., Grannemann, B., Soyombo, A., et al. (2017). Can C-reactive protein inform antidepressant medication selection in depressed outpatients? Findings from the CO-MED trial. *Psychoneuroendocrinology, 78,* 105–113.
21. Raison, C. L., Rutherford, R. E., Woolwine, B. J., Shuo, C., Schettler, P., Drake, D. F., et al. (2013). A randomized controlled trial of the tumor necrosis factor antagonist infliximab for treatment-resistant depression: the role of baseline inflammatory biomarkers. *JAMA Psychiatry, 70*(1), 31–41.
22. Fessel, W. J. (1963). The "antibrain" factors in psychiatric patients' sera. I. Further studies with a hemagglutination technique. *Arch Gen Psychiatry, 8,* 614–621.
23. Knight, J. G. (1982). Dopamine-receptor-stimulating autoantibodies: a possible cause of schizophrenia. *Lancet, 2*(8307), 1073–1076.
24. Swedo, S. E., Leonard, H. L., Garvey, M., Mittleman, B., Allen, A. J., Perlmutter, S., et al. (1998). Pediatric autoimmune neuropsychiatric disorders associated with streptococcal infections: clinical description of the first 50 cases. *Am J Psychiatry, 155*(2), 264–271.
25. Williams, K. A., & Swedo, S. E. (2015). Post-infectious autoimmune disorders: Sydenham's chorea, PANDAS and beyond. *Brain Res, 1617,* 144–154.
26. Teixeira, A. L., Rodrigues, D. H., Marques, A. H., Miguel, E. C., & Fontenelle, L. F. (2014). Searching for the immune basis of obsessive-compulsive disorder. *Neuroimmunomodulation, 21*(2–3), 152–158.
27. Titulaer, M. J., McCracken, L., Gabilondo, I., Armangue, T., Glaser, C., Iizuka, T., et al. (2013). Treatment and prognostic factors for long-term outcome in patients with anti-NMDA receptor encephalitis: an observational cohort study. *Lancet Neurol, 12*(2), 157–165.
28. Dalmau, J., Gleichman, A. J., Hughes, E. G., Rossi, J. E., Peng, X., Lai, M., et al. (2008). Anti-NMDA-receptor encephalitis: case series and analysis of the effects of antibodies. *Lancet Neurol, 7*(12), 1091–1098.
29. Pollak, T. A., McCormack, R., Peakman, M., Nicholson, T. R., & David, A. S. (2014). Prevalence of anti-N-methyl-D-aspartate (NMDA) receptor [corrected] antibodies in patients with schizophrenia and related psychoses: a systematic review and meta-analysis. *Psychol Med, 44*(12), 2475–2487.
30. Teixeira, A. L., Rocha, N. P., & Zhang, X. (2017). Anti-NMDAR antibodies as a new piece in schizophrenia's puzzle. *Future Sci OA, 3*(2), FSO178.
31. Rupprecht, R., Papadopoulos, V., Rammes, G., Baghai, T. C., Fan, J., Akula, N., et al. (2010). Translocator protein (18 kDa) (TSPO) as a therapeutic target for neurological and psychiatric disorders. *Nat Rev Drug Discov, 9*(12), 971–988.
32. van Berckel, B. N., Bossong, M. G., Boellaard, R., Kloet, R., Schuitemaker, A., Caspers, E., et al. (2008). Microglia activation in recent-onset schizophrenia: a quantitative (R)-[11C]PK11195 positron emission tomography study. *Biol Psychiatry, 64*(9), 820–822.

33. Doorduin, J., de Vries, E. F., Willemsen, A. T., de Groot, J. C., Dierckx, R. A., & Klein, H. C. (2009). Neuroinflammation in schizophrenia-related psychosis: a PET study. *J Nucl Med, 50*(11), 1801–1807.
34. Van Camp, N., Boisgard, R., Kuhnast, B., Thézé, B., Viel, T., Grégoire, M. C., et al. (2010). In vivo imaging of neuroinflammation: a comparative study between [(18) F]PBR111, [(11)C]CLINME and [(11)C]PK11195 in an acute rodent model. *Eur J Nucl Med Mol Imaging, 37*(5), 962–972.
35. Kreisl, W. C., Fujita, M., Fujimura, Y., Kimura, N., Jenko, K. J., Kannan, P., et al. (2010). Comparison of [(11)C]-(R)-PK 11195 and [(11)C]PBR28, two radioligands for translocator protein (18 kDa) in human and monkey: implications for positron emission tomographic imaging of this inflammation biomarker. *Neuroimage, 49*(4), 2924–2932.
36. Ottoy, J., De Picker, L., Verhaeghe, J., Deleye, S., Wyffels, L., Kosten, L., et al. (2018). [18F]PBR111 PET Imaging in Healthy Controls and Schizophrenia: Test - Retest Reproducibility and Quantification of Neuroinflammation. *J Nucl Med, 59,* 1267–1274.
37. Bloomfield, P. S., Selvaraj, S., Veronese, M., Rizzo, G., Bertoldo, A., Owen, D. R., et al. (2016). Microglial activity in people at ultra high risk of psychosis and in schizophrenia: an [(11)C]PBR28 PET brain imaging study. *Am J Psychiatry, 173*(1), 44–52.
38. Collste, K., Plavén-Sigray, P., Fatouros-Bergman, H., Victorsson, P., Schain, M., Forsberg, A., et al. (2017). Lower levels of the glial cell marker TSPO in drug-naïve first-episode psychosis patients as measured using PET and [(11)C]PBR28. *Mol Psychiatry, 22*(6), 850–856.
39. Coughlin, J. M., Wang, Y., Ambinder, E. B., Ward, R. E., Minn, I., Vranesic, M., et al. (2016). In vivo markers of inflammatory response in recent-onset schizophrenia: a combined study using [(11)C]DPA-713 PET and analysis of CSF and plasma. *Transl Psychiatry, 6,* e777.
40. Hafizi, S., Tseng, H. H., Rao, N., Selvanathan, T., Kenk, M., Bazinet, R. P., et al. (2017). Imaging microglial activation in untreated first-episode psychosis: a PET study with [(18)F]FEPPA. *Am J Psychiatry, 174*(2), 118–124.
41. Holmes, S. E., Hinz, R., Drake, R. J., Gregory, C. J., Conen, S., Matthews, J. C., et al. (2016). In vivo imaging of brain microglial activity in antipsychotic-free and medicated schizophrenia: a [(11)C](R)-PK11195 positron emission tomography study. *Mol Psychiatry, 21*(12), 1672–1679.
42. Kenk, M., Selvanathan, T., Rao, N., Suridjan, I., Rusjan, P., Remington, G., et al. (2015). Imaging neuroinflammation in gray and white matter in schizophrenia: an in-vivo PET study with [18F]-FEPPA. *Schizophr Bull, 41*(1), 85–93.
43. Turkheimer, F. E., Rizzo, G., Bloomfield, P. S., Howes, O., Zanotti-Fregonara, P., Bertoldo, A., et al. (2015). The methodology of TSPO imaging with positron emission tomography. *Biochem Soc Trans, 43*(4), 586–592.
44. Setiawan, E., Wilson, A. A., Mizrahi, R., Rusjan, P. M., Miler, L., Rajkowska, G., et al. (2015). Role of translocator protein density, a marker of neuroinflammation, in the brain during major depressive episodes. *JAMA Psychiatry, 72*(3), 268–275.
45. Holmes, S. E., Hinz, R., Conen, S., Gregory, C. J., Matthews, J. C., Anton-Rodriguez, J. M., et al. (2018). Elevated translocator protein in anterior cingulate in major

depression and a role for inflammation in suicidal thinking: a positron emission tomography study. *Biol Psychiatry, 83*(1), 61–69.

46. Hannestad, J., DellaGioia, N., Gallezot, J. D., Lim, K., Nabulsi, N., Esterlis, I., et al. (2013). The neuroinflammation marker translocator protein is not elevated in individuals with mild-to-moderate depression: a [^{11}C]PBR28 PET study. *Brain Behav Immun, 33,* 131–138.
47. Haarman, B. C., Riemersma-Van der Lek, R. F., de Groot, J. C., Ruhé, H. G., Klein, H. C., Zandstra, T. E., et al. (2014). Neuroinflammation in bipolar disorder—a [(11) C]-(R)-PK11195 positron emission tomography study. *Brain Behav Immun, 40,* 219–225.
48. Attwells, S., Setiawan, E., Wilson, A. A., Rusjan, P. M., Mizrahi, R., Miler, L., et al. (2017). Inflammation in the neurocircuitry of obsessive-compulsive disorder. *JAMA Psychiatry, 74*(8), 833–840.
49. Szymanski, S., Ashtari, M., Zito, J., Degreef, G., Bogerts, B., & Lieberman, J. (1991). Gadolinium-DTPA enhanced gradient echo magnetic resonance scans in first episode of psychosis and chronic schizophrenic patients. *Psychiatry Res, 40*(3), 203–207.
50. Sollini, M., Berchiolli, R., Kirienko, M., Rossi, A., Glaudemans, A., Slart, R., et al. (2018). *PET/MRI in infection and inflammation. Semin Nucl Med, 48*(3), 225–241.
51. Caetano, S. C., Olvera, R. L., Hatch, J. P., Sanches, M., Chen, H. H., Nicoletti, M., et al. (2011). Lower N-acetyl-aspartate levels in prefrontal cortices in pediatric bipolar disorder: a 1H magnetic resonance spectroscopy study. *J Am Acad Child Adolesc Psychiatry, 50*(1), 85–94.
52. Yüksel, C., & Öngür, D. (2010). Magnetic resonance spectroscopy studies of glutamate-related abnormalities in mood disorders. *Biol Psychiatry, 68*(9), 785–794.
53. Iwata, Y., Nakajima, S., Plitman, E., Mihashi, Y., Caravaggio, F., Chung, J. K., et al. (2018). Neurometabolite levels in antipsychotic-naïve/free patients with schizophrenia: a systematic review and meta-analysis of 1 H-MRS studies. *Prog Neuro-Psychopharmacol Biol Psychiatry, 86,* 340–352.
54. Zahr, N. M., Mayer, D., Rohlfing, T., Sullivan, E. V., & Pfefferbaum, A. (2014). Imaging neuroinflammation? A perspective from MR spectroscopy. *Brain Pathol, 24*(6), 654–664.
55. Gajamange, S., Raffelt, D., Dhollander, T., Lui, E., van der Walt, A., Kilpatrick, T., et al. (2018). Fibre-specific white matter changes in multiple sclerosis patients with optic neuritis. *NeuroImage: Clinical, 17,* 60–68.
56. van Geest, Q., Boeschoten, R. E., Keijzer, M. J., Steenwijk, M. D., Pouwels, P. J., Twisk, J. W., et al. (2018). Fronto-limbic disconnection in patients with multiple sclerosis and depression. *Mult Scler,* 1352458518767051. doi:10.1177/1352458518767051. [Epub ahead of print].
57. Repple, J., Meinert, S., Grotegerd, D., Kugel, H., Redlich, R., Dohm, K., et al. (2017). A voxel-based diffusion tensor imaging study in unipolar and bipolar depression. *Bipolar Disord, 19*(1), 23–31.

58. Bauer, I. E., Wu, M.-J., Meyer, T. D., Mwangi, B., Ouyang, A., Spiker, D., et al. (2016). The role of white matter in personality traits and affective processing in bipolar disorder. *J Psychiatr Res, 80*, 64–72.
59. Mwangi, B., Wu, M.-J., Bauer, I. E., Modi, H., Zeni, C. P., Zunta-Soares, G. B., et al. (2015). Predictive classification of pediatric bipolar disorder using atlas-based diffusion weighted imaging and support vector machines. *Psychiatry Res: Neuroimaging, 234*(2), 265–271.
60. Manogaran, P., Walker-Egger, C., Samardzija, M., Waschkies, C., Grimm, C., Rudin, M., et al. (2018). Exploring experimental autoimmune optic neuritis using multimodal imaging. *NeuroImage, 175*, 327–339.

8

Immune Dysfunction

A Common Feature of Major Psychiatric Disorders?

JOSÉ OLIVEIRA, LAURE TABOUY, AND MARION LEBOYER ■

INTRODUCTION

Psychiatric disorders affect nearly one-third of individuals during their lifetimes and are a leading cause of years lived with disability worldwide.[1,2] Health disparities are worrisome in individuals diagnosed with a mental illness, as they did not share the improvements in life expectancy observed in the overall population in the last decades. Excess mortality even seems to be increasing, with a median reduction in life expectancy estimated to be of 10 years, with 67.3% of deaths due to natural causes: i.e., cardiovascular disorders, pulmonary diseases, and infection.[3] A population-based cross-sectional survey including 17 countries identified an association between mental disorders and the subsequent onset of a chronic medical condition, in a dose–response manner, with odds ratios (OR) ranging from 1.2 (95% CI: 1.0–1.5) to 3.6 (95% CI: 2.0–6.6).[4] The elevated rates and burden of comorbidities in mental disorders highlight the need to better understand this association to help design preventive and therapeutic strategies.

Based on epidemiological and biological data, current views of most of the severe psychiatric disorders show them as multisystem disorders, evolving against a background of immune dysfunction, at least in a subset of patients. Inflammation is part of the complex physiological response to harmful stimuli such as external stressors, which have been associated with psychiatric disorders, i.e., psychosocial trauma and infections.[5,6] Immune dysfunction, resulting in chronic low-grade inflammation, metabolic syndrome, and autoinflammation/autoimmunity, is commonly interpreted as common feature associating psychiatric disorders and general medical comorbidities. Moreover, it is supposed that shared genetic factors, environmental insults (including alterations of the composition of gut microbiota), and socioeconomic milieux may be in play. Immune dysregulation in psychiatric disorders has clear transnosological features, which is a characteristic of several complex and chronic human disorders.

CHRONIC IMMUNE DYSFUNCTION IN PSYCHIATRIC DISORDERS

There are several lines of evidence indicating significant changes in immune components (cells and molecules) and mechanisms that can be implicated in the pathophysiology of major psychiatric disorders.

Chronic Low-Grade Inflammation

Immune dysfunction as an integral component of psychiatric disorders is highlighted by altered peripheral and central markers such as acute-phase proteins, cytokines and other immune mediators, auto-antibodies, gut microbiota, and microglia activation, associated with changes in neurotrophins, oxidative stress, and mitochondrial dysfunction. Peripheral cytokines and other soluble immune mediators are the main focus of most studies, because they characterize the systemic nature of psychiatric disorders, constitute potential trait and state biomarkers, offer new therapeutic interventions, and are easily assessed in peripheral samples.

Cytokine levels and their receptors and other immune mediators are altered in schizophrenia (SZ), bipolar disorder (BD), major depressive disorder (MDD), obsessive-compulsive disorder (OCD), autism spectrum disorders (ASD), and other psychiatric disorders, both peripherally and in the CNS. A meta-analysis by Potvin and collaborators (2008) included 62 studies (2,298 SZ patients and 1,858 healthy controls) and showed an association with increased levels of IL-1 receptor antagonist (IL-1RA), soluble IL-2R (sIL-2R), and IL-6.[7] Another meta-analysis found that IL-12, interferon (IFN)-γ, tumor necrosis factor (TNF)-α, and IL-2R are potential trait markers, while IL-1β, IL-6, and transforming growth factor (TGF)-β were found to be good markers of treatment response, increased in acute illness but reduced after treatment in SZ.[8] Fernandes et al. (2016) in a meta-analysis of 26 cross-sectional or longitudinal studies including 85,000 individuals found a moderate increase in peripheral CRP levels in SZ, regardless of antipsychotic use, and a positive correlation with severity of positive symptoms.[9] Antipsychotic treatment was associated with an increase in plasma levels of sIL-2R and reduced plasma levels of IL-1β and IFN-γ.[10]

A recent meta-analysis of 28 studies found preliminary evidence for similarities for cerebrospinal fluid (CSF) and alterations of tryptophan catabolite across psychiatric disorders.[11] The authors found that CSF levels of IL-1β and kynurenic acid were elevated in SZ and BD patients as compared to healthy controls.[11] In fact, single studies have reported increased levels of IL-6, but evidence remains poor, as differences were not found when comparing patients and controls in a study by Katila et al. (1994).[12–14] Also in the CSF, T cell subsets were found to be activated in acute SZ.[15–17]

In BD, a meta-analysis of 30 studies involving 2,599 individuals (1,351 diagnosed with BD, and 1,248 healthy controls) concluded that peripheral IL-4, IL-6, IL-10, sIL-2R, sIL-6R, TNF-α, sTNFR1, and IL-1RA increased as compared to controls.[18] Phasic differences have also been documented for TNF-α, sTNFR1, sIL-2R, IL-6, and IL-1RA.[18] Another meta-analysis, including 11 studies on peripheral CRP levels, found elevated CRP in BD when compared to controls, specifically in manic and euthymic patients, but not in depression, unrelated to lithium or antipsychotic medication use.[19] Similar to what has been documented in SZ, inflammation markers may act as trait or state markers. Also in BD, central markers of inflammation have been described, namely increased levels of IL-1β in the CSF, increased protein and mRNA levels of IL-1β, MyD88, and NF-κB p50 and p65 subunits, in postmortem prefrontal cortex, and decreased levels of TGF-β1, an anti-inflammatory cytokine, in postmortem frontal cortex.[20,21] Activation of T cell and of the monocyte-macrophage systems has also been reported in BD.[22,23]

A systematic review and meta-analysis of community and clinical samples on the association between peripheral CRP, IL-1, and IL-6 concluded there was a positive association between these markers and depression. Moreover, the authors suggested that body mass index (BMI) acts as a mediator/moderator, also showing an association of depression with CRP and IL-6 among patients with cardiac disease or cancer, indicating that comorbidity and behavioral aspects should be taken into consideration.[24] Moreover, another meta-analysis of longitudinal studies found a small but significant association between peripheral CRP and IL-6 levels and the subsequent development of depressive symptoms.[25] Several cytokines significantly decrease with antidepressant treatment, as shown by recent meta-analyses, but data on respondents versus non-respondents remain scarce.[26,27]

In addition, changes in intestinal permeability may also modulate the inflammation observed in depression. In individuals diagnosed with depression, the diversity of gut microbiota is decreased, with a lower number of *Bifidobacterium* and *Lactobacillus* spp. The development of the HPA axis is influenced by the gut microbiota and contributes to inflammation by increasing intestinal barrier permeability and facilitating bacterial translocation into the systemic circulation.

Some studies have shown higher levels of Immunoglobulin (Ig) A and (Ig M)-mediated immune responses directed against lipopolysaccharides (LPS) of gram-negative bacteria from the microbiota in patients with depression, and bacterial DNA was detected in the serum concomitantly with increased Toll-like receptor (TLR) 4 (TLR4) expression. Some studies also highlighted that the gut microbiota influence hippocampal neurogenesis, itself altered in the context of chronic low-grade inflammation.

Cytokine alterations have also been reported in several studies in OCD, autism, and other psychiatric disorders. Details on the particular aspects of markers of inflammation on individual psychiatric disorders will be developed in the subsequent this book.

Oxidative and Nitrosative Stress

Work on the biological underpinnings of psychiatric disorders has additionally focused on changes in oxidative and nitrosative stress (O&NS) system. A meta-analysis on oxidative stress markers in SZ included 44 studies and concluded that alterations are already present in first-episode psychosis, an effect that seems to be independent of medication use.[28] Moreover, total antioxidant status (TAS), red blood cell (RBC) catalase, and plasma nitrate are potential markers of acute exacerbations, while RBC superoxide dismutase may act as a trait marker.[28] In another meta-analysis in BD patients, patients had relatively high nitric oxide (NO) level, a molecule abundantly produced by stimulated immune cells, among others (e.g., vascular endothelial cells, neurons) with concomitantly occurring signs of oxidative damage; namely, DNA and RNA damage and lipid peroxidation.[29] Interestingly, a study comparing the integrity of white matter between BD patient and controls found that serum lipid hydroperoxides measures explained 59% and 51% of the fractional anisotropy and radial diffusivity variances, respectively.[30] Furthermore, increased O&NS, mitochondrial dysfunction, and lipid peroxidation products, 4-hydroxy-2-nonenal (4-HNE) and 8-isoprostane (8-Iso), are evident in BD prefrontal cortex.[31]

A recent meta-analysis on oxidative stress markers in ADHD included six studies and a total of 231 ADHD patients and 207 controls.[32] The authors found normal levels of antioxidants but insufficient response to oxidative stress, although results are considered preliminary because the information available is very scarce.[32] Also in depression, a meta-analysis including 115 articles found lower TAS in acute episodes, and that antioxidants were lower in depressed patients when compared to controls. Moreover, oxidative damage byproducts were higher (malondialdehyde and 8-F2-isoprostane) in depression than in controls, which seems to be partially reversed by antidepressant medication use.[33]

Metabolic Syndrome and Related Disorders

A dysregulated immune system in psychiatric patients potentially contributes to the development of several psychiatric and other medical comorbidities, including metabolic syndrome (MetS) and related disorders or autoimmune disorders.[34–37] A recent meta-analysis showed that the prevalence of MetS was 31.3% (95% CI: 27.3–35.5%) in individuals diagnosed with MDD, with similar prevalence in BD [31.7% (95% CI: 27.3–36.3%)] and SZ [33.4% (95% CI: 30.8–36.0%)] and other related psychotic disorders [34.6% (95% CI: 29.3–40.0%)].[38] Relative risks were of 1.57 (95% CI: 1.38–1.79), 1.58 (95% CI: 1.24–2.03), and 1.87 (95% CI: 1.53–2.29) in MDD, BD, and SZ and related psychotic disorders, respectively.[38] An overlap of pathophysiological processes is suspected to be in play. A prospective follow-up study of non-depressed individuals showed that those meeting criteria for MetS were twice as likely to develop depressive symptoms.[44] Moreover, higher plasma glucose levels after an oral glucose load were observed in naïve first-episode SZ, BD, and MDD patients.[45] In fact, a meta-analysis of 16 case-control studies comprising 15 samples showed that glucose homeostasis is already impaired in antipsychotic-naïve first-episode SZ patients.[46]

Preliminary findings suggest that inflammation may contribute to abnormal metabolic pathways as greater increases in hsCRP in first-onset psychosis patients were associated with dyslipidemia independently of weight-gain and baseline inflammation.[47] More preliminary evidence shows that adolescent-onset BD patients present high cardiovascular risk CRP and have higher rates of preexisting obesity (OR: 1.58) and hypertension (OR: 2.93).[48,49] Importantly, higher serum IL-6 in childhood was associated with risk of depression and psychotic experiences.[50] Early-life immune/inflammatory malfunctioning may contribute to both mental illness and increased burden of medical and other psychiatric comorbidities such as suicide.[51] Increased peripheral *in vivo* IL-1β as IL-6 and decreased *in vitro* IL-2 were shown in suicidal patients suffering from major psychiatric disorders, when compared with non-suicidal patients or healthy controls.[52]

Autoantibodies and Autoimmune Diseases

Individuals diagnosed with an autoimmune disease often present psychiatric symptoms, such as is the case with systemic lupus erythematosus, antiphospholipid syndrome, or multiple sclerosis.[53–55] A recent study using self-reported information in a population cohort including 8,174 individuals, concluded there is a significant comorbidity between depression and autoimmune diseases (OR: 1.66; 95% CI: 1.27–2.15).[56] In this study, autoimmune diseases were used as a pooled autoimmune phenotype including 23 ICD-10 (*International Classification of Diseases, 10th Ed.*) diagnoses. A temporal relationship was strongly suggested, as autoimmune disease onset was associated with subsequent hazard of depression onset by 1.39 (95% CI: 1.11–1.74), and depression increased the subsequent hazard of autoimmune disease onset by 1.40 (95% CI: 1.09–1.80). A population-based study including approximately 1.1 million individuals (Danish registries) found that depression was associated with an increased risk of autoimmune diseases (15 pooled ICD-10 diagnoses) (incidence rate ratio [IRR]: 1.25; 95% CI: 1.19–1.31), when compared to individuals without a history of depression.[57] Eaton et al. (2013) found that a prior hospital contact because of autoimmune disease increased the risk of a subsequent mood disorder (MDD or BD) by 45% (IRR: 1.45; 95% CI: 1.39–1.52). Autoimmunity and previous infections interacted synergistically, further increasing the risk 2.35 times (IRR: 2.35; 95% CI: 2.25–2.46).[58]

In SZ, a large population-based study also in the Danish population revealed that previous hospital contacts due to an autoimmune disease, or previous hospitalization for an infection, increased the risk of a subsequent diagnosis of SZ by 45% and 60%, respectively.[59] Both factors synergistically raised the risk 2.25 times.[59] In this study, a dose–response relationship has been suggested. Having three or more infections plus an autoimmune disease diagnosis was associated with a rate ratio of 3.40.[59]

However, the non-specific nature of the autoimmune phenomena is also worthy of remark, as studies on markers of central and peripheral autoimmunity (neuronal cell surface autoantibodies and other general antibodies [e.g., antinuclear antibodies, ANAs; anti-parietal cell antibodies, APCAs; anti-mitochondrial antibodies, AMAs; anti-smooth muscle antibodies, ASMAs], respectively) suggest that they may be the consequence of a general immune dysfunction producing diffuse brain autoimmune involvement.[60,61] Anti-brain antibodies recently identified by Margari and collaborators (2015) include autoantibodies to hypothalamus, hippocampus, and cerebellum, and anti-nuclear autoantibodies.[61] N-methyl-D-aspartate (NMDA) receptor autoantibodies, observed in anti-NMDA receptor encephalitis, may be an exception, as they have been linked with psychiatric symptoms, such as mood swings, aggressive behavior, and psychosis, with a very plausible explanatory mechanism and, in some cases, with isolated psychiatric presentations.[62–66]

In the particular case of BD, autoimmune thyroiditis remains the most studied immune dysfunction in BD. These disorders have been described as comorbid, independently of lithium treatment, and associations with microbial influence have been proposed in a preliminary study; namely, for parvovirus B19 and comorbid BD and autoimmune thyroid diseases in women.[67–69] Interestingly, a genetic vulnerability to thyroid diseases among BD patients has been identified in the TLR4 pathogen-receptor (*TLR4 rs*4986790 G and *TLR4 rs*4986791 T alleles, independently of lithium treatment).[70] Autoimmunity in BD is suggested to be inherited as a common trait, as significantly higher prevalence of autoimmune thyroiditis was observed among female offspring of BD parents.[71,72]

In OCD, a link between streptococcal infection and the development of CNS immune dysregulation (including autoimmune phenomena) was strongly suggested after the description of pediatric autoimmune neuropsychiatric disorders associated with streptococcal infections (PANDAS).[73] Family studies found higher rates of autoimmune disorders in first-degree relatives of OCD patients.[74] Notwithstanding, streptococcal throat infection has been recently associated with an 18% increased risk of any mental illness (IRR: 1.18; 95% CI: 1.15–1.21), particularly with OCD/tic disorders, increasing the risk by 51% (IRR: 1.51; 95% CI: 1.28–1.77) and 35% (IRR: 1.35; 95% CI: 1.21–1.50), respectively.[75] In addition, a meta-analysis that pooled data on serum and CSF showed that seropositivity for anti-basal ganglia antibodies was associated with primary OCD (without autoimmune or neurological comorbidity) with an OR of 4.97 (95% CI: 2.88–8.55).[76] Interestingly, a recent small-size case-control study using brain PET showed inflammation-related activation of microglia is involved in the neurocircuitry of OCD.[77]

A personal and maternal history of autoimmune disorders increased the risk of ADHD by 24% (IRR: 1.24; 95% CI: 1.10–1.40) and 12% (IRR: 1.12; 95% CI: 1.06–1.19), respectively.[78] In particular, maternal (not paternal) history of thyrotoxicosis, type 1 diabetes, autoimmune hepatitis, psoriasis, and ankylosing spondylitis was found to be associated.[78]

Increased Permeability of the Digestive Barrier and Dysbiosis

In psychiatry, cytokines are probably not simply blameless markers of inflammation, as external administration of cytokines or induction of their production after endotoxin infusion in humans are associated with the development of depressive symptoms.[79] One of the suspected culprits in fueling chronic pro-inflammatory states is gut dysbiosis and increased gut permeability, but studies on this issue are still inconclusive.

Plasma levels of IgA and/or IgM directed against commensal bacteria LPS, a major constituent of the bacterial cell wall, are increased in BD, suggestive of intestinal bacterial translocation.[80] Besides the conceivable existence of a "leaky gut" contribution to systemic inflammation in mood disorders, systemic inflammation has also been suggested to increase intestinal permeability in humans, possibly through the facilitation of paracellular mechanisms.[81] The pleiotropy and wide expression of innate immune receptors such as TLR4, present in brain structures rich in vasculature and lacking a normal BBB like the circumventricular organs, place them as potential transductors of gut to brain immune-inflammatory axis. TLRs' expression in other leaky structures such as choroid plexus and leptomeninges in endothelial and perivascular cells of the BBB have also been proposed in neurons, astroglia, and microglia.[82] Gárate and collaborators (2013), in a murine model, using antibiotic intestinal decontamination, suggested a role for intestinal bacterial translocation for the upregulation of TLR4 expression in mice after stress exposure.[83] The crucial role of the microbiota in the development and maturation of the innate and adaptive immune systems at the gut level, particularly the gut-associated lymphoid tissue (GALT) in the early years of life, is worth remark.[84] Fecal samples from depressed patients transferred to microbiota-deficient mice induced depressive-like behavioral and physiological features; namely, anhedonia and anxiety-like behaviors, as well as alterations in tryptophan metabolism, not transmitted by their non-depressed counterparts.[85] Interestingly, depressed patients have decreased gut microbiota richness and diversity, and over-representation of some subpopulations such as *Enterobacteriaceae* and *Alistipes*.[86,87]

Once again, the influence of commensals has been suggested to start very early in life. O'Neil and colleagues (2016), in a large population-based Swedish cohort, observed an association between elective birth by caesarean section, and any psychosis, but also specifically with affective psychoses (Hazard Ratio [HR]: 1.17; 95% CI: 1.05–1.31).[88] This study replicated a previous nested case-control study in the Finish population, showing a 2.5-fold increased risk of BD in those born by elective caesarean section (95% CI: 1.32–4.78; $P < 0.01$).[89] Caesarean section may disturb the bacterial colonization of the gut in the initial months of life, as children born by elective caesarean are initially colonized from the hospital environment and maternal skin instead of microbes from the birth canal.

Human Endogenous Retroviruses

Human endogenous retroviruses (HERVs) are constituents of the human genomic DNA, and have been proposed to be a "missing link" between infections, chronic immune dysfunction, and risk of psychiatric disorders.[90] HERVs belong to the superfamily of transposable elements resulting from the integration of infectious retroviruses' genetic elements along the evolution of humans.[91] Although epigenetic silencing mechanisms, as well as the predominance of defective or inactive copies, prevent the expression of HERVs,[92] HERVs may also be responsive to environmental stressors and thus be reactivated.[90] This has been shown for influenza and *Herpes simplex* type 1 viruses, which act as potent transactivators of HERV-W elements' expression.[93,94] The reactivation of HERV-W causes the production of Envelope (Env) protein (Nellåker et al., 2006), which is a TLR4 agonist that thus activates inflammation and causes neurotoxic effects through the activation of this pattern recognition receptor.[93,95] DNA methylation of HERV-W promoter regions is in its levels early in life, but infection-induced disturbances in the epigenetic control of HERV-W genetic transcription may create a HERV-related genetic vulnerability to additional/secondary exposures later in life and potentially cause neurodevelopmental issues with lifelong implications for neuropsychiatric disorders.[92,96]

Some studies have documented the presence of HERV-W expression in the brain, CSF, and serum of patients diagnosed with SZ, but the presence and influence of HERV-W transcriptional profile and antigenic potential (e.g., Env and Gag [capsid/matrix] proteins) remains poorly understood in patients diagnosed with other psychiatric disorders.[97–100] One study involving 45 patients diagnosed with SZ and 91 patients diagnosed with BD and 73 healthy controls found HERV-W Env transcription to be increased in both psychiatric diagnoses when compared with the control group.[101]

ENVIRONMENTAL RISK FACTORS ARE SHARED IN MOST PSYCHIATRIC DIAGNOSES

While different environmental factors are implicated in the pathogenesis of psychiatric disorders, immune mechanisms may be the driven force underlying this association.

Childhood Adversities

Adverse and traumatic events are among the most studied environmental stressors in association with psychiatric disorders, especially those occurring early in life.[102,103] Childhood adversities have been associated with the manifestation of several psychiatric disorders such as MDD, BD, or SZ.[104] Childhood trauma is associated with an earlier age at onset and a severe clinical presentation; namely, a greater prevalence of suicide attempts, rapid cycling, greater proneness towards depression, and more cannabis use.[103,105–111] These data support a neurodevelopmental model in which early psychosocial stress may have a spectrum of effects on mental well-being. Interestingly, such events are also associated with several general medical conditions such as diabetes, obesity, cardiovascular diseases, autoimmunity, cancer, and neurodegeneration in adults with a history of childhood maltreatment.[112–116] Moreover, these individuals die younger due to chronic diseases.[117] One of the proposed mechanisms underlying these associations is the induction of maladaptive chronic inflammatory responses.

Stressful life events are known to induce acute and chronic alterations in human inflammatory responses, including increased CRP and pro-inflammatory cytokine levels, such as IL-6 and TNF-α.[104,118–120] Maltreated children present with increased markers of inflammation when compared with those who were not maltreated, and this association persists through adulthood.[120] In a large-scale longitudinal prospective study (birth cohort, $n = 1{,}037$), the association between depression and high CRP levels was significantly attenuated and no longer significant when the effect of childhood maltreatment was taken into account.[121] More than 10% of adult low-grade inflammation (CRP) was attributable to maltreatment.[120]

The influences of stress may start very early in life. In this regard, newborns exposed to maternal stress *in utero* had higher *ex vivo* IL-8 levels after a microbial challenge, while teenagers who were raised in harsh family environments had increasingly higher *ex vivo* inflammatory cytokine increases to LPS stimulation during follow-up analysis when compared to those coming from supportive family environments.[122,123] In this sense, adults raised in low socioeconomic status environments as children had higher levels of TLR-4 mRNA and displayed larger *ex vivo* cytokine production to microbial challenges when compared to their higher socioeconomic status counterparts.[124,125] On the other hand, adults who were raised in environments of low socioeconomic status as children but who had mothers who were warm and caring, exhibited less TLR-stimulated production of IL-6 compared with individuals raised in similar environments but not experiencing maternal warmth.[126] Together, these observations suggest that immune responses to other kinds of "stressors" (e.g., infections) may be persistently disrupted and inadequate in individuals exposed to childhood trauma, and that this damage is potentially preventable.

Interestingly, gut microbiota are also implicated in the establishment of stress-related psychiatric disorders. Maternal stress decreases the abundance of some bacteria, like *Lactobacillus spp*, in the maternal vaginal canal, resulting in decreased transmission of these bacteria to offspring. In ADHD, for example, there is a link between microbiota from the mother, dietary components modulating the gut, and vaginal microbiota during pregnancy, and the development of some symptoms of ADHD in offspring. It is also well established that the HPA axis, stress, and the immune system are interconnected, and that the gut microbiota mediate this interaction as reshaping of microbiota composition in the gut, increasing permeability of intestinal barrier, and bacterial translocation in systemic circulation.

Infection

Infection has been described in BD, SZ, and autism potentially acting during periods of susceptibility, mostly early in life, fitting a multiple-hit model. Indeed, several environmental challenges seem to cluster during prenatal/perinatal periods; chiefly maternal influenza infection, but also maternal stress, nutritional deficiencies, and obstetric complications.[142–148] Other risk factors may diminish the threshold for a first mood or psychotic episode, for example. The influences of stress and trauma during childhood are relevant, as well as during adolescence.[149–153]

In autism, the congenital infection association is one the most important research hypotheses.[154] Any maternal inpatient diagnosis of infection during pregnancy was associated with approximately 30% increase in autism risk, particularly with intellectual disability, in all trimesters of gestation.[155] Furthermore, rubella, measles and mumps, cytomegalovirus, polyomaviruses, and influenza have been associated with autism.[156–162]

Infectious insults are also well described in BD and SZ. Although pathogen-specific characteristics may be in play, several lines of evidence also suggest the resulting immune response is implicated in the subsequent development of a psychiatric disorder.[5] In fact, infectious insults are strong inducers of inflammatory responses and potentially of autoimmunity.[163–165] Several infectious agents have been associated with BD and SZ, the most studied ones being *Toxoplasma gondii*, influenza, cytomegalovirus, or herpes-simplex virus type 1, the latter also associated with worse cognitive functioning.[166–168] Sutterland and collaborators (2015), in a meta-analysis, suggest that *Toxoplasma gondii* infection seems to precede SZ.[168] However, infection is also suggested to immediately precede the instigation of a mood episode or psychotic symptoms. In this sense, a significantly higher prevalence of signs of urinary tract infection has been reported in acutely admitted patients with non-affective and affective psychoses when compared with control subjects.[169] Moreover, anti–*Toxoplasma gondii* IgM antibodies, suggestive of a first contact, were significantly higher in mania at hospital admission compared with healthy controls.[170] Epidemiological data based on nationwide (Denmark) registers showed that any history of hospitalization for infection increased the risk of SZ by 60% (IRR: 1.60; 95% CI: 1.56–1.64), of BD by 61% (IRR: 1.61; 95% CI: 1.55–1.68), and of MDD by 63% (IRR: 1.63; 95% CI: 1.61–1.66).[58,59] The authors also reported a synergistic interaction between hospital contacts for autoimmune diseases and hospitalizations for infectious diseases, increasing the risk of SZ and of mood disorders (BD or MDD) even further, 2.25 (IRR: 2.25; 95% CI: 2.04–2.46) and 2.35 times (IRR: 2.35; 95% CI: 2.25–2.46), respectively.[58,59] Interestingly, toxoplasmosis in mice induces gastrointestinal inflammation, a decrease of microbiota diversity, and increase of gut microbiota translocation to the circulation. A dysregulation in the composition of the gut microbiota following the infection of *Toxoplasma gondii* could thus contribute to psychiatric disorders through stimulation of the immune system or production of bacterial metabolites that have mood destabilizing or pro-psychotic effects.

CONCLUSIONS

Research in the field of immunopsychiatry clearly demonstrates the presence of immune dysfunction in psychiatric disorders. Major psychiatric disorders are often associated with alterations of peripheral cytokines, chemokines, and other immune mediators, and of immune cell subpopulations, over-representation of autoimmune and autoinflammatory phenomena, microglia activation, and markers of inflammation in the CNS. Moreover, the comorbidity with metabolic abnormalities and with general medical disorders, signs of gut dysbiosis, and increased gut permeability provide further evidence linking psychiatry disorders with inflammation. Immune dysregulation observed across psychiatric disorders is persistent and may be a consequence of: (1) the impact of mental illness on behavior, (2) the deterioration of physical health and/or, (3) the interacting genetic and environmental risk factors early in life, thus preceding the onset of psychiatric syndromes. In whatever case, inflammation may constitute an opportunity for the development of novel preventive strategies in those at risk, and of therapeutic strategies alleviating the burden of mental illness in those affected.

The rapid growth of immunopsychiatry reflects the alarming association of immune dysfunction with psychiatric disorders. Until now, most research has focused on general aspects of inflammation, documenting its presence and association with clinical characteristics. Current knowledge should be consolidated and guide future research on the clarification of the mechanisms linking inflammation and physical illnesses to specific psychiatric phenotypes/dimensions, and inspire mental health policies in further managing mental illness as a multisystem condition.

REFERENCES

1. Steel Z, Marnane C, Iranpour C, et al. The global prevalence of common mental disorders: a systematic review and meta-analysis 1980–2013. *Int J Epidemiol.* 2014;43(2):476–493. doi:10.1093/ije/dyu038.
2. Vigo D, Thornicroft G, Atun R. Estimating the true global burden of mental illness. *Lancet Psychiatry.* 2016;3(2):171–178. doi:10.1016/S2215-0366(15)00505-2.
3. Walker ER, McGee RE, Druss BG. Mortality in mental disorders and global disease burden implications: a systematic review and meta-analysis. *JAMA Psychiatry.* 2015;72(4):334–341. doi:10.1001/jamapsychiatry.2014.2502.
4. Scott KM, Lim C, Al-Hamzawi A, et al. Association of mental disorders with subsequent chronic physical conditions: world mental health surveys from 17 countries. *JAMA Psychiatry.* 2016;73(2):150–158. doi:10.1001/jamapsychiatry.2015.2688.
5. Leboyer M, Oliveira J, Tamouza R, Groc L. Is it time for immunopsychiatry in psychotic disorders? *Psychopharmacology (Berl).* 2016;233(9):1651–1660. doi:10.1007/s00213-016-4266-1.

6. Danese A, Caspi A, Williams B, et al. Biological embedding of stress through inflammation processes in childhood. *Mol Psychiatry.* 2011;16(3):244–246. doi:10.1038/mp.2010.5.
7. Potvin S, Stip E, Sepehry AA, Gendron A, Bah R, Kouassi E. Inflammatory cytokine alterations in schizophrenia: a systematic quantitative review. *Biol Psychiatry.* 2008;63(8):801–808. doi:10.1016/j.biopsych.2007.09.024.
8. Miller BJ, Buckley P, Seabolt W, Mellor A, Kirkpatrick B. Meta-analysis of cytokine alterations in schizophrenia: clinical status and antipsychotic effects. *Biol Psychiatry.* 2011;70(7):663–671. doi:10.1016/j.biopsych.2011.04.013.
9. Fernandes BS, Steiner J, Bernstein H-G, et al. C-reactive protein is increased in schizophrenia but is not altered by antipsychotics: meta-analysis and implications. *Mol Psychiatry.* 2016;21(4):554–564. doi:10.1038/mp.2015.87.
10. Tourjman V, Kouassi É, Koué M-È, et al. Antipsychotics' effects on blood levels of cytokines in schizophrenia: a meta-analysis. *Schizophr Res.* 2013;151(1–3):43–47. doi:10.1016/j.schres.2013.10.011.
11. Wang AK, Miller BJ. Meta-analysis of cerebrospinal fluid cytokine and tryptophan catabolite alterations in psychiatric patients: comparisons between schizophrenia, bipolar disorder, and depression. *Schizophr Bull.* 2018 Jan 13;44(1):75–83. doi:10.1093/schbul/sbx035.
12. Sasayama D, Hattori K, Wakabayashi C, et al. Increased cerebrospinal fluid interleukin-6 levels in patients with schizophrenia and those with major depressive disorder. *J Psychiatr Res.* 2013;47(3):401–406. doi:10.1016/j.jpsychires.2012.12.001.
13. Schwieler L, Larsson MK, Skogh E, et al. Increased levels of IL-6 in the cerebrospinal fluid of patients with chronic schizophrenia—significance for activation of the kynurenine pathway. *J Psychiatry Neurosci JPN.* 2015;40(2):126–133.
14. Katila H, Hurme M, Wahlbeck K, Appelberg B, Rimón R. Plasma and cerebrospinal fluid interleukin-1 beta and interleukin-6 in hospitalized schizophrenic patients. *Neuropsychobiology.* 1994;30(1):20–23.
15. Nikkilä HV, Müller K, Ahokas A, Rimón R, Andersson LC. Increased frequency of activated lymphocytes in the cerebrospinal fluid of patients with acute schizophrenia. *Schizophr Res.* 2001;49(1–2):99–105.
16. Miller BJ, Gassama B, Sebastian D, Buckley P, Mellor A. Meta-analysis of lymphocytes in schizophrenia: clinical status and antipsychotic effects. *Biol Psychiatry.* 2013;73(10):993–999. doi:10.1016/j.biopsych.2012.09.007.
17. Upthegrove R, Manzanares-Teson N, Barnes NM. Cytokine function in medication-naïve first episode psychosis: a systematic review and meta-analysis. *Schizophr Res.* 2014;155(1–3):101–108. doi:10.1016/j.schres.2014.03.005.
18. Modabbernia A, Taslimi S, Brietzke E, Ashrafi M. Cytokine alterations in bipolar disorder: a meta-analysis of 30 studies. *Biol Psychiatry.* 2013;74(1):15–25. doi:10.1016/j.biopsych.2013.01.007.
19. Dargél AA, Godin O, Kapczinski F, Kupfer DJ, Leboyer M. C-reactive protein alterations in bipolar disorder: a meta-analysis. *J Clin Psychiatry.* 2015;76(2):142–150. doi:10.4088/JCP.14r09007.
20. Rao JS, Harry GJ, Rapoport SI, Kim HW. Increased excitotoxicity and neuroinflammatory markers in postmortem frontal cortex from bipolar disorder patients. *Mol Psychiatry.* 2010;15(4):384–392. doi:10.1038/mp.2009.47.

21. Söderlund J, Olsson SK, Samuelsson M, et al. Elevation of cerebrospinal fluid interleukin-1ß in bipolar disorder. *J Psychiatry Neurosci JPN*. 2011;36(2):114–118. doi:10.1503/jpn.100080.
22. Breunis MN, Kupka RW, Nolen WA, et al. High numbers of circulating activated T cells and raised levels of serum IL-2 receptor in bipolar disorder. *Biol Psychiatry*. 2003;53(2):157–165.
23. Drexhage RC, Hoogenboezem TH, Versnel MA, Berghout A, Nolen WA, Drexhage HA. The activation of monocyte and T cell networks in patients with bipolar disorder. *Brain Behav Immun*. 2011;25(6):1206–1213. doi:10.1016/j.bbi.2011.03.013.
24. Howren MB, Lamkin DM, Suls J. Associations of depression with C-reactive protein, IL-1, and IL-6: a meta-analysis. *Psychosom Med*. 2009;71(2):171–186. doi:10.1097/PSY.0b013e3181907c1b.
25. Valkanova V, Ebmeier KP, Allan CL. CRP, IL-6 and depression: a systematic review and meta-analysis of longitudinal studies. *J Affect Disord*. 2013;150(3):736–744. doi:10.1016/j.jad.2013.06.004.
26. Köhler CA, Freitas TH, Stubbs B, et al. Peripheral alterations in cytokine and chemokine levels after antidepressant drug treatment for major depressive disorder: systematic review and meta-analysis. *Mol Neurobiol*. 2018 May;55(5):4195–4206. doi:10.1007/s12035-017-0632-1.
27. Więdłocha M, Marcinowicz P, Krupa R, et al. Effect of antidepressant treatment on peripheral inflammation markers—A meta-analysis. *Prog Neuropsychopharmacol Biol Psychiatry*. 2018 Jan 3;80(Pt C):217–226. doi:10.1016/j.pnpbp.2017.04.026.
28. Flatow J, Buckley P, Miller BJ. Meta-analysis of oxidative stress in schizophrenia. *Biol Psychiatry*. 2013;74(6):400–409. doi:10.1016/j.biopsych.2013.03.018.
29. Brown NC, Andreazza AC, Young LT. An updated meta-analysis of oxidative stress markers in bipolar disorder. *Psychiatry Res*. 2014;218(1–2):61–68. doi:10.1016/j.psychres.2014.04.005.
30. Versace A, Andreazza AC, Young LT, et al. Elevated serum measures of lipid peroxidation and abnormal prefrontal white matter in euthymic bipolar adults: toward peripheral biomarkers of bipolar disorder. *Mol Psychiatry*. 2014;19(2):200–208. doi:10.1038/mp.2012.188.
31. Andreazza AC, Wang J-F, Salmasi F, Shao L, Young LT. Specific subcellular changes in oxidative stress in prefrontal cortex from patients with bipolar disorder. *J Neurochem*. 2013;127(4):552–561. doi:10.1111/jnc.12316.
32. Joseph N, Zhang-James Y, Perl A, Faraone SV. Oxidative stress and ADHD: a meta-analysis. *J Atten Disord*. 2015;19(11):915–924. doi:10.1177/1087054713510354.
33. Liu T, Zhong S, Liao X, et al. A meta-analysis of oxidative stress markers in depression. *PLoS One*. 2015;10(10):e0138904. doi:10.1371/journal.pone.0138904.
34. Leonard BE, Schwarz M, Myint AM. The metabolic syndrome in schizophrenia: is inflammation a contributing cause? *J Psychopharmacol Oxf Engl*. 2012;26(5 Suppl):33–41. doi:10.1177/0269881111431622.
35. Leboyer M, Soreca I, Scott J, et al. Can bipolar disorder be viewed as a multi-system inflammatory disease? *J Affect Disord*. 2012;141(1):1–10. doi:10.1016/j.jad.2011.12.049.

36. Halaris A. Inflammation-associated co-morbidity between depression and cardiovascular disease. *Curr Top Behav Neurosci.* 2017;31:45–70. doi:10.1007/7854_2016_28.
37. Witthauer C, Gloster AT, Meyer AH, Lieb R. Physical diseases among persons with obsessive compulsive symptoms and disorder: a general population study. *Soc Psychiatry Psychiatr Epidemiol.* 2014;49(12):2013–2022. doi:10.1007/s00127-014-0895-z.
38. Vancampfort D, Stubbs B, Mitchell AJ, et al. Risk of metabolic syndrome and its components in people with schizophrenia and related psychotic disorders, bipolar disorder and major depressive disorder: a systematic review and meta-analysis. *World Psychiatry.* 2015;14(3):339–347. doi:10.1002/wps.20252.
39. Wang PS, Angermeyer M, Borges G, et al. Delay and failure in treatment seeking after first onset of mental disorders in the World Health Organization's World Mental Health Survey Initiative. *World Psychiatry.* 2007;6(3):177–185.
40. Levin JB, Krivenko A, Howland M, Schlachet R, Sajatovic M. Medication adherence in patients with bipolar disorder: a comprehensive review. *CNS Drugs.* 2016;30(9):819–835. doi:10.1007/s40263-016-0368-x.
41. McIntyre RS, Rasgon NL, Kemp DE, et al. Metabolic syndrome and major depressive disorder: co-occurrence and pathophysiologic overlap. *Curr Diab Rep.* 2009;9(1):51–59.
42. Godin O, Etain B, Henry C, et al. Metabolic syndrome in a French cohort of patients with bipolar disorder: results from the FACE-BD cohort. *J Clin Psychiatry.* 2014;75(10):1078–1085; quiz 1085. doi:10.4088/JCP.14m09038.
43. Godin O, Leboyer M, Gaman A, et al. Metabolic syndrome, abdominal obesity and hyperuricemia in schizophrenia: Results from the FACE-SZ cohort. *Schizophr Res.* 2015;168(1–2):388–394. doi:10.1016/j.schres.2015.07.047.
44. Koponen H, Jokelainen J, Keinänen-Kiukaanniemi S, Kumpusalo E, Vanhala M. Metabolic syndrome predisposes to depressive symptoms: a population-based 7-year follow-up study. *J Clin Psychiatry.* 2008;69(2):178–182.
45. Garcia-Rizo C, Kirkpatrick B, Fernandez-Egea E, Oliveira C, Bernardo M. Abnormal glycemic homeostasis at the onset of serious mental illnesses: a common pathway. *Psychoneuroendocrinology.* 2016;67:70–75. doi:10.1016/j.psyneuen.2016.02.001.
46. Pillinger T, Beck K, Gobjila C, Donocik JG, Jauhar S, Howes OD. Impaired glucose homeostasis in first-episode schizophrenia: a systematic review and meta-analysis. *JAMA Psychiatry.* 2017;74(3):261–269. doi:10.1001/jamapsychiatry.2016.3803.
47. Russell A, Ciufolini S, Gardner-Sood P, et al. Inflammation and metabolic changes in first episode psychosis: preliminary results from a longitudinal study. *Brain Behav Immun.* 2015;49:25–29. doi:10.1016/j.bbi.2015.06.004.
48. Goldstein BI, Collinger KA, Lotrich F, et al. Preliminary findings regarding proinflammatory markers and brain-derived neurotrophic factor among adolescents with bipolar spectrum disorders. *J Child Adolesc Psychopharmacol.* 2011;21(5):479–484. doi:10.1089/cap.2011.0009.
49. Jerrell JM, McIntyre RS, Tripathi A. A cohort study of the prevalence and impact of comorbid medical conditions in pediatric bipolar disorder. *J Clin Psychiatry.* 2010;71(11):1518–1525. doi:10.4088/JCP.09m05585ora.

50. Khandaker GM, Pearson RM, Zammit S, Lewis G, Jones PB. Association of serum interleukin 6 and C-reactive protein in childhood with depression and psychosis in young adult life: a population-based longitudinal study. *JAMA Psychiatry.* 2014;71(10):1121–1128. doi:10.1001/jamapsychiatry.2014.1332.
51. Courtet P, Giner L, Seneque M, Guillaume S, Olie E, Ducasse D. Neuroinflammation in suicide: Toward a comprehensive model. *World J Biol Psychiatry.* July 2015:1–23. doi:10.3109/15622975.2015.1054879.
52. Black C, Miller BJ. Meta-analysis of cytokines and chemokines in suicidality: distinguishing suicidal versus nonsuicidal patients. *Biol Psychiatry.* 2015;78(1):28–37. doi:10.1016/j.biopsych.2014.10.014.
53. Lyketsos CG, Kozauer N, Rabins PV. Psychiatric manifestations of neurologic disease: where are we headed? *Dialog Clin Neurosci.* 2007;9(2):111–124.
54. Meszaros ZS, Perl A, Faraone SV. Psychiatric symptoms in systemic lupus erythematosus: a systematic review. *J Clin Psychiatry.* 2012;73(7):993–1001. doi:10.4088/JCP.11r07425.
55. Gris J-C, Nobile B, Bouvier S. Neuropsychiatric presentations of antiphospholipid antibodies. *Thromb Res.* 2015;135(Suppl 1):S56–S59. doi:10.1016/S0049-3848(15)50445-3.
56. Euesden J, Danese A, Lewis CM, Maughan B. A bidirectional relationship between depression and the autoimmune disorders—new perspectives from the National Child Development Study. *PLoS One.* 2017;12(3):e0173015. doi:10.1371/journal.pone.0173015. eCollection 2017.
57. Andersson NW, Gustafsson LN, Okkels N, et al. Depression and the risk of autoimmune disease: a nationally representative, prospective longitudinal study. *Psychol Med.* 2015;45(16):3559–3569. doi:10.1017/S0033291715001488.
58. Benros ME, Waltoft BL, Nordentoft M, et al. Autoimmune diseases and severe infections as risk factors for mood disorders: a nationwide study. *JAMA Psychiatry.* 2013;70(8):812–820. doi:10.1001/jamapsychiatry.2013.1111.
59. Benros ME, Nielsen PR, Nordentoft M, Eaton WW, Dalton SO, Mortensen PB. Autoimmune diseases and severe infections as risk factors for schizophrenia: a 30-year population-based register study. *Am J Psychiatry.* 2011;168(12):1303–1310. doi:10.1176/appi.ajp.2011.11030516.
60. Chiaie RD, Caronti B, Macrì F, et al. Anti-purkinje cell and natural autoantibodies in a group of psychiatric patients. Evidences for a correlation with the psychopathological status. *Clin Pract Epidemiol Ment Health CP EMH.* 2012;8:81–90. doi:10.2174/1745017901208010081.
61. Margari F, Petruzzelli MG, Mianulli R, Campa MG, Pastore A, Tampoia M. Circulating anti-brain autoantibodies in schizophrenia and mood disorders. *Psychiatry Res.* 2015;230(2):704–708. doi:10.1016/j.psychres.2015.10.029.
62. Bard L, Groc L. Glutamate receptor dynamics and protein interaction: lessons from the NMDA receptor. *Mol Cell Neurosci.* 2011;48(4):298–307. doi:10.1016/j.mcn.2011.05.009.
63. Kantrowitz JT, Javitt DC. N-methyl-d-aspartate (NMDA) receptor dysfunction or dysregulation: the final common pathway on the road to schizophrenia? *Brain Res Bull.* 2010;83(3–4):108–121. doi:10.1016/j.brainresbull.2010.04.006.

64. Kayser MS, Titulaer MJ, Gresa-Arribas N, Dalmau J. Frequency and characteristics of isolated psychiatric episodes in anti–N-methyl-d-aspartate receptor encephalitis. *JAMA Neurol.* 2013;70(9):1133–1139. doi:10.1001/jamaneurol.2013.3216.
65. Kuo YL, Tsai HF, Lai MC, Lin CH, Yang YK. Anti-NMDA receptor encephalitis with the initial presentation of psychotic mania. *J Clin Neurosci Off J Neurosurg Soc Australas.* 2012;19(6):896–898. doi:10.1016/j.jocn.2011.10.006.
66. Ladepeche L, Dupuis JP, Bouchet D, et al. Single-molecule imaging of the functional crosstalk between surface NMDA and dopamine D1 receptors. *Proc Natl Acad Sci U S A.* 2013;110(44):18005–18010. doi:10.1073/pnas.1310145110.
67. Kupka RW, Nolen WA, Post RM, et al. High rate of autoimmune thyroiditis in bipolar disorder: lack of association with lithium exposure. *Biol Psychiatry.* 2002;51(4):305–311.
68. Padmos RC, Bekris L, Knijff EM, et al. A high prevalence of organ-specific autoimmunity in patients with bipolar disorder. *Biol Psychiatry.* 2004;56(7):476–482. doi:10.1016/j.biopsych.2004.07.003.
69. Hammond CJ, Hobbs JA. Parvovirus B19 infection of brain: possible role of gender in determining mental illness and autoimmune thyroid disorders. *Med Hypotheses.* 2007;69(1):113–116. doi:10.1016/j.mehy.2006.11.023.
70. Oliveira J, Busson M, Etain B, et al. Polymorphism of Toll-like receptor 4 gene in bipolar disorder. *J Affect Disord.* 2014;152–154: 395–402. doi:10.1016/j.jad.2013.09.043.
71. Hillegers MHJ, Reichart CG, Wals M, et al. Signs of a higher prevalence of autoimmune thyroiditis in female offspring of bipolar parents. *Eur Neuropsychopharmacol.* 2007;17(6–7):394–399. doi:10.1016/j.euroneuro.2006.10.005.
72. Bocchetta A, Traccis F, Mosca E, Serra A, Tamburini G, Loviselli A. Bipolar disorder and antithyroid antibodies: review and case series. *Int J Bipolar Disord.* 2016;4(1):5. doi:10.1186/s40345-016-0046-4.
73. Swedo SE. Pediatric autoimmune neuropsychiatric disorders associated with streptococcal infections (PANDAS). *Mol Psychiatry.* 2002;7(Suppl 2):S24–S25. doi:10.1038/sj.mp.4001170.
74. Pérez-Vigil A, Fernández de la Cruz L, Brander G, Isomura K, Gromark C, Mataix-Cols D. The link between autoimmune diseases and obsessive-compulsive and tic disorders: a systematic review. *Neurosci Biobehav Rev.* 2016;71:542–562. doi:10.1016/j.neubiorev.2016.09.025.
75. Orlovska S, Vestergaard CH, Bech BH, Nordentoft M, Vestergaard M, Benros ME. Association of streptococcal throat infection with mental disorders: testing key aspects of the PANDAS hypothesis in a nationwide study. *JAMA Psychiatry.* 2017;74(7):740–746. doi:10.1001/jamapsychiatry.2017.0995.
76. Pearlman DM, Vora HS, Marquis BG, Najjar S, Dudley LA. Anti-basal ganglia antibodies in primary obsessive-compulsive disorder: systematic review and meta-analysis. *Br J Psychiatry J Ment Sci.* 2014;205(1):8–16. doi:10.1192/bjp.bp.113.137018.
77. Attwells S, Setiawan E, Wilson AA, et al. Inflammation in the neurocircuitry of obsessive-compulsive disorder. *JAMA Psychiatry.* 2017;74(8):833–840. doi:10.1001/jamapsychiatry.2017.1567.

78. Nielsen PR, Benros ME, Dalsgaard S. Associations between autoimmune diseases and attention-deficit/hyperactivity disorder: a nationwide study. *J Am Acad Child Adolesc Psychiatry*. 2017;56(3):234–240.e1. doi:10.1016/j.jaac.2016.12.010.
79. Berk M, Williams LJ, Jacka FN, et al. So depression is an inflammatory disease, but where does the inflammation come from? *BMC Med*. 2013;11:200. doi:10.1186/1741–7015-11-200.
80. Maes M, Kubera M, Leunis J-C, Berk M. Increased IgA and IgM responses against gut commensals in chronic depression: further evidence for increased bacterial translocation or leaky gut. *J Affect Disord*. 2012;141(1):55–62. doi:10.1016/j.jad.2012.02.023.
81. Hietbrink F, Besselink MGH, Renooij W, et al. Systemic inflammation increases intestinal permeability during experimental human endotoxemia. *Shock Augusta Ga*. 2009;32(4):374–378. doi:10.1097/SHK.0b013e3181a2bcd6.
82. García Bueno B, Caso JR, Madrigal JLM, Leza JC. Innate immune receptor Toll-like receptor 4 signalling in neuropsychiatric diseases. *Neurosci Biobehav Rev*. 2016;64:134–147. doi:10.1016/j.neubiorev.2016.02.013.
83. Gárate I, Garcia-Bueno B, Madrigal JLM, et al. Stress-induced neuroinflammation: role of the Toll-like receptor-4 pathway. *Biol Psychiatry*. 2013;73(1):32–43. doi:10.1016/j.biopsych.2012.07.005.
84. Mangiola F, Ianiro G, Franceschi F, Fagiuoli S, Gasbarrini G, Gasbarrini A. Gut microbiota in autism and mood disorders. *World J Gastroenterol*. 2016;22(1):361–368. doi:10.3748/wjg.v22.i1.361.
85. Kelly JR, Borre Y, O' Brien C, et al. Transferring the blues: depression-associated gut microbiota induces neurobehavioural changes in the rat. *J Psychiatr Res*. 2016;82:109–118. doi:10.1016/j.jpsychires.2016.07.019.
86. Jiang H, Ling Z, Zhang Y, et al. Altered fecal microbiota composition in patients with major depressive disorder. *Brain Behav Immun*. 2015;48:186–194. doi:10.1016/j.bbi.2015.03.016.
87. Naseribafrouei A, Hestad K, Avershina E, et al. Correlation between the human fecal microbiota and depression. *Neurogastroenterol Motil*. 2014;26(8):1155–1162. doi:10.1111/nmo.12378.
88. O'Neill SM, Curran EA, Dalman C, et al. Birth by caesarean section and the risk of adult psychosis: a population-based cohort study. *Schizophr Bull*. 2016;42(3):633–641. doi:10.1093/schbul/sbv152.
89. Chudal R, Sourander A, Polo-Kantola P, et al. Perinatal factors and the risk of bipolar disorder in Finland. *J Affect Disord*. 2014;155:75–80. doi:10.1016/j.jad.2013.10.026.
90. Leboyer M, Tamouza R, Charron D, Faucard R, Perron H. Human endogenous retrovirus type W (HERV-W) in schizophrenia: a new avenue of research at the gene-environment interface. *World J Biol Psychiatry*. 2013;14(2):80–90. doi:10.3109/15622975.2010.601760.
91. Griffiths DJ. Endogenous retroviruses in the human genome sequence. *Genome Biol*. 2001;2(6): reviews1017.1–reviews1017.5.
92. Schulz WA, Steinhoff C, Florl AR. Methylation of endogenous human retroelements in health and disease. *Curr Top Microbiol Immunol*. 2006;310:211–250.

93. Nellåker C, Yao Y, Jones-Brando L, Mallet F, Yolken RH, Karlsson H. Transactivation of elements in the human endogenous retrovirus W family by viral infection. *Retrovirology*. 2006;3:44. doi:10.1186/1742-4690-3-44.
94. Perron H, Lang A. The human endogenous retrovirus link between genes and environment in multiple sclerosis and in multifactorial diseases associating neuroinflammation. *Clin Rev Allergy Immunol.* 2010;39(1):51–61. doi:10.1007/s12016-009-8170-x.
95. Rolland A, Jouvin-Marche E, Viret C, Faure M, Perron H, Marche PN. The envelope protein of a human endogenous retrovirus-W family activates innate immunity through CD14/TLR4 and promotes Th1-like responses. *J Immunol Baltim Md 1950*. 2006;176(12):7636–7644.
96. Lees-Murdock DJ, Walsh CP. DNA methylation reprogramming in the germ line. *Epigenetics*. 2008;3(1):5–13.
97. Huang W-J, Liu Z-C, Wei W, Wang G-H, Wu J-G, Zhu F. Human endogenous retroviral pol RNA and protein detected and identified in the blood of individuals with schizophrenia. *Schizophr Res.* 2006;83(2–3):193–199. doi:10.1016/j.schres.2006.01.007.
98. Karlsson H, Schröder J, Bachmann S, Bottmer C, Yolken RH. HERV-W-related RNA detected in plasma from individuals with recent-onset schizophrenia or schizoaffective disorder. *Mol Psychiatry*. 2004;9(1):12–13. doi:10.1038/sj.mp.4001439.
99. Karlsson H, Bachmann S, Schröder J, McArthur J, Torrey EF, Yolken RH. Retroviral RNA identified in the cerebrospinal fluids and brains of individuals with schizophrenia. *Proc Natl Acad Sci U S A*. 2001;98(8):4634–4639. doi:10.1073/pnas.061021998.
100. Yao Y, Schröder J, Nellåker C, et al. Elevated levels of human endogenous retrovirus-W transcripts in blood cells from patients with first episode schizophrenia. *Genes Brain Behav*. 2008;7(1):103–112. doi:10.1111/j.1601-183X.2007.00334.x.
101. Perron H, Hamdani N, Faucard R, et al. Molecular characteristics of human endogenous retrovirus type-W in schizophrenia and bipolar disorder. *Transl Psychiatry*. 2012;2:e201. doi:10.1038/tp.2012.125.
102. Middlebrooks JS, Audage NC. The effects of childhood stress on health across the lifespan. 2008. http://health-equity.pitt.edu/932/. Accessed February 11, 2015.
103. Daruy-Filho L, Brietzke E, Lafer B, Grassi-Oliveira R. Childhood maltreatment and clinical outcomes of bipolar disorder. *Acta Psychiatr Scand*. 2011;124(6):427–434. doi:10.1111/j.1600-0447.2011.01756.x.
104. Coelho R, Viola TW, Walss-Bass C, Brietzke E, Grassi-Oliveira R. Childhood maltreatment and inflammatory markers: a systematic review. *Acta Psychiatr Scand*. 2014;129(3):180–192. doi:10.1111/acps.12217.
105. Etain B, Aas M, Andreassen OA, et al. Childhood trauma is associated with severe clinical characteristics of bipolar disorders. *J Clin Psychiatry*. 2013;74(10):991–998. doi:10.4088/JCP.13m08353.
106. Etain B, Henry C, Bellivier F, Mathieu F, Leboyer M. Beyond genetics: childhood affective trauma in bipolar disorder. *Bipolar Disord*. 2008;10(8):867–876. doi:10.1111/j.1399-5618.2008.00635.x.

107. Garno JL, Goldberg JF, Ramirez PM, Ritzler BA. Impact of childhood abuse on the clinical course of bipolar disorder. *Br J Psychiatry J Ment Sci.* 2005;186:121–125. doi:10.1192/bjp.186.2.121.
108. Cremniter D, Jamain S, Kollenbach K, et al. CSF 5-HIAA levels are lower in impulsive as compared to nonimpulsive violent suicide attempters and control subjects. *Biol Psychiatry.* 1999;45(12):1572–1579.
109. Mandelli L, Petrelli C, Serretti A. The role of specific early trauma in adult depression: a meta-analysis of published literature. Childhood trauma and adult depression. *Eur Psychiatry J Assoc Eur Psychiatr.* 2015;30(6):665–680. doi:10.1016/j.eurpsy.2015.04.007.
110. Infurna MR, Reichl C, Parzer P, Schimmenti A, Bifulco A, Kaess M. Associations between depression and specific childhood experiences of abuse and neglect: a meta-analysis. *J Affect Disord.* 2016;190:47–55. doi:10.1016/j.jad.2015.09.006.
111. Tatham EL, Ramasubbu R, Gaxiola-Valdez I, et al. White matter integrity in major depressive disorder: Implications of childhood trauma, 5-HTTLPR and BDNF polymorphisms. *Psychiatry Res.* 2016;253:15–25. doi:10.1016/j.pscychresns.2016.04.014.
112. Kelly-Irving M, Lepage B, Dedieu D, et al. Childhood adversity as a risk for cancer: findings from the 1958 British birth cohort study. *BMC Public Health.* 2013;13(1):767. doi:10.1186/1471-2458-13-767.
113. Miller GE, Chen E, Parker KJ. Psychological stress in childhood and susceptibility to the chronic diseases of aging: moving toward a model of behavioral and biological mechanisms. *Psychol Bull.* 2011;137(6):959–997. doi:10.1037/a0024768.
114. Rich-Edwards JW, Spiegelman D, Lividoti Hibert EN, et al. Abuse in childhood and adolescence as a predictor of type 2 diabetes in adult women. *Am J Prev Med.* 2010;39(6):529–536. doi:10.1016/j.amepre.2010.09.007.
115. Shonkoff JP, Garner AS, Committee on psychosocial aspects of child and family health, committee on early childhood, adoption, and dependent care, section on developmental and behavioral pediatrics. The lifelong effects of early childhood adversity and toxic stress. *Pediatrics.* 2012;129(1):e232–e246. doi:10.1542/peds.2011-2663.
116. Slopen N, Koenen KC, Kubzansky LD. Childhood adversity and immune and inflammatory biomarkers associated with cardiovascular risk in youth: a systematic review. *Brain Behav Immun.* 2012;26(2):239–250. doi:10.1016/j.bbi.2011.11.003.
117. Kelly-Irving M, Lepage B, Dedieu D, et al. Adverse childhood experiences and premature all-cause mortality. *Eur J Epidemiol.* 2013;28(9):721–734. doi:10.1007/s10654-013-9832-9.
118. Carpenter LL, Gawuga CE, Tyrka AR, Lee JK, Anderson GM, Price LH. Association between plasma IL-6 response to acute stress and early-life adversity in healthy adults. *Neuropsychopharmacol.* 2010;35(13):2617–2623. doi:10.1038/npp.2010.159.
119. Copeland WE, Wolke D, Lereya ST, Shanahan L, Worthman C, Costello EJ. Childhood bullying involvement predicts low-grade systemic inflammation into adulthood. *Proc Natl Acad Sci U S A.* 2014;111(21):7570–7575. doi:10.1073/pnas.1323641111.
120. Danese A, Pariante CM, Caspi A, Taylor A, Poulton R. Childhood maltreatment predicts adult inflammation in a life-course study. *Proc Natl Acad Sci U S A.* 2007;104(4):1319–1324. doi:10.1073/pnas.0610362104.

121. Danese A, Moffitt TE, Pariante CM, Ambler A, Poulton R, Caspi A. Elevated inflammation levels in depressed adults with a history of childhood maltreatment. *Arch Gen Psychiatry.* 2008;65(4):409–415. doi:10.1001/archpsyc.65.4.409.
122. Miller GE, Chen E. Harsh family climate in early life presages the emergence of a proinflammatory phenotype in adolescence. *Psychol Sci.* 2010;21(6):848–856. doi:10.1177/0956797610370161.
123. Wright RJ, Visness CM, Calatroni A, et al. Prenatal maternal stress and cord blood innate and adaptive cytokine responses in an inner-city cohort. *Am J Respir Crit Care Med.* 2010;182(1):25–33. doi:10.1164/rccm.200904-0637OC.
124. Miller G, Chen E. Unfavorable socioeconomic conditions in early life presage expression of proinflammatory phenotype in adolescence. *Psychosom Med.* 2007;69(5):402–409. doi:10.1097/PSY.0b013e318068fcf9.
125. Miller GE, Chen E, Fok AK, et al. Low early-life social class leaves a biological residue manifested by decreased glucocorticoid and increased proinflammatory signaling. *Proc Natl Acad Sci U S A.* 2009;106(34):14716–14721. doi:10.1073/pnas.0902971106.
126. Chen E, Miller GE, Kobor MS, Cole SW. Maternal warmth buffers the effects of low early-life socioeconomic status on pro-inflammatory signaling in adulthood. *Mol Psychiatry.* 2011;16(7):729–737. doi:10.1038/mp.2010.53.
127. Dougherty LR, Klein DN, Davila J. A growth curve analysis of the course of dysthymic disorder: the effects of chronic stress and moderation by adverse parent-child relationships and family history. *J Consult Clin Psychol.* 2004;72(6):1012–1021. doi:10.1037/0022-006X.72.6.1012.
128. Heim C, Newport DJ, Heit S, et al. Pituitary-adrenal and autonomic responses to stress in women after sexual and physical abuse in childhood. *JAMA.* 2000;284(5):592–597.
129. Fries GR, Vasconcelos-Moreno MP, Gubert C, et al. Hypothalamic-pituitary-adrenal axis dysfunction and illness progression in bipolar disorder. *Int J Neuropsychopharmacol.* 2014;18(1). doi:10.1093/ijnp/pyu043.
130. Binder EB, Bradley RG, Liu W, et al. Association of FKBP5 polymorphisms and childhood abuse with risk of posttraumatic stress disorder symptoms in adults. *JAMA.* 2008;299(11):1291–1305. doi:10.1001/jama.299.11.1291.
131. Caspi A, Sugden K, Moffitt TE, et al. Influence of life stress on depression: moderation by a polymorphism in the 5-HTT gene. *Science.* 2003;301(5631):386–389. doi:10.1126/science.1083968.
132. Klengel T, Mehta D, Anacker C, et al. Allele-specific FKBP5 DNA demethylation mediates gene-childhood trauma interactions. *Nat Neurosci.* 2013;16(1):33–41. doi:10.1038/nn.3275.
133. Miller S, Hallmayer J, Wang PW, Hill SJ, Johnson SL, Ketter TA. Brain-derived neurotrophic factor val66met genotype and early life stress effects upon bipolar course. *J Psychiatr Res.* 2013;47(2):252–258. doi:10.1016/j.jpsychires.2012.10.015.
134. Perroud N, Courtet P, Vincze I, et al. Interaction between BDNF Val66Met and childhood trauma on adult's violent suicide attempt. *Genes Brain Behav.* 2008;7(3):314–322. doi:10.1111/j.1601-183X.2007.00354.x.
135. Perroud N, Jaussent I, Guillaume S, et al. *COMT* but not serotonin-related genes modulates the influence of childhood abuse on anger traits. *Genes Brain Behav.* 2010;9(2):193–202. doi:10.1111/j.1601-183X.2009.00547.x.

136. Romens SE, McDonald J, Svaren J, Pollak SD. Associations between early life stress and gene methylation in children. *Child Dev*. 2015;86(1):303–309. doi:10.1111/cdev.12270.
137. Roth TL, Lubin FD, Funk AJ, Sweatt JD. Lasting epigenetic influence of early-life adversity on the BDNF gene. *Biol Psychiatry*. 2009;65(9):760–769. doi:10.1016/j.biopsych.2008.11.028.
138. Oliveira J, Etain B, Lajnef M, et al. Combined effect of TLR2 gene polymorphism and early life stress on the age at onset of bipolar disorders. *PLoS One*. 2015;10(3):e0119702. doi:10.1371/journal.pone.0119702.
139. Belvederi Murri M, Prestia D, Mondelli V, et al. The HPA axis in bipolar disorder: systematic review and meta-analysis. *Psychoneuroendocrinology*. 2016;63:327–342. doi:10.1016/j.psyneuen.2015.10.014.
140. Giovanoli S, Engler H, Engler A, et al. Stress in puberty unmasks latent neuropathological consequences of prenatal immune activation in mice. *Science*. 2013;339(6123):1095–1099. doi:10.1126/science.1228261.
141. Giovanoli S, Engler H, Engler A, et al. Preventive effects of minocycline in a neurodevelopmental two-hit model with relevance to schizophrenia. *Transl Psychiatry*. 2016;6:e772. doi:10.1038/tp.2016.38.
142. Susser ES, Lin SP. Schizophrenia after prenatal exposure to the Dutch Hunger Winter of 1944–1945. *Arch Gen Psychiatry*. 1992;49(12):983–988.
143. Torrey EF, Miller J, Rawlings R, Yolken RH. Seasonality of births in schizophrenia and bipolar disorder: a review of the literature. *Schizophr Res*. 1997;28(1):1–38.
144. Brown AS, van Os J, Driessens C, Hoek HW, Susser ES. Further evidence of relation between prenatal famine and major affective disorder. *Am J Psychiatry*. 2000;157(2):190–195.
145. Cannon M, Jones PB, Murray RM. Obstetric complications and schizophrenia: historical and meta-analytic review. *Am J Psychiatry*. 2002;159(7):1080–1092.
146. Malaspina D, Corcoran C, Kleinhaus KR, et al. Acute maternal stress in pregnancy and schizophrenia in offspring: a cohort prospective study. *BMC Psychiatry*. 2008;8:71. doi:10.1186/1471-244X-8-71.
147. Kleinhaus K, Harlap S, Perrin M, et al. Prenatal stress and affective disorders in a population birth cohort. *Bipolar Disord*. 2013;15(1):92–99. doi:10.1111/bdi.12015.
148. Canetta SE, Bao Y, Co MDT, et al. Serological documentation of maternal influenza exposure and bipolar disorder in adult offspring. *Am J Psychiatry*. 2014;171(5):557–563. doi:10.1176/appi.ajp.2013.13070943.
149. Mortensen PB, Pedersen CB, Melbye M, Mors O, Ewald H. Individual and familial risk factors for bipolar affective disorders in Denmark. *Arch Gen Psychiatry*. 2003;60(12):1209–1215. doi:10.1001/archpsyc.60.12.1209.
150. Etain B, Mathieu F, Henry C, et al. Preferential association between childhood emotional abuse and bipolar disorder. *J Trauma Stress*. 2010;23(3):376–383. doi:10.1002/jts.20532.
151. Valipour G, Saneei P, Esmaillzadeh A. Serum vitamin D levels in relation to schizophrenia: a systematic review and meta-analysis of observational studies. *J Clin Endocrinol Metab*. 2014;99(10):3863–3872. doi:10.1210/jc.2014-1887.

152. Di Forti M, Sallis H, Allegri F, et al. Daily use, especially of high-potency cannabis, drives the earlier onset of psychosis in cannabis users. *Schizophr Bull.* 2014;40(6):1509–1517. doi:10.1093/schbul/sbt181.
153. Geoffroy PA, Scott J, Boudebesse C, et al. Sleep in patients with remitted bipolar disorders: a meta-analysis of actigraphy studies. *Acta Psychiatr Scand.* 2015;131(2):89–99. doi:10.1111/acps.12367.
154. Meltzer A, Van de Water J. The role of the immune system in autism spectrum disorder. *Neuropsychopharmacol.* 2017;42(1):284–298. doi:10.1038/npp.2016.158.
155. Lee BK, Magnusson C, Gardner RM, et al. Maternal hospitalization with infection during pregnancy and risk of autism spectrum disorders. *Brain Behav Immun.* 2015;44:100–105. doi:10.1016/j.bbi.2014.09.001.
156. Deykin EY, MacMahon B. Viral exposure and autism. *Am J Epidemiol.* 1979;109(6):628–638.
157. Libbey JE, Sweeten TL, McMahon WM, Fujinami RS. Autistic disorder and viral infections. *J Neurovirol.* 2005;11(1):1–10. doi:10.1080/13550280590900553.
158. Lintas C, Altieri L, Lombardi F, Sacco R, Persico AM. Association of autism with polyomavirus infection in postmortem brains. *J Neurovirol.* 2010;16(2):141–149. doi:10.3109/13550281003685839.
159. Atladóttir HÓ, Henriksen TB, Schendel DE, Parner ET. Autism after infection, febrile episodes, and antibiotic use during pregnancy: an exploratory study. *Pediatrics.* 2012;130(6):e1447–1454. doi:10.1542/peds.2012-1107.
160. Shi L, Fatemi SH, Sidwell RW, Patterson PH. Maternal influenza infection causes marked behavioral and pharmacological changes in the offspring. *J Neurosci.* 2003;23(1):297–302.
161. Zhang X, Lv C-C, Tian J, et al. Prenatal and perinatal risk factors for autism in China. *J Autism Dev Disord.* 2010;40(11):1311–1321. doi:10.1007/s10803-010-0992-0.
162. Zerbo O, Iosif A-M, Walker C, Ozonoff S, Hansen RL, Hertz-Picciotto I. Is maternal influenza or fever during pregnancy associated with autism or developmental delays? Results from the CHARGE (CHildhood Autism Risks from Genetics and Environment) study. *J Autism Dev Disord.* 2013;43(1):25–33. doi:10.1007/s10803-012-1540-x.
163. Yolken RH, Torrey EF. Are some cases of psychosis caused by microbial agents? A review of the evidence. *Mol Psychiatry.* 2008;13(5):470–479. doi:10.1038/mp.2008.5.
164. Pásztói M, Misják P, György B, et al. Infection and autoimmunity: lessons of animal models. *Eur J Microbiol Immunol.* 2011;1(3):198–207. doi:10.1556/EuJMI.1.2011.3.3.
165. Jung S, Schickel J-N, Kern A, et al. Chronic bacterial infection activates autoreactive B cells and induces isotype switching and autoantigen-driven mutations. *Eur J Immunol.* 2016;46(1):131–146. doi:10.1002/eji.201545810.
166. Hamdani N, Daban-Huard C, Godin O, et al. Effects of cumulative Herpesviridae and *Toxoplasma gondii* infections on cognitive function in healthy, bipolar, and schizophrenia subjects. *J Clin Psychiatry.* 2017;78(1):e18–e27. doi:10.4088/JCP.15m10133.

167. Oliveira J, Oliveira-Maia AJ, Tamouza R, Brown AS, Leboyer M. Infectious and immunogenetic factors in bipolar disorder. *Acta Psychiatr Scand.* 2017;136(4):409–423. doi:10.1111/acps.12791.
168. Sutterland AL, Fond G, Kuin A, et al. Beyond the association. *Toxoplasma gondii* in schizophrenia, bipolar disorder, and addiction: systematic review and meta-analysis. *Acta Psychiatr Scand.* 2015;132(3):161–179. doi:10.1111/acps.12423.
169. Graham KL, Carson CM, Ezeoke A, Buckley PF, Miller BJ. Urinary tract infections in acute psychosis. *J Clin Psychiatry.* 2014;75(4):379–385. doi:10.4088/JCP.13m08469.
170. Dickerson F, Stallings C, Origoni A, et al. Antibodies to *Toxoplasma gondii* in individuals with mania. *Bipolar Disord.* 2014;16(2):129–136. doi:10.1111/bdi.12123.

9

Immunology of Substance Use Disorders

HAITHAM SALEM, SCOTT D. LANE, AND ANTONIO L. TEIXEIRA ■

INTRODUCTION

Addiction is a complex condition that leads to a series of medical and socioeconomic consequences, exacting a huge toll from society. According to the National Institute on Drug Abuse (NIDA), 40 million Americans ages 12 and older abuse or are addicted to nicotine, alcohol, or illicit drugs.[1] Nine out of ten people who abuse drugs began abusing substances before they were 18 years old, and those who began using addictive substances before age 15 are nearly seven times more likely to develop a substance problem than those who delay first use until age 21 or older.[1]

Historically viewed as a character flaw, addiction has been re-conceptualized as a chronic illness with a myriad of symptoms, and this understanding of addiction as a medical condition is important for its prevention and treatment.[2] Indeed, a wealth of evidence has shown that addiction to alcohol and other drugs is associated with significant changes in the brain. Furthermore, several medical conditions are frequently comorbid with addiction, such as diabetes, asthma, and hypertension. Both in addiction and in other psychiatric disorders, the role of the immune system is gathering increasing attention as an important mechanism. The goal of this chapter is to describe the neurobiological framework underlying addiction, and the emerging understanding of the role played by immune mechanisms and related clinical implications.

NEUROBIOLOGY OF ADDICTION

There are three key brain areas that form networks intimately implicated in the development and persistence of substance use disorders: the basal ganglia, the extended amygdala, and the prefrontal cortex (Table 9.1). The basal ganglia is associated with rewarding or pleasurable effects of substance use, and the formation of habitual substance taking. The extended amygdala is involved in stress and the negative emotional states that typically follow substance withdrawal. The prefrontal cortex exerts control over substance taking and is part of the reward circuitry activated by drugs of abuse.[3]

Addiction can be described as a repeating cycle with three stages. Each stage is particularly associated with one of these aforementioned brain regions.[4] This three-stage model is well supported by human and animal research and provides a useful heuristic to depict the processes of addiction, thus informing prevention and treatment efforts.[5]

Table 9.1 Neurobiology of Addiction and Addiction Cycle

Addiction Cycle	Neural Substrates	Features
Binge/ Intoxication	Ventral tegmental area (a main source of dopamine) and basal ganglia (ventral striatal or *nucleus accumbens* and dorsal striatum)	Activation of the reward system generates the pleasurable feelings associated with substance use, and engenders changes in the response to drug-related stimuli (e.g., people, places). Over time, these stimuli activate the system on their own, triggering powerful urges for substance intake called "incentive salience."
Withdrawal/ Negative Affect	Extended amygdala	In the absence of the abused substance, the person experiences negative emotions and, sometimes, symptoms of physical illness. Two main mechanisms are associated with the negative feelings: (1) de-activation of the reward circuit of the basal ganglia (reward deficit), and (2) activation of the brain stress system in the extended amygdala (stress surfeit).
Preoccupation/ Anticipation	Prefrontal cortex	Experiencing preoccupation with substance use (craving), and drug-seeking. Due to executive dysfunction, the person may experience difficulty with decision making, including problems suppressing/inhibiting drug-seeking behavior.

In the binge/intoxication phase, dopamine receptors are activated, either directly or indirectly, by addictive substances, particularly stimulants such as cocaine, amphetamines, and nicotine.[6] The opioid system of the brain—which includes endorphins, enkephalins, and dynorphins, and their receptors mu, delta, and kappa—mediates the reward system of other substances such as opioids and alcohol. Activation of the opioid system stimulates the *nucleus accumbens* through the dopaminergic system. Cannabinoids (THC) act on the brain endogenous cannabinoid system that also contributes to the reward by modulating the function of dopamine neurons.

The overlap of reward and habit formation neurocircuits facilitates the intense desire/craving for the substance and the compulsive substance seeking behavior that occurs when individuals with addiction are exposed to drug cues.

During withdrawal periods, individuals with addiction experience anhedonia, i.e., an overall reduction in the sensitivity of the brain reward system both to addictive substances and also to natural reinforcers, such as food and sex. Human brain imaging studies in addiction have consistently shown long-lasting deficits in D2 dopamine receptors compared with non-addicted individuals.[7,8] This general loss of reward sensitivity may also account for the compulsive escalation of substance use as addicted individuals attempt to regain the pleasurable feelings the reward system once provided.[9]

A second phenomenon that occurs during this stage is the activation of stress neurotransmitters in the extended amygdala, most notably in the hypothalamic-pituitary-adrenal (HPA) axis. These stress neurotransmitters include corticotrophin-releasing factor (CRF), norepinephrine, and dynorphins. Studies have shown that blocking the activation of stress receptors in the brain reduced alcohol consumption in both alcohol-dependent rats and humans with alcohol use disorder.[10] Accordingly, an additional motivation for drug and alcohol seeking behaviors among individuals with substance use disorders is possibly related to the suppression of an overactive brain stress system that produces negative emotions and feelings.[11]

After a period of abstinence, the addicted person starts exhibiting craving, or a preoccupation with substance use. This stage involves the prefrontal cortex, which is responsible for executive functioning, helping people making decisions, i.e., selecting strategies to follow ("Go system") while inhibiting others ("Stop system").[12]

When substance-seeking behavior is triggered by environmental cues, there is an increased activity in the prefrontal cortex that signals the *nucleus accumbens* to release glutamate, promoting incentive salience, which creates a powerful urge to use the substance. The prefrontal cortex activation also engages habit formation in the dorsal striatum, contributing to the impulsivity associated with substance abuse.[13] Notably, people with alcohol, cocaine, or opioid use disorders usually show impairment in executive functions, including disruption of decision-making and behavioral inhibition. As people progress through the addiction cycle and the intensity with which they experience each of the stages, the addiction cycle tends to escalate over time.[14]

IMMUNOLOGY OF ADDICTION

Due in part to the activation of stress responses during drug withdrawal, immune/inflammatory changes are expected in this phase. It has also been proposed that the innate immune system might play a role in the development of addictions, especially in the context of stressful situations.

Alcohol

In animal models, alcohol exposure exerts a significant impact on the peripheral and the neuro-immune systems, i.e., the immune cells and molecules in the central nervous system. Binge and chronic alcohol exposure induces neuroimmune activation through Toll-like receptors (TLR) and HMGB1,[15,16] while TLR4 and CD14 play an important role in the acute ethanol effects on GABAergic transmission in the central amygdala.[17] Interestingly, targeted disruption of TLR4 in the central amygdala reduces alcohol consumption.[18] Pro-inflammatory cytokines facilitate alcohol withdrawal-induced anxiety via the CRF signaling in the central amygdala.[19,20]

In humans, polymorphisms of genes encoding the cytokines IL-1β and IL-10 are associated with the susceptibility to alcoholism.[21,22] The levels of the pro-inflammatory cytokines IL-8 and IL-1β were positively correlated with alcohol consumption and craving, while the levels of the anti-inflammatory cytokine IL-10 were negatively correlated with them, suggesting that activation of innate immune signaling may modulate alcohol craving and consumption.[23,24]

Early studies of humans with alcohol use disorders showed that chronic alcohol abuse for 12 to 15 years resulted in reduced numbers of peripheral T cells.[25] This finding was confirmed in people who engaged in a short period of binge drinking and in individuals who drank heavily for six months. Interestingly, abstinence for 30 days was sufficient to restore lymphocyte numbers to physiological levels.[26] Actually, chronic alcohol exposure disrupts T cell homeostasis.[27] Chronic alcohol abusers exhibited a decreased frequency of naïve CD4 and CD8 T cells (CD45RA+), but an increased frequency of memory T cells (CD45RO+).[28] Together, these observations suggest that chronic alcohol consumption results in lymphopenia, which, through compensatory mechanisms, accelerates conversion of naïve T cells into memory T cells.[29]

Chronic alcohol abusers with liver disease showed increased activation of B cells leading to high titers of circulating antibodies directed against liver-specific autoantigens.[30] Antibodies against the alcohol dehydrogenase enzyme have been found in 50% of chronic abusers with alcoholic hepatitis, while anti-phospholipid antibodies were detected in almost 80% of patients with alcoholic hepatitis or cirrhosis.[31]

Cocaine

In an early preclinical study of the effects of cocaine on the immune system, Watson and colleagues described a suppression of the immune system following cocaine administration in mice, as measured by ear swelling and plaque-forming cell splenic assay. This finding was attributed to the cocaine effect on the σ_1 receptor in a wide range of leukocyte populations, including T cells, B cells, and natural killer (NK) cells.[32]

Besides direct effects of cocaine on immune cells, there is a generalized enhanced activity of the stress system, including the HPA axis, during cocaine dependence that also influences immune cells.[33] The immune system undergoes changes during early-protracted cocaine withdrawal in the form of upregulation of TNF-α.[34] Also after cocaine withdrawal, Kubera and colleagues documented an increase of TNF-α and a reduction of the production of IL-10 by splenocytes in mice following exposure to a conditioned stimuli previously paired with cocaine administration.[35]

In human studies, in line with the concept of a chronic stress state, cocaine users showed decreased blood levels of IL-10 and significant elevations of TNF-α when exposed to stress compared with social drinkers.[36] In a case-control study nested in a cross-sectional population-based survey, serum levels of the pro-inflammatory cytokine IL-6 were also significantly increased, while IL-10 levels were decreased among cocaine users.[37] In an outpatient population–based study, plasma levels of the pro-inflammatory cytokines IL-1β, CX3CL1/fractalkine, and CXCL12/SDF-1 were positively correlated with cocaine symptom severity. Interestingly, IL-1β was increased in users with psychiatric disorders in comparison to the users with no psychiatric diagnosis.[38] Peripheral TNF-α levels were able to predict crack cocaine acute withdrawal symptom severity.[39] These results suggest that abstinence is associated with higher levels of peripheral pro-inflammatory cytokines, and this may play a role in the behavioral negative reinforcing effects of cocaine use disorder. Moreover, peripheral cytokines may be regarded as putative biomarkers for cocaine users, although further studies must confirm that.

Cocaine metabolites, specifically benzoylecgonine, have been shown to covalently modify endogenous proteins (e.g., albumin) present in the plasma through the acylation of the ε-amino group of the protein lysine residues through a nucleophilic attack.[40] These modified proteins have the potential to be recognized as "foreign" antigens through binding to major histocompatibility complexes (MHC) located on antigen-presenting cells (e.g., macrophages, dendritic cells), and ultimately affect the functions of T and B lymphocytes. This finding fostered the concept of using immune modulatory strategies in addiction treatment (e.g., cocaine vaccine).[41]

Opiates

Opiates' interaction with the neuroimmune system can enhance the rewarding properties of opiates such as morphine.[42] Glia activation results in the release of pro-inflammatory cytokines, which can affect glia-neuronal signaling, including dopamine signaling, ultimately influencing behavioral outcomes after opioid exposure. More specifically, TLRs have been shown to be important for modulating glia-neuronal signaling and the reinforcing properties of opiates.[43–45]

In humans, a key pro-inflammatory cytokine, IL-1β, has been shown to be upregulated following morphine exposure[46] and single-nucleotide polymorphisms implicated in increased IL-1β production have been associated with risk of opioid dependence.[47]

Amphetamines

Evidence from animal models has shown that central nervous system exposure to methamphetamine results in the activation of microglia, which has been linked to damage to striatal dopaminergic terminals and reduction in striatal dopamine through inflammatory processes.[48–51] Pharmacological agents (e.g., AV411, Ibudilast) that reduce glial activation have been shown to attenuate methamphetamine-induced relapse.[52] In humans, neuroimaging data revealed significant increase in microglial cells in the midbrain, striatum, thalamus, orbitofrontal cortex, and insular cortex of methamphetamine addicted individuals, further linking methamphetamine exposure to microglia activation.[53]

CLINICAL IMPLICATIONS

Drug-mediated neuroimmune modulation may contribute to addiction through different mechanisms. Importantly, those signaling pathways could provide targets for the development of biomarkers and therapeutic strategies.

Biomarkers

Biomarkers for addiction stand to aid in the classification of patients into categories or subgroups that may have predictive validity, thus providing targets for treatment, prediction of response, and improvement in outcomes. Another possibility is to identify biomarkers of vulnerability.[54] Taking into account promising results of some inflammatory biomarkers, such as C-reactive protein and IL-6 in the context of mood disorders, it would be interesting to test whether these markers have any categorizing and/or predictive value in identifying groups of patients with substance use disorders and comorbid mood disorders.

Vaccines and Other Immune-Based Strategies

Several trials were carried out in a quest to develop a vaccine to treat addiction. The first effort at an anti-drug vaccine was against morphine/heroin in the 1970s.[55] However, this interest lessened when methadone became available for heroin addiction.[56] The interest in anti-drug vaccines reawakened in the 1990s when Bagasra and colleagues reported an anti-cocaine vaccine.[57]

The first anti-cocaine vaccine (GNC) was developed by Jada and colleagues in 1995. When GNC-vaccinated rats were dosed with cocaine, there was suppression of locomotor activity as compared with non-vaccinated animals, and lower levels of cocaine were found in the brain.[58] In 2001, Fox and Kantak developed a vaccine made from succinyl norcocaine conjugated to cholera toxin B (dubbed TA-CD), and showed that the vaccine was effective in eliciting levels of antibodies enough to antagonize self-administration of cocaine in rats.[59,60]

The first clinical trial of an anti-cocaine vaccine (TA-CD) was conducted in 2002 on 24 subjects recruited from a residential treatment program. Three different doses were given three times at monthly intervals. Anti-cocaine antibodies appeared after the second injection, peaked at three months, and fell to baseline by one year. The level of antibodies correlated with the vaccine dose.[61] Another study recruited cocaine-dependent subjects from an outpatient treatment center, and evaluated the effects of two different doses of the vaccine, showing that the subjects who received the higher dose had higher levels of antibodies in their serum and fewer positive urine samples indicating cocaine use. They also reported less euphoric effects if they had taken cocaine. Antibodies did not appear in the serum before two weeks, and waned by six months, but booster vaccinations in a few subjects increased the antibody titers.[62] It must be noted that in both trials, 25–30% of subjects produced relatively low levels of antibodies.

In 2009, a large trial of an anti-cocaine vaccine involving 115 subjects from an outpatient maintenance program measured the antibody levels monthly after the first two weeks, and tested urine samples three times a week for the entire study period. Like the previous studies, the level of antibodies varied substantially among individuals. Only 38% of the subjects in this trial attained the target concentration of antibodies, but this group had significantly more cocaine-free urine samples than either the low antibody group or the placebo group.[63]

More recently, dAd5GNE prepared from human adenovirus has been developed as anti-cocaine vaccine and showed promising results in mice, rats, and nonhuman primates demonstrating that dAd5GNE induces high levels of high-affinity anti-cocaine antibodies that sequester systemically administered cocaine in the blood, preventing cocaine-induced hyperlocomotor activity and sensitization.[64–67] In 2016, dAd5GNE was approved for its first phase I clinical trial, which is in progress.

Methamphetamine vaccines have also been used in rodents with positive results.[68,69] In those trials, immune sera were able to bind to free methamphetamine, but this binding capacity was lost over time.[69]

In preclinical studies, minocycline attenuated the development of behavioral sensitization to amphetamines[70] and cocaine.[71] The exact mechanism of action of minocycline has to be determined. Two main mechanisms have been proposed: an anti-inflammatory effect via inhibition of activated microglia, and/or enhancing glutamate release via stabilization of the NMDA receptors.[72–74]

In a double-blind placebo-controlled crossover study, 15 healthy human volunteers were given oral dextroamphetamine capsules to simulate amphetamines' effect on the body, then they were treated with 200 mg/day minocycline. Results showed attenuation of some of the subjective-rewarding responses to amphetamines and reduced cortisol levels, thus supporting its possible use in the treatment of stimulant addiction.[75]

FUTURE PERSPECTIVES AND CONCLUSION

Addiction treatment is very challenging. Biomarker development is a frontier that needs further exploration, and immune biomarkers are promising in this regard. Another potentially interesting area is addiction immunotherapy with the development of strategies (e.g., vaccines) able to block or attenuate the addiction cycle. However, there are several factors that need to be improved before anti-drug vaccines and immune-related strategies can achieve their full potential. For instance, it is not understood why there is so much variability in antibody responses to drugs among human subjects.

Also of paramount importance is the acknowledgement that vaccines or immune-related strategies will not be the "end of the game" for treating substance abuse, but rather can serve as one component in a diverse treatment platform that includes pharmacological and psychosocial approaches.

REFERENCES

1. Ahrnsbrak, R., Bose, J., Hedden, S. L., Lipari, R. N., & Park-Lee, E. (2017). Substance Abuse and Mental Health Services Administration. Key substance use and mental health indicators in the United States: Results from the 2016 National Survey on Drug Use and Health *(HHS Publication No. SMA 17-5044, NSDUH Series H-52).* Rockville, MD: Center for Behavioral Health Statistics and Quality, Substance Abuse and Mental Health Services Administration. Retrieved from https://www.samhsa.gov/data/.
2. McLellan, A. T., Lewis, D. C., O'Brien, C. P., & Kleber, H. D. (2000). Drug dependence, a chronic medical illness: Implications for treatment, insurance, and outcomes evaluation. *Journal of the American Medical Association, 284*(13), 1689–1695.
3. Kalivas, P. W., & Volkow, N. D. (2005). The neural basis of addiction: A pathology of motivation and choice. *The American Journal of Psychiatry, 162*(8), 1403–1413.
4. Koob, G. F., & Le Moal, M. (1997). Drug abuse: Hedonic homeostatic dysregulation. *Science, 278*(5335), 52–58.
5. Koob, G. F., & Volkow, N. D. (2010). Neurocircuitry of addiction. *Neuropsychopharmacology, 35*(1), 217–238.
6. Nestler, E. J. (2005). Is there a common molecular pathway for addiction? *Nature Neuroscience, 8*(11), 1445–1449.
7. Volkow, N. D., Tomasi, D., Wang, G. J., et al. (2014). Stimulant-induced dopamine increases are markedly blunted in active cocaine abusers. *Molecular Psychiatry, 19*(9), 1037–1043.
8. Volkow, N. D., Wang, G. J., Fowler, J. S., et al. (1997). Decreased striatal dopaminergic responsiveness in detoxified cocaine-dependent subjects. *Nature, 386*(6627), 830–833.
9. Koob, G. F., & Le Moal, M. (2001). Drug addiction, dysregulation of reward, and allostasis. *Neuropsychopharmacology, 24*(2), 97–129.
10. Vendruscolo, L. F., Estey, D., Goodell, V., et al. (2015). Glucocorticoid receptor antagonism decreases alcohol seeking in alcohol-dependent individuals. *The Journal of Clinical Investigation, 125*(8), 3193–3197.
11. Parsons, L. H., & Hurd, Y. L. (2015). Endocannabinoid signalling in reward and addiction. *Nature Reviews Neuroscience, 16*(10), 579–594.

12. Koob, G. F., Arends, M. A., & Le Moal, M. (2014). *Drugs, addiction, and the brain.* Waltham, MA: Academic Press.
13. Goldstein, R. Z., & Volkow, N. D. (2011). Dysfunction of the prefrontal cortex in addiction: Neuroimaging findings and clinical implications. *Nature Reviews Neuroscience, 12*(11), 652–669.
14. Davis, M., Walker, D. L., Miles, L., & Grillon, C. (2010). Phasic vs sustained fear in rats and humans: Role of the extended amygdala in fear vs anxiety. *Neuropsychopharmacology, 35*(1), 105–135.
15. Alfonso-Loeches, S., Pascual-Lucas, M., Blanco, A. M., Sanchez-Vera, I., & Guerri, C. (2010). Pivotal role of TLR4 receptors in alcohol-induced neuroinflammation and brain damage. *The Journal of Neuroscience, 30*, 8285–8295.
16. Crews, F. T., Qin, L., Sheedy, D., Vetreno, R. P., & Zou, J. (2013). High mobility group box 1/Toll-like receptor danger signaling increases brain neuroimmune activation in alcohol dependence. *Biological Psychiatry, 73*, 602–612.
17. Bajo, M., Madamba, S. G., Roberto, M., et al. (2014). Innate immune factors modulate ethanol interaction with GABAergic transmission in mouse central amygdala. *Brain, Behavior, and Immunity, 40*, 191–202.
18. Liu, J., Yang, A. R., Kelly, T., et al. (2011). Binge alcohol drinking is associated with GABAA alpha2-regulated Toll-like receptor 4 (TLR4) expression in the central amygdala. *Proceedings of the National Academy of Sciences of the United States of America, 108*, 4465–4470.
19. Knapp, D. J., Whitman, B. A., Wills, T. A., et al. (2011). Cytokine involvement in stress may depend on corticotrophin releasing factor to sensitize ethanol withdrawal anxiety. *Brain, Behavior, and Immunity, 25*(Suppl. 1), S146–S154.
20. Whitman, B. A., Knapp, D. J., Werner, D. F., Crews, F. T., & Breese, G. R. (2013). The cytokine mRNA increase induced by withdrawal from chronic ethanol in the sterile environment of brain is mediated by CRF and HMGB1 release. *Alcoholism, Clinical and Experimental Research, 37*, 2086–2097.
21. Marcos, M., Pastor, I., Gonzalez-Sarmiento, R., & Laso, F. J. (2008). Interleukin-10 gene polymorphism is associated with alcoholism but not with alcoholic liver disease. *Alcohol and Alcoholism, 43*, 523–528.
22. Pastor, I. J., Laso, F. J., Romero, A., & Gonzalez-Sarmiento, R. (2005). Interleukin-1 gene cluster polymorphisms and alcoholism in Spanish men. *Alcohol and Alcoholism, 40*, 181–186.
23. Leclercq, S., Cani, P. D., Neyrinck, A. M., et al. (2012). Role of intestinal permeability and inflammation in the biological and behavioral control of alcohol-dependent subjects. *Brain, Behavior, and Immunity, 26*, 911–918.
24. Leclercq, S., De Saeger, C., Delzenne, N., de Timary, P., & Starkel, P. (2014). Role of inflammatory pathways, blood mononuclear cells, and Gut-derived bacterial products in alcohol dependence. *Biological Psychiatry, 76*(9), 725–733. doi:10.1016/j.biopsych.2014.02.003. Epub 2014 Feb 14.
25. Liu, Y. K. (1973). Leukopenia in alcoholics. *American Journal of Medicine, 54*(5), 605–610.
26. Tønnesen, H., Andersen, J. R., Pedersen, A. E., & Kaiser, A. H. (1990). Lymphopenia in heavy drinkers—reversibility and relation to the duration of drinking episodes. *Annals of Medicine, 22*(4), 229–231.

27. Song, K., Coleman, R. A., Zhu, X., et al. (2002). Chronic ethanol consumption by mice results in activated splenic T cells. *Journal of Leukocyte Biology, 72*(6), 1109–1116.
28. Zhang, H., & Meadows, G. G. (2005). Chronic alcohol consumption in mice increases the proportion of peripheral memory T cells by homeostatic proliferation. *Journal of Leukocyte Biology, 78*(5), 1070–1080.
29. Cho, B. K., Rao, V. P., Ge, Q., Eisen, H. N., & Chen, J. (2000). Homeostasis-stimulated proliferation drives naive T cells to differentiate directly into memory T cells. *Journal of Experimental Medicine, 192*(4), 549–556.
30. McFarlane, I. G. (2000). Autoantibodies in alcoholic liver disease. *Addiction Biology, 5*(2), 141–151.
31. Pasala, S., Barr, T., & Messaoudi, I. (2015). Impact of alcohol abuse on the adaptive immune system. *Alcohol Research: Current Reviews, 37*(2), 185–197.
32. Watson, E. S., Murphy, J. C., ElSohly, H. N., ElSohly, M. A., & Turner, C. E. (1983). Effects of the administration of coca alkaloids on the primary immune responses of mice: Interaction with delta 9-tetrahydrocannabinol and ethanol. *Toxicology and Applied Pharmacology, 71*(1), 1–13.
33. Sinha, R. (2001). How does stress increase risk of drug abuse and relapse? *Psychopharmacology (Berlin), 158*(4), 343–359.
34. Wang, Y., Huang, D. S., & Watson, R. R. (1994). In vivo and in vitro cocaine modulation on production of cytokines in C57BL/6 mice. *Life Sciences, 54*(6), 401–411.
35. Kubera, M., Filip, M., Budziszewska, B., et al. (2008). Immunosuppression induced by a conditioned stimulus associated with cocaine self-administration. *Journal of Pharmacological Sciences, 107*(4), 361–369.
36. Fox, H. C., D'Sa, C., Kimmerling, A., et al. (2012). Immune system inflammation in cocaine dependent individuals: Implications for medications development. *Human Psychopharmacology, 27*(2), 156–166. doi:10.1002/hup.1251.
37. Moreira, F. P., Medeiros, J. R., Lhullier, A. C., et al. (2016). Cocaine abuse and effects in the serum levels of cytokines IL-6 and IL-10. *Drug & Alcohol Dependence, 158*, 181–185. doi:10.1016/j.drugalcdep.2015.11.024.
38. Araos, P., Pedraz, M., Serrano, A., et al. (2015). Plasma profile of pro-inflammatory cytokines and chemokines in cocaine users under outpatient treatment: Influence of cocaine symptom severity and psychiatric co-morbidity. *Addiction Biology, 20*(4), 756–772. doi:10.1111/adb.12156.
39. Levandowski, M. L., Viola, T. W., Wearick-Silva, L. E., et al. (2014). Early life stress and tumor necrosis factor superfamily in crack cocaine withdrawal. *Journal of Psychiatric Research, 53*, 180–186. doi:10.1016/j.jpsychires.2014.02.017.
40. Deng, S. X., Bharat, N., Fischman, M. C., & Landry, D. W. (2002). Covalent modification of proteins by cocaine. *Proceedings of the National Academy of Sciences of the United States of America, 99*(6), 3412–3416.
41. Marasco, C. C., Goodwin, C. R., Winder, D., Schramm-Sapyta, N., McLean, J. A., & Wikswo, J. P. (2014). Systems-level view of cocaine addiction: The interconnection of the immune and nervous systems. *Experimental Biology and Medicine (Maywood, NJ), 239*(11), 1433–1442. doi:10.1177/1535370214537747.
42. Bland, S. T., Hutchinson, M. R., Maier, S. F., Watkins, L. R., & Johnson, K. W. (2009). The glial activation inhibitor AV411 reduces morphine-induced nucleus accumbens dopamine release. *Brain, Behavior, and Immunity, 23*, 492–497.

43. Gay, N. J., Symmons, M. F., Gangloff, M., & Bryant, C. E. (2014). Assembly and localization of Toll-like receptor signalling complexes. *Nature Reviews Immunology, 14*(8), 546–558.
44. Trotta, T., Porro, C., Calvello, R., & Panaro, M. A. (2014). Biological role of Toll-like receptor-4 in the brain. *Journal of Neuroimmunology, 268*(1–2), 1–12.
45. Terashvili, M., Wu, H. E., Schwasinger, E. T., Hung, K. C., Hong, J. S., & Tseng, L. F. (2008). (+)-Morphine attenuates the (−)-morphine-produced conditioned place preference and the mu-opioid receptor-mediated dopamine increase in the posterior nucleus accumbens of the rat. *European Journal of Pharmacology, 587*, 147–154.
46. Raghavendra, V., Tanga, F. Y., & DeLeo, J. A. (2004). Attenuation of morphine tolerance, withdrawal-induced hyperalgesia, and associated spinal inflammatory immune responses by propentofylline in rats. *Neuropsychopharmacology, 29*, 327–334.
47. Liu, L., Hutchinson, M. R., White, J. M., Somogyi, A. A., & Coller, J. K. (2009). Association of IL-1B genetic polymorphisms with an increased risk of opioid and alcohol dependence. *Pharmacogenetics and Genomics, 19*, 869–876.
48. Ehrlich, L. C., Hu, S., Sheng, W. S., et al. (1998). Cytokine regulation of human microglial cell IL-8 production. *Journal of Immunology, 160*, 1944–1948.
49. Gadient, R. A., & Otten, U. H. (1997). Interleukin-6 (IL-6) A molecule with both beneficial and destructive potentials. *Progress in Neurobiology, 52*, 379–390.
50. McGuire, S. O., Ling, Z. D., Lipton, J. W., Sortwell, C. E., Collier, T. J., & Carvey, P. M. (2001). Tumor necrosis factor alpha is toxic to embryonic mesencephalic dopamine neurons. *Experimental Neurology, 169*, 219–230.
51. Thomas, D. M., & Kuhn, D. M. (2005). Attenuated microglial activation mediates tolerance to the neurotoxic effects of methamphetamine. *Journal of Neurochemistry, 92*, 790–797.
52. Beardsley, P. M., Shelton, K. L., Hendrick, E., & Johnson, K. W. (2010). The glial cell modulator and phosphodiesterase inhibitor, AV411 (Ibudilast), attenuates prime- and stress-induced methamphetamine relapse. *European Journal of Pharmacology, 637*, 102–108.
53. Sekine, Y., Ouchi, Y., Sugihara, G., et al. (2008). Methamphetamine causes microglial activation in the brains of human abusers. *The Journal of Neuroscience, 28*, 5756–5761.
54. Volkow, N. D., Baler, R. D., & Goldstein, R. Z. (2011). Addiction: Pulling at the neural threads of social behaviors. *Neuron, 69*(4), 599–602.
55. Spector, S., Berkowitz, B., Flynn, E. J., & Peskar, B. (1973). Antibodies to morphine, barbiturates, and serotonin. *Pharmacology Review,25*, 281–291.
56. Kreek, M. J., Borg, L., Ducat, E., & Ray, B. (2010). Pharmacotherapy in the treatment of addiction: Methadone. *Journal of Addictive Diseases, 29*(2), 200–216.
57. Bagasra, O., Forman, L. J., Howeedy, A., & Whittle, P. (1992). A potential vaccine for cocaine abuse prophylaxis. *Immunopharmacology, 23*(3), 173–179.
58. Carrera, M. R., Ashley, J. A., Parsons, L. H., Wirsching, P., Koob, G. F., & Janda, K. D. (1995). Suppression of psychoactive effects of cocaine by active immunization. *Nature, 378*(6558), 727–730.
59. Kantak, K. M., Collins, S. L., Lipman, E. G., Bond, J., Giovanoni, K., & Fox, B. S. (2000). Evaluation of anti-cocaine antibodies and a cocaine vaccine in a rat self-administration model. *Psychopharmacology (Berlin), 148*(3), 251–262.

60. Kantak, K. M., Collins, S. L., Bond, J., & Fox, B. S. (2001). Time course of changes in cocaine self-administration behavior in rats during immunization with the cocaine vaccine IPC-1010. *Psychopharmacology (Berlin), 153*(3), 334–340.
61. Kosten, T. R., & Biegel, D. (2002). Therapeutic vaccines for substance dependence. *Expert Review of Vaccines, 1*(3), 363–371.
62. Martell, B. A., Mitchell, E., Poling, J., Gonsai, K., & Kosten, T. R. (2005). Vaccine pharmacotherapy for the treatment of cocaine dependence. *Biological Psychiatry, 58*(2), 158–164.
63. Martell, B. A., Orson, F. M., Poling, J., Mitchell, E., Rossen, R. D., Gardner, T., & Kosten, T. R. (2009). Cocaine vaccine for the treatment of cocaine dependence in methadone-maintained patients: A randomized, double-blind, placebo-controlled efficacy trial. *Archives of General Psychiatry, 66*(10), 1116–1123.
64. Hicks, M. J., De, B. P., Rosenberg, J. B., et al. (2011). Cocaine analog coupled to disrupted adenovirus: A vaccine strategy to evoke high-titer immunity against addictive drugs. *Molecular Therapy, 19*(3), 612–619.
65. Wee, S., Hicks, M. J., De, B. P., et al. (2012). Novel cocaine vaccine linked to a disrupted adenovirus gene transfer vector blocks cocaine psychostimulant and reinforcing effects. *Neuropsychopharmacology, 37*(5), 1083–1091.
66. Maoz, A., Hicks, M. J., & Vallabhjosula, S. (2013). Adenovirus capsid-based anti-cocaine vaccine prevents cocaine from binding to the nonhuman primate CNS dopamine transporter. *Neuropsychopharmacology, 38*(11), 2170–2178.
67. Hicks, M. J., Kaminsky, S. M., De, B. P., et al. (2014). Fate of systemically administered cocaine in nonhuman primates treated with the dAd5GNE anticocaine vaccine. *Human Gene Therapy Clinical Development, 25*(1), 40–49. doi:10.1089/humc.2013.23).
68. Duryee, M. J., Bevins, R. A., Reichel, C. M., et al. (2009). Immune responses to methamphetamine by active immunization with peptide-based, molecular adjuvant-containing vaccines. *Vaccine, 27*(22), 2981–2988.
69. Gentry, W. B., Rüedi-Bettschen, D., Owens, SM. (2009). Development of active and passive human vaccines to treat methamphetamine addiction. *Human Vaccine, 5*(4), 206–213.
70. Zhang, L., Kitaichi, K., Fujimoto, Y., et al. (2006). Protective effects of minocycline on behavioral changes and neurotoxicity in mice after administration of methamphetamine. *Progress in Neuro-Psychopharmacology & Biological Psychiatry, 30*(8), 1381–1393.
71. Chen, H., Uz, T., & Manev, H. (2009). Minocycline affects cocaine sensitization in mice. *Neuroscience Letters, 452*(3), 258–261.
72. Lisiecka, D. M., Suckling, J., Barnes, T. R., et al. (2015). the benefit of micocycline on negative symptoms in early-phase psychosis in addition to standard care—extent and mechanism (BeneMin), Study protocol for randomized controlled trial. *Trials, 16*, 71. doi:10.1186/s13063-015-0580-x.
73. Fujita, Y., Kunitachi, S., Iyo, M., & Hashimoto, K. (2012). The antibiotic minocycline prevents methamphetamine-induced rewarding effects in mice. *Pharmacology Biochemistry & Behavior, 101*(2), 303–306. doi:10.1016/j.pbb.2012.01.005. Epub 2012 Jan 12.
74. Attarzadeh, Y. G., Arezoomandan, R., & Haghparast, A. (2014). Minocycline, an antibiotic with inhibitory effect on microglial activation, attenuates the maintenance

and reinstatement of methamphetamine-seeking behavior in rat. *Progress in Neuro-Psychopharmacology & Biological Psychiatry, 53*, 142–148. doi:10.1016/g.pnpbp.2014.04.008.

75. Sofuoglu, M., Mooney, M., Kosten, T., Waters, A., & Hashimoto, K. (2011). Minocycline attenuates subjective-rewarding effects of dextroamphetamine in humans. *Psychopharmacology, 213*(1), 61–68. doi:10.1007/s00213-010-2014-5.

10

Immune System Dysregulation and Schizophrenia

LAURA STERTZ AND CONSUELO WALSS-BASS ■

INTRODUCTION

Schizophrenia is a complex illness that affects around 1% of the population. This psychiatric disorder often has an unpredictable course and is characterized by symptoms of psychosis (delusions, hallucinations, and disordered thought), blunted affect, and cognitive abnormalities (impaired memory, attention, and executive function). Schizophrenia is considered a complex familial disorder with multiple susceptibility genes of small effect. Different individuals may carry different sets of genes that, in combination with particular environmental factors, will determine the overall expression pattern and outcome severity of the illness.

Although several hypotheses regarding the causes of schizophrenia have been proposed across the years, its precise etiopathogenesis is still unknown. Some major examples are the dopaminergic and glutamatergic hypotheses, which propose that schizophrenia is primarily associated with dysfunction of dopamine/glutamate signaling pathways,[1,2] and the neurodevelopment hypothesis,[3] which focuses on aberrant genes that cause early neural hazards. Another line of studies suggests that schizophrenia is associated with conditions of enhanced innate immune response and overproduction of pro-inflammatory cytokines, leading to the immune dysregulation hypothesis of schizophrenia that will be discussed in this chapter.

IMMUNE ACTIVATION AND SCHIZOPHRENIA

The integrity of the blood-brain barrier (BBB) is essential to reduce the entry of inflammatory cells and antibodies in the brain. In an intact brain, only a few mononuclear cells, such as activated T cells and macrophages, migrate into the central nervous system (CNS).[4] Under certain environmental conditions, such as exposure to infectious agents or ischemic conditions, macrophages produce inflammatory cytokines such as interleukin 1-beta (IL-1β), tumor necrosis factor-α (TNF-α), and interleukin 6 (IL-6).[5] These inflammatory molecules can signal through the endothelial cells to change the tight junction structure, leading to increased permeability of the BBB and alterations in brain structure and function.[6] Importantly, levels of these molecules have been repeatedly shown to be elevated in schizophrenia[5–7] and a recent meta-analysis support elevated IL-6 levels in schizophrenia.[8] Another meta-analysis revealed that some markers such as IL-6, TNFα, soluble IL-2 receptor (sIL-2R), and IL-1 receptor antagonist (IL-1RA) show trans-diagnostic validity in different major psychiatric disorders.[9] A comparison of cerebrospinal fluid (CSF) samples from patients with schizophrenia and controls revealed that the proportion of mononuclear cells is significantly higher in patients as well.[5]

The importance of macrophages and T lymphocytes, and the cytokines produced by them, has been highlighted in the macrophage–T cell theory of schizophrenia.[10,11] According to this hypothesis, chronically activated macrophages and T cells synthesize inflammatory compounds that destabilize the brain and lead to schizophrenic symptoms. Monocytes expressing CD54 receptors bind to intracellular adhesion molecules (ICAMs), which are important factors for the transmigration of monocytes through the endothelial cells into the brain. Soluble ICAMs levels have been shown to be increased in schizophrenia.[12]

IMMUNE T CELL ACTIVATION AND SCHIZOPHRENIA: FOCUS ON IL-17

Immune activation and/or dysregulation have been consistently reported in schizophrenia.[13] The risk factors leading to this inflammatory state in schizophrenia include genetic vulnerability interacting with environmental stressors such as infectious agents, trauma, and toxins.[14] Researchers have hypothesized that some mediators of inflammation are related to disease manifestation and symptom severity in schizophrenia. For instance, we reported positive correlation between the levels of circulating cytokines and severity of symptoms, as assessed by the Positive and Negative Syndrome Scale (PANSS), in patients with schizophrenia, and showed these cytokines to be part of the IL-17 pathway.[15]

Th17 cells, a subset of T helper cells, appear to have evolved as cells bridging the innate and the adaptive immunity, and are specialized for enhanced cell protection from microorganisms that are not well guarded against the Th1 and Th2 immunity. In this context, Th17 cells have emerged as crucial players in mucosal defense against infections.[16] Th17 responses are very important in host defense but also in promoting chronic inflammation and autoimmunity.[17] IL-17 is linked to atopic, inflammatory, and autoimmune diseases. IL-17 induces expression of several cytokines known to contain the nuclear factor kappa B (NF-κB) binding sites in their promoter regions. NF-κB is the principal transcription factor in the initiation of the inflammatory response. In response to IL-17, neutrophil-specific chemokines such as IL-8/CXCL8 are generated, as well as granulopoietic cytokines such as granulocyte-colony stimulating factor (G-CSF) and granulocyte-macrophage colony-stimulating factor (GM-CSF). Also, by acting on macrophages, IL-17 stimulates the release of TNF-α.[18]

It has been proposed that fetal programming of cellular immune components leads to the generation of long-lasting effector/memory Th17 cells, which are reactivated upon a second "hit." Activated Th17 cells can infiltrate the CNS by disrupting the BBB, thereby activating microglia, leading to a neuroinflammatory state.[19] Th17 cells lead to increased inflammatory mediators' release from microglia.[20] For instance, in experimental autoimmune encephalomyelitis (EAE), a mouse model for multiple sclerosis, Th17 cells and CD4(+) T cells that produce both interferon gamma (IFN-γ) and IL-17, infiltrate the brain prior to the development of clinical symptoms of EAE. This coincides with the activation of CD11b(+) microglia and the local production of IL-1β, TNF-α, and IL-6 in the CNS.[21]

In contrast to the autoimmunity-promoting Th17 cells, thymus-derived natural regulatory cells (Tregs) represent a unique population of cells that inhibit T cell proliferation and autoimmune processes.[7] Tregs are a component of the immune system that suppresses immune responses of other T cells. This is an important "self-check" built into the immune system to prevent excessive or pathological reactions. These cells are involved in shutting down immune responses after they have successfully eliminated invading organisms, and also in preventing autoimmunity. Accordingly, Th17 and Treg cells play opposite roles in immune responses. While Treg cells use the Foxp3 transcription factor, the process of differentiation of Th17 requires retinoic acid receptor-related orphan receptor gamma t isoform (RORγt) and signal transducer and activator of transcription 3 (STAT3) transcription factors. At the same time, both cell subsets require transforming growth factor beta (TGF-β) for their development, but there is a reciprocal regulation in the generation of these cells. While TGF-β induces Foxp3 expression (Tregs), in the presence of IL-6 and IL-23, TGF-β will instead induce Th17 differentiation.[22] Tregs are known to have a neuroprotective effect, attenuating microglia-mediated inflammation.[23] Overall, these studies suggest a well-organized T cell activity pattern, through coordinated expression of Th17- and Treg-related factors. There is an intensive crosstalk between infiltrating immune cells/mediators and the resident neural cells, which can be significantly affected when the T cell pattern becomes imbalanced. Although not yet specifically investigated in schizophrenia, these mechanisms pose an interesting model that may be relevant for its pathogenesis, particularly regarding the evidence of immune activation and the role of infectious agents in schizophrenia, as discussed later.

BRAIN IMMUNE CELL ACTIVATION AND SCHIZOPHRENIA

Cerebral inflammation and immune activation in schizophrenia have been evidenced by reports of increased microglial activity and elevated levels of two markers: serine protein inhibitor 3 (SERPINA3), an acute-phase protein, and interferon-induced transmembrane protein (IFITM), an immune-related protein involved in viral replication, in postmortem brains of schizophrenia patients.[24] Postmortem investigation of brain tissue has also found elevated microglial cell density (with a hypertrophic morphology) in schizophrenia compared with controls,[25] particularly in the frontal and temporal lobes,[26] although some studies did not confirm that.[27]

Microglial cells seems to be activated during active psychosis. Cells visualized with a positron emission tomography (PET) tracer (PK11195), an indicator of microglia activation that binds to benzodiazepine receptors,[28] were found to have greater receptor expression in patients with recent-onset psychosis/schizophrenia.[29] Activated microglia stimulate astrocytes to produce S100B, a marker of inflammation that is considered to be the equivalent of C-reactive protein (CRP) in the brain.[30] S100B serum levels are elevated in schizophrenia patients, and antipsychotics such as haloperidol and clozapine have been shown to decrease S100B release from glial cells.[31]

When activated, microglia undergo morphological changes and also express the 18kDa translocator protein (TSPO), a mitochondrial membrane protein that is usually expressed at low levels in glial cells of healthy brains.[32] Upregulation of TSPO expression has been proposed as an *in vivo* biomarker of microglial activation during neuroinflammation.[33–37] Elevations in microglial activity can be measured *in vivo* with PET using radioligands specific for TSPO.[35] Investigations have revealed increases in TSPO binding in medicated patients with schizophrenia when compared to healthy controls.[29] The first investigation of microglia using PET in schizophrenia, using the first-generation TSPO radioligand (R)-[11C]PK11195, found elevated (R)-[11C]PK11195 binding in total brain gray matter in a cohort of 10 patients with schizophrenia within five years of diagnosis.[38] Another (R)-[11C]PK11195 study found an elevation in hippocampal binding potential and a non-statistically significant 30% increase in total gray matter binding potential in seven chronically medicated patients with schizophrenia.[39] Results from second-generation TSPO radiotracers have been so far mixed. Takano et al. and Kenk et al. used [11C]DAA1106 and [18F]-FEPPA, respectively, and did not find any difference in schizophrenia compared with healthy controls. However, Takano et al. found positive correlations between [11C]DAA1106 and the severity of symptoms according to the PANSS in patients.[40] A more recent TSPO ligand has now been developed, [11C]PBR28, with higher brain uptake and improved binding specificity compared to [11C]-(R)-PK11195.[41,42] In non-human primates, inflammation-induced microglial activity caused marked increases in [11C]PBR28 signal, a result that was confirmed by postmortem analysis.[43]

Caution needs to be used when interpreting data from TSPO studies due to methodological differences in measures of TSPO availability, considering that TSPO can be expressed in astrocytes in addition to microglia, and more recently, the suggestion that TSPO may, in fact, be part of an anti-inflammatory response.[44]

It is noteworthy that transcriptome data from the hippocampus of schizophrenia patients indicated that abnormally expressed molecules related to immune and inflammation pathways were more likely to be expressed in endothelial cells of blood vessels, in blood monocytes within the blood vessels, and in perivascular astrocytes in patients, than in lymphocytes or microglia. As discussed before, the BBB may be abnormal in schizophrenia, or cells proximal to the BBB may be responding to factors in the blood of patients that then cause abnormal function in neurons and microglia.[45] Accordingly, a brain structure that would be interesting to analyze, and is normally overlooked, is the choroid plexus.[46] This is a plexus of cells that produces CSF into the ventricles of the brain and is important for communication between the brain and the periphery of the body. Building co-expression networks from the combined gene expression data of two different tissue collections, researchers found a module related to immune/inflammation response that was over-expressed in schizophrenia.[47] Taking into account the function of upregulated genes, the findings suggested that the choroid plexus of patients is responding to peripheral stimuli by upregulating genes related to immune function and inflammation.

STRESS AND INFLAMMATION

Stress and immune system reactivity play a major role in the development and exacerbation of psychotic symptoms in schizophrenia.[48–50] First-time outbreaks of psychosis and schizophrenia are more frequent in individuals exposed to high levels of stress, such as newly recruited military personnel.[51] Furthermore, evidence suggests that with increasing life events or situational stress, psychotic symptoms increase,[48] possibly due to over-activation of the immune system and over-expression of inflammatory cytokines.[52] Stress induces over-activation of the NMDA receptor, caused by excessive cytokine-induced glutamate release from astrocytes. Increased inflammation in the CNS also leads to impaired function of oligodendrocytes, which are damaged due to overexposure to cytokines such as TNF-α, leading to impaired myelination.[53] These processes have been shown to be altered in schizophrenia.

Rodent models of immune challenge have shown that maternal viral and bacterial infections during pregnancy lead to behaviors in the offspring that are considered rodent analogues of schizophrenia-like symptoms: deficits in prepulse inhibition, novelty-induced hyperactivity, and cognitive impairments.[54] These behavioral changes are thought to be due to imbalance in pro- and anti-inflammatory cytokines.[54,55] Supporting these animal models, prospective birth cohort studies with serologically documented gestational infection and immune biomarkers in humans suggest that gestational exposure to specific infections (such as influenza [H1N1] and *Toxoplasma gondii*) during pregnancy significantly increases the risk of schizophrenia in the offspring.[56] There is additional extensive literature supporting the role of infectious agents in schizophrenia.[57,58] In this context, Carter discussed how some strong schizophrenia candidate genes, including immune system genes, play direct roles in regulating the life cycle of several pathogens, and how understanding the interplay between gene function and pathogen exposure may help explain the role of gene–environment interactions in schizophrenia development.[59] Altogether, the evidence suggests that exposure to stress/immune challenges (either through stressful life events or infections) produces a host of biological processes that, in genetically susceptible individuals, can lead to the development of schizophrenia.

Inflammation is inextricably and cyclically linked to mitochondrial dysfunction and production of reactive oxygen species, i.e., oxidative stress, with both processes reciprocally inducing each other in a positive feedback loop, leading to disturbances in energy metabolism. Abnormalities in mitochondria function have been reported in schizophrenia, including decreased mitochondrial respiration and changes in mitochondrial morphology.[60] Of interest, mitochondrial diseases are characterized by psychotic and affective symptoms as well as cognitive decline. Several studies performing transcriptome profiling of schizophrenia postmortem brains have identified downregulation of genes involved in energy metabolism.[61–63]

Alterations in immune and energy metabolism pathways, via exposure to stressful events or infectious agents, can have an impact on the regulation of brain processes such as synaptic plasticity, neurogenesis, mood, and cognition. Therefore, persistent alterations in immune processes could potentially have long-term consequences leading to the onset and/or progression of psychiatric disorders. A study measuring gene expression profiles in postmortem brain of schizophrenic patients at different stages of the illness (defined by illness length) found both distinct as well as overlapping networks. Importantly, pathways of inflammation and immune response were associated with long-term illness, suggesting that this pathway may be related to disease progression in schizophrenia.[64] An increased inflammatory state may be either the result of progression of the primary underlying process, or a consequence of failure of adaptive homeostatic mechanisms occurring as part of illness progression.

ASSOCIATION OF IMMUNE SYSTEM GENES WITH SCHIZOPHRENIA

Schizophrenia is a very heterogeneous disorder with a wide array of symptomatology and extraordinary phenotypical variety among individuals. The impingement of different environmental factors in genetically susceptible individuals probably determines the overall expression pattern and outcome severity of the illness. Furthermore, schizophrenia is most likely not caused by mutations in a few specific genes, but by the disruption of several genes involved in key developmental signaling pathways, such as the immune system pathway. Understanding how genetic make-up, in combination with environmental stimuli, modulates expression of immune system genes and regulates brain function and behavior, and how this could lead to the occurrence of psychiatric disorders, is an area of intense research. As discussed later, several studies demonstrating associations of genetic variants in immune system genes with schizophrenia have been reported in the past few years, further supporting the hypothesis that immune system dysregulation plays an important role in the pathogenesis of schizophrenia (Table 10.1).

Two independent large studies of schizophrenia patients and controls found significant associations with variants mapping to the region of chromosome 6 that forms part of the major histocompatibility complex (MHC) and increased risk for schizophrenia.[65,66] The MHC has been associated with an increased risk of autoimmune diseases.[67,68] Importantly, a large population-based study (7,704 patients with schizophrenia and 192,590 control subjects) in the Danish

Table 10.1 EVIDENCE SUPPORTING THE ASSOCIATION OF IMMUNE SYSTEM GENES WITH SCHIZOPHRENIA

Gene/Gene Product	Reference	Comments
Tumor necrosis factor (TNF)	Saviouk et al. (2005)[82]	Association of a haplotype in the promoter region of TNF
Colony stimulating factor, receptor 2 alpha (CSF2RA); interleukin 3 receptor alpha (IL3RA)	Lencz et al. (2007)[75]	Evidence from genome-wide association, followed by gene sequencing in an independent population
Neuregulin 1 (NRG1), IL-1β	Hanninen et al. (2007)[84]	Synergistic interaction between NRG1 and IL-1b increases risk for schizophrenia
Interleukin 18 (IL-18 pathway)	Shirts et al. (2008)[79]	Genes in the IL-18 pathway and HSV seropositivity are associated with schizophrenia
Interleukin 12-p40 (IL-12-p40)	Ozbey et al. (2008)[80]	Association of gene promoter variants in a Turkish population
IL3RA	Sun et al. (2008, 2009)[76,77]	Association studies in the Han Chinese population
Interleukin 10 (IL-10)	Ozbey et al. (2009)[81]	SNP in the IL-10 promoter is associated with schizophrenia
Major histocompatibility complex (MHC), transcription factor 4 (TCF4)	Stefansson et al. (2009)[65]	Association with markers spanning the MHC region on 6p21.3-22.1 and a marker in intron 4 of TCF4
Interleukin 1 (IL-1) genes: IL-1α, IL-1β, IL-1RA	Xu and He (2010)[83]	Convergent evidence of association of IL-1 gene complex
NRG1	Marballi et al. (2010)[86]	A NRG1 transmembrane mutation is associated with increased levels of pro-inflammatory cytokines
MHC, TCF4	Li et al. (2010)[66]	Common variants associated in the Han Chinese population
Interferon gamma (IFNγ)	Paul-Samojedny et al. (2010)[78]	SNP associated with increased risk for schizophrenia in males
Inter-alpha-trypsin-inhibitor family of genes (NEK4-ITIH1-ITIH3-ITIH4)	Psychiatric GWAS Consortium Bipolar Disorder Working Group (2011)[72]	Association with SNPs in the region of NEK4-ITIH1-ITIH3-ITIH4

Table 10.1 CONTINUED

Gene/Gene Product	Reference	Comments
MHC	Schizophrenia Working Group of the Psychiatric Genomics Consortium (2014)[14]	Association with alleles in the MHC locus
Complement component 4 (C4)	Sekar et al. (2016)[70]	Association with structurally diverse alleles of the C4 gene

population found that a family history of any autoimmune disease is associated with a 45% increase in the risk for schizophrenia development.[69] The association of the MHC with schizophrenia was more recently validated in the largest genome-wide association study (GWAS) of schizophrenia to date (36,989 cases and 113,075 controls) in which, among the 108 loci that met criteria for genome-wide significance, the locus on chromosome 6 harboring MHC genes was the most significant.[14] A follow-up study pinpointed the MHC association to alleles of the complement component 4 (C4) gene.[70] C4 activates complement component 3 (C3), marking the synapse for phagocytosis, and the complement receptor 3 (CR3) drives synaptic pruning by microglia. The finding of C4 alleles associated with schizophrenia indicates an important genetic influence on the extent of synaptic pruning. Recently, a RNA sequencing (RNAseq)-based transcriptomic profiling study in lymphoblastoid cell lines of a European ancestry sample (529 schizophrenia cases and 660 controls) found differential expression in components of the complement system (CR1, CR2, CD55, and C3), though not for C4A.[71]

A combined GWAS analysis of schizophrenia and bipolar disorder found evidence of association of ITIH3-ITIH4.[72] These genes are part of the inter-alpha-trypsin-inhibitor family of genes that are involved in conferring stability to the extracellular matrix and play an important role in regulation of inflammation.[73,74] In another GWAS, Lencz et al. found evidence that two immune system genes—colony stimulating factor, receptor 2 alpha (CSF2RA) and interleukin 3 receptor alpha (IL3RA)—are associated with schizophrenia.[75] Two later studies further support the finding of association of IL3RA with schizophrenia.[76,77] A single-nucleotide polymorphism (SNP) in the first intron of the IFN-γ gene was found to be associated with a 1.66-fold higher risk of paranoid schizophrenia in males, but not in females.[78] A comprehensive evaluation of genes in the IL-18 pathway revealed associations with schizophrenia and herpes virus seropositivity.[79] Significant differences in both genotype and allele frequencies of IL12-p40 gene promoter variants were found between schizophrenia patients and controls in a Turkish population.[80] The same group found evidence of association of an IL-10 gene promoter polymorphism with schizophrenia in the Turkish population.[81] A haplotype in the tumor necrosis factor promoter has also been associated with schizophrenia.[82] Evidence of association of interleukin-1 genes (IL-1 alpha, IL-1 beta, and IL-1 receptor antagonist) with susceptibility to schizophrenia was reported.[83] Polymorphisms in neuregulin 1 (NRG1), a strong schizophrenia candidate gene, have been shown to interact synergistically with IL-1β polymorphisms to increase risk for schizophrenia.[84] Our group reported the first evidence of association of NRG1 with immune dysregulation. A valine-to-leucine mutation in the transmembrane region of NRG1, previously implicated in schizophrenia,[85] was found to be associated with increased levels of autoimmune antibodies and pro-inflammatory cytokines in plasma, and greater production of pro-inflammatory cytokines in lymphoblastoid cell lines, from individuals carrying the mutation compared to non-carriers.[86]

IMPLICATIONS AND FUTURE RESEARCH

Detection of specific inflammatory fingerprints may help to identify schizophrenia patients who will benefit from adjunctive anti-inflammatory or immunomodulation therapy. It has been reported that immunomodulation therapy can change aberrant monocytes/macrophages causing inflammation and over-production of cytokines by restoring them to their neuroprotective state, thus reducing inflammation and normalizing the cellular environment of critical nerve cells. For example, promising results have been demonstrated by clinical trials using nonsteroidal anti-inflammatory drugs (NSAIDs) in addition to antipsychotic treatment. Particularly, aspirin has been shown to successfully decrease the severity of symptoms in patients.[87] More recently, in Japan, a 24-year-old male with treatment-resistant schizophrenia with predominant severe delusions and hallucinations, who received bone marrow transplantation for acute myeloid leukemia, showed a remarkable reduction in psychotic symptoms without administration of neuroleptics. The patient also showed significant improvement in social functioning, even after four years.[88] These studies are a source of optimism that immunomodulation may open up an alternative method for treatment of schizophrenia.

In conclusion, there are many different theories that have been proposed to explain the wide range of symptomatology in schizophrenia, including alterations of the dopaminergic and/or glutamatergic systems,[1,2] abnormal neurodevelopment,[3] and the theory of immune system imbalance discussed here. While all of these individual theories have a strong body of evidence to support them, they are not mutually exclusive, and in fact are interconnected. For example, recent studies indicate that the kynurenine pathway, by which tryptophan is metabolized into kynurenic and quinolinic acid, is activated by pro-inflammatory cytokines. It has been shown that these acids regulate NMDA receptor activity and may also be involved in dopamine regulation. Therefore, although the precise interrelationship between the different pathways remains to be demonstrated in schizophrenia, a unifying theory that explains all the different lines of evidence may soon start to emerge.

REFERENCES

1. Carlsson, A. (1988). The current status of the dopamine hypothesis of schizophrenia. *Neuropsychopharmacology, 1*(3), 179–186. doi:10.1016/0893-133x(88)90012-7
2. Kim, J. S., Kornhuber, H. H., Schmid-Burgk, W., & Holzmüller, B. (1980). Low cerebrospinal fluid glutamate in schizophrenic patients and a new hypothesis on schizophrenia. *Neurosci Lett, 20*(3), 379–382. doi:10.1016/0304-3940(80)90178-0
3. Weinberger, D. R. (1987). Implications of normal brain development for the pathogenesis of schizophrenia. *Arch Gen Psychiatry, 44*(7), 660. doi:10.1001/archpsyc.1987.01800190080012
4. Silverman, A. J., Sutherland, A. K., Wilhelm, M., & Silver, R. (2000). Mast cells migrate from blood to brain. *J Neurosci, 20*(1), 401–408.
5. Nikkila, H. V., Muller, K., Ahokas, A., Miettinen, K., Rimon, R., & Andersson, L. C. (1999). Accumulation of macrophages in the CSF of schizophrenic patients during acute psychotic episodes. *Am J Psychiatry, 156*(11), 1725–1729. doi:10.1176/ajp.156.11.1725
6. Ek, C. J., Dziegielewska, K. M., Stolp, H., & Saunders, N. R. (2006). Functional effectiveness of the blood-brain barrier to small water-soluble molecules in developing and adult opossum (*Monodelphis domestica*). *J Comp Neurol, 496*(1), 13–26. doi:10.1002/cne.20885
7. Strous, R. D., & Shoenfeld, Y. (2006). Schizophrenia, autoimmunity and immune system dysregulation: A comprehensive model updated and revisited. *J Autoimmun, 27*(2), 71–80. doi:10.1016/j.jaut.2006.07.006
8. Wang, A. K., & Miller, B. J. (2017). Meta-analysis of cerebrospinal fluid cytokine and tryptophan catabolite alterations in psychiatric patients: Comparisons between schizophrenia, bipolar disorder, and depression. *Schizophr Bull.* doi:10.1093/schbul/sbx035
9. Goldsmith, D. R., Rapaport, M. H., & Miller, B. J. (2016). A meta-analysis of blood cytokine network alterations in psychiatric patients: Comparisons between schizophrenia, bipolar disorder and depression. *Mol Psychiatry, 21*(12), 1696–1709. doi:10.1038/mp.2016.3
10. Drexhage, R. C., Knijff, E. M., Padmos, R. C., Heul-Nieuwenhuijzen, L. V. D., Beumer, W., Versnel, M. A., & Drexhage, H. A. (2010). The mononuclear phagocyte system and its cytokine inflammatory networks in schizophrenia and bipolar disorder. *Expert Rev Neurother, 10*(1), 59–76. doi:10.1586/ern.09.144

11. Smith, R. S. (1992). A comprehensive macrophage-T lymphocyte theory of schizophrenia. *Med Hypotheses, 39*(3), 248–257. doi:10.1016/0306-9877(92)90117-u
12. Lawson, C., & Wolf, S. (2009). ICAM-1 signaling in endothelial cells. *Pharmacol Rep, 61*(1), 22–32. doi:10.1016/s1734-1140(09)70004-0
13. Bergink, V., Gibney, S. M., & Drexhage, H. A. (2014). Autoimmunity, inflammation, and psychosis: A search for peripheral markers. *Biol Psychiatry, 75*(4), 324–331. doi:10.1016/j.biopsych.2013.09.037
14. Consortium, S. W. G. o. t. P. G. (2014). Biological insights from 108 schizophrenia-associated genetic loci. *Nature, 511*(7510), 421–427. doi:10.1038/nature13595
15. Dimitrov, D. H., Lee, S., Yantis, J., Valdez, C., Paredes, R. M., Braida, N., . . . Walss-Bass, C. (2013). Differential correlations between inflammatory cytokines and psychopathology in veterans with schizophrenia: Potential role for IL-17 pathway. *Schizophr Res, 151*(1–3), 29–35. doi:10.1016/j.schres.2013.10.019
16. Weaver, C. T., Hatton, R. D., Mangan, P. R., & Harrington, L. E. (2007). IL-17 family cytokines and the expanding diversity of effector T cell lineages. *Annu Rev Immunol, 25*, 821–852. doi:10.1146/annurev.immunol.25.022106.141557
17. Korn, T., Bettelli, E., Oukka, M., & Kuchroo, V. K. (2009). IL-17 and Th17 Cells. *Annu Rev Immunol, 27*, 485–517. doi:10.1146/annurev.immunol.021908.132710
18. Witowski, J., Ksiazek, K., & Jorres, A. (2004). Interleukin-17: A mediator of inflammatory responses. *Cell Mol Life Sci, 61*(5), 567–579. doi:10.1007/s00018-003-3228-z
19. Debnath, M., & Berk, M. (2014). Th17 pathway-mediated immunopathogenesis of schizophrenia: Mechanisms and implications. *Schizophr Bull, 40*(6), 1412–1421. doi:10.1093/schbul/sbu049
20. Holley, M. M., & Kielian, T. (2012). Th1 and Th17 cells regulate innate immune responses and bacterial clearance during central nervous system infection. *J Immunol, 188*(3), 1360–1370. doi:10.4049/jimmunol.1101660
21. Murphy, A. C., Lalor, S. J., Lynch, M. A., & Mills, K. H. (2010). Infiltration of Th1 and Th17 cells and activation of microglia in the CNS during the course of experimental autoimmune encephalomyelitis. *Brain Behav Immun, 24*(4), 641–651. doi:10.1016/j.bbi.2010.01.014
22. Yamazaki, T., Yang, X. O., Chung, Y., Fukunaga, A., Nurieva, R., Pappu, B., . . . Dong, C. (2008). CCR6 regulates the migration of inflammatory and regulatory T cells. *J Immunol, 181*(12), 8391–8401.
23. Frick, L. R., Williams, K., & Pittenger, C. (2013). Microglial dysregulation in psychiatric disease. *Clin Dev Immunol, 2013*, 608654. doi:10.1155/2013/608654
24. Trepanier, M. O., Hopperton, K. E., Mizrahi, R., Mechawar, N., & Bazinet, R. P. (2016). Postmortem evidence of cerebral inflammation in schizophrenia: A systematic review. *Mol Psychiatry, 21*(8), 1009–1026. doi:10.1038/mp.2016.90
25. Bayer, T. A., Buslei, R., Havas, L., & Falkai, P. (1999). Evidence for activation of microglia in patients with psychiatric illnesses. *Neurosci Lett, 271*(2), 126–128.
26. Radewicz, K., Garey, L. J., Gentleman, S. M., & Reynolds, R. (2000). Increase in HLA-DR immunoreactive microglia in frontal and temporal cortex of chronic schizophrenics. *J Neuropathol Exp Neurol, 59*(2), 137–150.
27. Steiner, J., Mawrin, C., Ziegeler, A., Bielau, H., Ullrich, O., Bernstein, H. G., & Bogerts, B. (2006). Distribution of HLA-DR-positive microglia in schizophrenia reflects

impaired cerebral lateralization. *Acta Neuropathol, 112*(3), 305–316. doi:10.1007/s00401-006-0090-8

28. Bessis, A., Béchade, C., Bernard, D., & Roumier, A. (2006). Microglial control of neuronal death and synaptic properties. *Glia, 55*(3), 233–238. doi:10.1002/glia.20459
29. van Berckel, B. N., Bossong, M. G., Boellaard, R., Kloet, R., Schuitemaker, A., Caspers, E., . . . Kahn, R. S. (2008). Microglia activation in recent-onset schizophrenia: A quantitative (R)-[11C]PK11195 positron emission tomography study. *Biol Psychiatry, 64*(9), 820–822. doi:10.1016/j.biopsych.2008.04.025
30. Sen, J., & Belli, A. (2007). S100B in neuropathologic states: The CRP of the brain? *J Neurosci Res, 85*(7), 1373–1380. doi:10.1002/jnr.21211
31. Zhang, X. Y., Xiu, M. H., Song, C., Chen, D. C., Wu, G. Y., Haile, C. N., . . . Kosten, T. R. (2010). Increased serum S100B in never-medicated and medicated schizophrenic patients. *J Psychiatr Res, 44*(16), 1236–1240. doi:10.1016/j.jpsychires.2010.04.023
32. Rupprecht, R., Papadopoulos, V., Rammes, G., Baghai, T. C., Fan, J., Akula, N., . . . Schumacher, M. (2010). Translocator protein (18 kDa) (TSPO) as a therapeutic target for neurological and psychiatric disorders. *Nat Rev Drug Discov, 9*(12), 971–988. doi:10.1038/nrd3295
33. Mattner, F., Staykova, M., Berghofer, P., Wong, H. J., Fordham, S., Callaghan, P., . . . Katsifis, A. (2013). Central nervous system expression and PET imaging of the translocator protein in relapsing-remitting experimental autoimmune encephalomyelitis. *J Nucl Med, 54*(2), 291–298. doi:10.2967/jnumed.112.108894
34. Abourbeh, G., Theze, B., Maroy, R., Dubois, A., Brulon, V., Fontyn, Y., . . . Boisgard, R. (2012). Imaging microglial/macrophage activation in spinal cords of experimental autoimmune encephalomyelitis rats by positron emission tomography using the mitochondrial 18 kDa translocator protein radioligand [(1)(8)F]DPA-714. *J Neurosci, 32*(17), 5728–5736. doi:10.1523/jneurosci.2900-11.2012
35. Karlstetter, M., Nothdurfter, C., Aslanidis, A., Moeller, K., Horn, F., Scholz, R., . . . Langmann, T. (2014). Translocator protein (18 kDa) (TSPO) is expressed in reactive retinal microglia and modulates microglial inflammation and phagocytosis. *J Neuroinflammat, 11*, 3. doi:10.1186/1742-2094-11-3
36. Lartey, F. M., Ahn, G. O., Shen, B., Cord, K. T., Smith, T., Chua, J. Y., . . . Loo, B. W., Jr. (2014). PET imaging of stroke-induced neuroinflammation in mice using [18F]PBR06. *Mol Imaging Biol, 16*(1), 109–117. doi:10.1007/s11307-013-0664-5
37. Dickens, A. M., Vainio, S., Marjamaki, P., Johansson, J., Lehtiniemi, P., Rokka, J., . . . Airas, L. (2014). Detection of microglial activation in an acute model of neuroinflammation using PET and radiotracers 11C-(R)-PK11195 and 18F-GE-180. *J Nucl Med, 55*(3), 466–472. doi:10.2967/jnumed.113.125625
38. van Berckel, B. N., Bossong, M. G., Boellaard, R., Kloet, R., Schuitemaker, A., Caspers, E., . . . Kahn, R. S. (2008). Microglia activation in recent-onset schizophrenia: A quantitative (R)-[11C]PK11195 positron emission tomography study. *Biol Psychiatry, 64*(9), 820–822. doi:10.1016/j.biopsych.2008.04.025
39. Doorduin, J., de Vries, E. F., Willemsen, A. T., de Groot, J. C., Dierckx, R. A., & Klein, H. C. (2009). Neuroinflammation in schizophrenia-related psychosis: A PET study. *J Nucl Med, 50*(11), 1801–1807. doi:10.2967/jnumed.109.066647

40. Takano, A., Arakawa, R., Ito, H., Tateno, A., Takahashi, H., Matsumoto, R., . . . Suhara, T. (2010). Peripheral benzodiazepine receptors in patients with chronic schizophrenia: A PET study with [11C]DAA1106. *Int J Neuropsychopharmacol, 13*(7), 943–950. doi:10.1017/S1461145710000313
41. Van Camp, N., Boisgard, R., Kuhnast, B., Theze, B., Viel, T., Gregoire, M. C., . . . Tavitian, B. (2010). In vivo imaging of neuroinflammation: A comparative study between [(18)F]PBR111, [(11)C]CLINME and [(11)C]PK11195 in an acute rodent model. *Eur J Nucl Med Mol Imaging, 37*(5), 962–972. doi:10.1007/s00259-009-1353-0
42. Kreisl, W. C., Fujita, M., Fujimura, Y., Kimura, N., Jenko, K. J., Kannan, P., . . . Innis, R. B. (2010). Comparison of [(11)C]-(R)-PK 11195 and [(11)C]PBR28, two radioligands for translocator protein (18 kDa) in human and monkey: Implications for positron emission tomographic imaging of this inflammation biomarker. *Neuroimage, 49*(4), 2924–2932. doi:10.1016/j.neuroimage.2009.11.056
43. Hannestad, J., Gallezot, J. D., Schafbauer, T., Lim, K., Kloczynski, T., Morris, E. D., . . . Cosgrove, K. P. (2012). Endotoxin-induced systemic inflammation activates microglia: [(1)(1)C]PBR28 positron emission tomography in nonhuman primates. *Neuroimage, 63*(1), 232–239. doi:10.1016/j.neuroimage.2012.06.055
44. Notter, T., Coughlin, J. M., Gschwind, T., Weber-Stadlbauer, U., Wang, Y., Kassiou, M., . . . Meyer, U. (2017). Translational evaluation of translocator protein as a marker of neuroinflammation in schizophrenia. *Mol Psychiatry*. doi:10.1038/mp.2016.248
45. Hwang, Y., Kim, J., Shin, J. Y., Kim, J. I., Seo, J. S., Webster, M. J., . . . Kim, S. (2013). Gene expression profiling by mRNA sequencing reveals increased expression of immune/inflammation-related genes in the hippocampus of individuals with schizophrenia. *Transl Psychiatry, 3*, e321. doi:10.1038/tp.2013.94
46. Cataldo, A. M., McPhie, D. L., Lange, N. T., Punzell, S., Elmiligy, S., Ye, N. Z., . . . Cohen, B. M. (2010). Abnormalities in mitochondrial structure in cells from patients with bipolar disorder. *Am J Pathol, 177*(2), 575–585. doi:S0002-9440(10)60116-2 [pii] 10.2353/ajpath.2010.081068
47. Kim, S., Hwang, Y., Lee, D., & Webster, M. J. (2016). Transcriptome sequencing of the choroid plexus in schizophrenia. *Transl Psychiatry, 6*(11), e964. doi:10.1038/tp.2016.229
48. van Winkel, R., Stefanis, N. C., & Myin-Germeys, I. (2008). Psychosocial stress and psychosis. A review of the neurobiological mechanisms and the evidence for gene-stress interaction. *Schizophr Bull, 34*(6), 1095–1105. doi:10.1093/schbul/sbn101
49. Myin-Germeys, I., & van Os, J. (2007). Stress-reactivity in psychosis: Evidence for an affective pathway to psychosis. *Clin Psychol Rev, 27*(4), 409–424. doi:10.1016/j.cpr.2006.09.005
50. Pariante, C. M. (2008). Pituitary volume in psychosis: The first review of the evidence. *J Psychopharmacol, 22*(2 Suppl), 76–81. doi:10.1177/0269881107084020
51. Knobler, H. Y. (2000). First psychotic episodes among Israeli youth during military service. *Mil Med, 165*(3), 169–172.
52. Dimitrov, D. H. (2011). Correlation or coincidence between monocytosis and worsening of psychotic symptoms in veterans with schizophrenia? *Schizophr Res, 126*(1–3), 306–307. doi:10.1016/j.schres.2010.06.003

53. Miller, A. H., Maletic, V., & Raison, C. L. (2009). Inflammation and its discontents: The role of cytokines in the pathophysiology of major depression. *Biol Psychiatry, 65*(9), 732–741. doi:10.1016/j.biopsych.2008.11.029
54. Patterson, P. (2003). Of mice and mental illness. *Nat Neurosci, 6*(4), 323. doi:10.1038/nn0403-323
55. Nawa, H., & Takei, N. (2006). Recent progress in animal modeling of immune inflammatory processes in schizophrenia: Implication of specific cytokines. *Neurosci Res, 56*(1), 2–13. doi:10.1016/j.neures.2006.06.002
56. Brown, A. S., & Derkits, E. J. (2010). Prenatal infection and schizophrenia: A review of epidemiologic and translational studies. *Am J Psychiatry, 167*(3), 261–280. doi:10.1176/appi.ajp.2009.09030361
57. Yolken, R. H., & Torrey, E. F. (1995). Viruses, schizophrenia, and bipolar disorder. *Clin Microbiol Rev, 8*(1), 131–145.
58. Torrey, E. F., Bartko, J. J., Lun, Z. R., & Yolken, R. H. (2007). Antibodies to *Toxoplasma gondii* in patients with schizophrenia: A meta-analysis. *Schizophr Bull,* 33(3), 729–736. doi:10.1093/schbul/sbl050
59. Carter, C. J. (2008). Schizophrenia susceptibility genes directly implicated in the life cycles of pathogens: Cytomegalovirus, influenza, *Herpes simplex,* rubella, and *Toxoplasma gondii. Schizophr Bull, 35*(6), 1163–1182. doi:10.1093/schbul/sbn054
60. Clay, H. B., Sillivan, S., & Konradi, C. (2011). Mitochondrial dysfunction and pathology in bipolar disorder and schizophrenia. *Int J Dev Neurosci, 29*(3), 311–324. doi:10.1016/j.ijdevneu.2010.08.007
61. Middleton, F. A., Mirnics, K., Pierri, J. N., Lewis, D. A., & Levitt, P. (2002). Gene expression profiling reveals alterations of specific metabolic pathways in schizophrenia. *J Neurosci, 22*(7), 2718–2729. doi:20026209
62. Iwamoto, K., Bundo, M., & Kato, T. (2005). Altered expression of mitochondria-related genes in postmortem brains of patients with bipolar disorder or schizophrenia, as revealed by large-scale DNA microarray analysis. *Hum Mol Genet, 14*(2), 241–253. doi:10.1093/hmg/ddi022
63. Konradi, C., Eaton, M., MacDonald, M. L., Walsh, J., Benes, F. M., & Heckers, S. (2004). Molecular evidence for mitochondrial dysfunction in bipolar disorder. *Arch Gen Psychiatry, 61*(3), 300–308. doi:10.1001/archpsyc.61.3.300
64. Narayan, S., Tang, B., Head, S. R., Gilmartin, T. J., Sutcliffe, J. G., Dean, B., & Thomas, E. A. (2008). Molecular profiles of schizophrenia in the CNS at different stages of illness. *Brain Res, 1239,* 235–248. doi:1.1016/j.brainres.2008.08.023
65. Stefansson, H., Ophoff, R. A., Steinberg, S., Andreassen, O. A., Cichon, S., Rujescu, D., . . . Collier, D. A. (2009). Common variants conferring risk of schizophrenia. *Nature, 460*(7256), 744–747. doi:10.1038/nature08186
66. Li, T., Li, Z., Chen, P., Zhao, Q., Wang, T., Huang, K., . . . Shi, Y. (2010). Common variants in major histocompatibility complex region and TCF4 gene are significantly associated with schizophrenia in Han Chinese. *Biol Psychiatry, 68*(7), 671–673. doi:10.1016/j.biopsych.2010.06.014
67. Harley, J. B., Alarcon-Riquelme, M. E., Criswell, L. A., Jacob, C. O., Kimberly, R. P., Moser, K. L., . . . Kelly, J. A. (2008). Genome-wide association scan in women with

systemic lupus erythematosus identifies susceptibility variants in ITGAM, PXK, KIAA1542 and other loci. *Nat Genet, 40*(2), 204–210. doi:10.1038/ng.81

68. van Heel, D. A., Franke, L., Hunt, K. A., Gwilliam, R., Zhernakova, A., Inouye, M., . . . Wijmenga, C. (2007). A genome-wide association study for celiac disease identifies risk variants in the region harboring IL2 and IL21. *Nat Genet, 39*(7), 827–829. doi:10.1038/ng2058
69. Eaton, W. W., Byrne, M., Ewald, H., Mors, O., Chen, C.-Y., Agerbo, E., & Mortensen, P. B. (2006). Association of schizophrenia and autoimmune diseases: Linkage of Danish national registers. *Am J Psychiatry, 163*(3), 521–528. doi:10.1176/appi.ajp.163.3.521
70. Sekar, A., Bialas, A. R., de Rivera, H., Davis, A., Hammond, T. R., Kamitaki, N., . . . McCarroll, S. A. (2016). Schizophrenia risk from complex variation of complement component 4. *Nature, 530*(7589), 177–183. doi:10.1038/nature16549
71. Sanders, A. R., Drigalenko, E. I., Duan, J., Moy, W., Freda, J., Goring, H. H. H., & Gejman, P. V. (2017). Transcriptome sequencing study implicates immune-related genes differentially expressed in schizophrenia: New data and a meta-analysis. *Transl Psychiatry*, 7(4), e1093. doi:10.1038/tp.2017.47
72. Psychiatric, G. C. B. D. W. G. (2011). Large-scale genome-wide association analysis of bipolar disorder identifies a new susceptibility locus near ODZ4. *Nat Genet, 43*(10), 977–983. doi:10.1038/ng.943
73. Hamm, A., Veeck, J., Bektas, N., Wild, P. J., Hartmann, A., Heindrichs, U., . . . Dahl, E. (2008). Frequent expression loss of inter-alpha-trypsin inhibitor heavy chain (ITIH) genes in multiple human solid tumors: A systematic expression analysis. *BMC Cancer, 8*, 25. doi:10.1186/1471-2407-8-25
74. Kashyap, R. S., Nayak, A. R., Deshpande, P. S., Kabra, D., Purohit, H. J., Taori, G. M., & Daginawala, H. F. (2009). Inter-alpha-trypsin inhibitor heavy chain 4 is a novel marker of acute ischemic stroke. *Clin Chim Acta, 402*(1–2), 160–163.
75. Lencz, T., Morgan, T. V., Athanasiou, M., Dain, B., Reed, C. R., Kane, J. M., . . . Malhotra, A. K. (2007). Converging evidence for a pseudoautosomal cytokine receptor gene locus in schizophrenia. *Mol Psychiatry, 12*(6), 572–580. doi:10.1038/sj.mp.4001983
76. Sun, S., Wang, F., Wei, J., Cao, L. Y., Wu, G. Y., Lu, L., . . . Zhang, X. Y. (2008). Association between interleukin-3 receptor alpha polymorphism and schizophrenia in the Chinese population. *Neurosci Lett, 440*(1), 35–37. doi:10.1016/j.neulet.2008.05.029
77. Sun, S., Wei, J., Li, H., Jin, S., Li, P., Ju, G., . . . Zhang, X. Y. (2009). A family-based study of the IL3RA gene on susceptibility to schizophrenia in a Chinese Han population. *Brain Res, 1268*, 13–16. doi:10.1016/j.brainres.2009.02.071
78. Paul-Samojedny, M., Owczarek, A., Suchanek, R., Kowalczyk, M., Fila-Danilow, A., Borkowska, P., . . . Kowalski, J. (2010). Association study of interferon gamma (IFN-γ) +874T/A gene polymorphism in patients with paranoid schizophrenia. *J Mol Neurosci, 43*(3), 309–315. doi:10.1007/s12031-010-9442-x
79. Shirts, B. H., Wood, J., Yolken, R. H., & Nimgaonkar, V. L. (2008). Comprehensive evaluation of positional candidates in the IL-18 pathway reveals suggestive associations with schizophrenia and herpes virus seropositivity. *Am J Med Genet B: Neuropsychiatr Genet, 147B*(3), 343–350. doi:10.1002/ajmg.b.30603

80. Ozbey, U., Tug, E., Kara, M., & Namli, M. (2008). The value of interleukin-12B (p40) gene promoter polymorphism in patients with schizophrenia in a region of East Turkey. *Psychiatry Clin Neurosci, 62*(3), 307–312. doi:10.1111/j.1440-1819.2008.01798.x
81. Ozbey, U., Tug, E., & Namli, M. (2009). Interleukin-10 gene promoter polymorphism in patients with schizophrenia in a region of East Turkey. *World J Biol Psychiatry, 10*(4–2), 461–468. doi:10.1080/15622970802626580
82. Saviouk, V., Chow, E. W. C., Bassett, A. S., & Brzustowicz, L. M. (2004). Tumor necrosis factor promoter haplotype associated with schizophrenia reveals a linked locus on 1q44. *Mol Psychiatry, 10*(4), 375–383. doi:10.1038/sj.mp.4001582
83. Xu, M., & He, L. (2010). Convergent evidence shows a positive association of interleukin-1 gene complex locus with susceptibility to schizophrenia in the Caucasian population. *Schizophr Res, 120*(1–3), 131–142. doi:10.1016/j.schres.2010.02.1031
84. Hänninen, K., Katila, H., Saarela, M., Rontu, R., Mattila, K. M., Fan, M., . . . Lehtimäki, T. (2007). Interleukin-1 beta gene polymorphism and its interactions with neuregulin-1 gene polymorphism are associated with schizophrenia. *Eur Arch Psychiatry Clin Neurosci, 258*(1), 10–15. doi:10.1007/s00406-007-0756-9
85. Walss-Bass, C., Liu, W., Lew, D. F., Villegas, R., Montero, P., Dassori, A., . . . Raventos, H. (2006). A novel missense mutation in the transmembrane domain of neuregulin 1 is associated with schizophrenia. *Biol Psychiatry, 60*(6), 548–553. doi:10.1016/j.biopsych.2006.03.017
86. Marballi, K., Quinones, M. P., Jimenez, F., Escamilla, M. A., Raventós, H., Soto-Bernardini, M. C., . . . Walss-Bass, C. (2010). In vivo and in vitro genetic evidence of involvement of neuregulin 1 in immune system dysregulation. *J Mol Med (Berl), 88*(11), 1133–1141. doi:10.1007/s00109-010-0653-y
87. Sommer, I. E., van Westrhenen, R., Begemann, M. J., de Witte, L. D., Leucht, S., & Kahn, R. S. (2014). Efficacy of anti-inflammatory agents to improve symptoms in patients with schizophrenia: An update. *Schizophr Bull, 40*(1), 181–191. doi:10.1093/schbul/sbt139
88. Miyaoka, T., Wake, R., Hashioka, S., Hayashida, M., Oh-Nishi, A., Azis, I. A., . . . Horiguchi, J. (2017). Remission of psychosis in treatment-resistant schizophrenia following bone marrow transplantation: A case report. *Front Psychiatry, 8*, 174. doi:10.3389/fpsyt.2017.00174

11

Depression as a Neuroinflammatory Condition

ALESSANDRA BORSINI AND PATRICIA A. ZUNSZAIN ■

INTRODUCTION

Major depressive disorder (MDD) is a multifactorial complex disorder characterized by persistent low mood or loss of interest in usual activities for a continuous period of at least two weeks. Feelings of low mood include sadness or emptiness, often accompanied by feelings of worthlessness, inappropriate guilt or regret, helplessness, and self-hatred. Patients also display poor memory, altered ability to think or concentrate, insomnia or hypersomnia, eating disturbances (including both weight loss or weight gain), a significant reduction in energy levels, fatigue, and in the most severe cases, recurrent suicidal ideation. These are the diagnostic criteria of depression that are formulated in the *Diagnostic and Statistical Manual of Mental Disorders, Fifth Edition* (DSM-5, 2013).

Psychiatrists, psychologists, or general practitioners can normally assess depressive symptoms. The aim of the diagnosis is to identify such symptomatology and the possible causes, and simultaneously to exclude other potential biological, psychological, or social factors. Indeed, such factors might not be involved in the definition of major depression as a mental illness, although affecting patients' mood. The standard Hamilton Depression Rating Scale (HAM-D)[1] and the Beck Inventory of Depression (BDI) Scale,[2] originally developed for assessing the severity of mood symptoms, have frequently been used for screening purposes. However, it is sometimes difficult to infer an accurate diagnosis being guided by such symptom-based tools, since the diagnostic criteria often overlap with other comorbidities reported by depressed patients, including anxiety, fatigue, or general illness.[3] The prevalence of those comorbidities in the vast majority of patients makes the interpretation of clinical studies difficult, especially of those that aim at identifying neurobiological mechanisms underlying the depressive condition.[4] Indeed, there is still no biological test that can confirm the presence of this disorder. Clinicians base their diagnosis on their expertise in identifying the symptomatology and other potential factors that may have affected the course of the disease, in order to provide the patient with the most indicated treatment option. However, despite patients' receiving comprehensive psychological and biological therapeutic approaches, there are still high levels of relapses and treatment-resistant cases.[5] Moreover, the absence of a valid biological confirmation of the disorder and the fact that antidepressants remain the main type of treatment for depression suggest the need for further research aiming at a better understanding of the distinct molecular mechanisms underlying depression.

CLINICAL EPIDEMIOLOGY OF MDD

Around 350 million people are affected with depression globally, with more than 75% of them experiencing several episodes throughout life. Over 50% have a second episode of depression within six months after the first one, unless they are under regular antidepressant treatment.[6] Furthermore, MDD can lead to death by suicide. It is reported that up to 50% of the total number of people who commit suicide per year worldwide are experiencing a depressive episode, and that around 15% of patients who never received treatment attempt suicide.[7] MDD is twice as prevalent in women as in men.

Furthermore, the problem of MDD extends far beyond this, and it can severely affect the quality of life and the general health of the individuals affected. Very many studies have reported an increase in the number of comorbidities associated with depression, including cardiovascular disease (CVD), stroke, and diabetes mellitus.[8] Moreover, evidence supports a link between both late-life and early-life depression and neurodegenerative diseases.[9,10] Indeed, recent evidence has shown that individuals having an early onset of depression have a higher risk of developing Alzheimer's disease,[9] and that individuals with late-life depression are twice as likely to develop different subtypes of dementia.[10] It is therefore clear that depression represents a major challenge for our society, and that further research in this field is needed.

INFLAMMATION IN MDD

Sub-chronic levels of inflammation represented by an upregulation of distinct inflammatory molecules, also known as cytokines, have been proposed as one of the potential mechanisms mediating the depressive condition. Cytokines are low-molecular-weight regulatory proteins or glycoproteins secreted by various cells in the body, including white blood cells, and participating in acute and chronic inflammation via a complex network of interactions.[11] Many cytokines are referred to as interleukins (ILs), a name indicating that they are secreted by some leukocytes and act on other similar cell types.[12] Two other important types of cytokines include interferons (IFNs), which have the ability to activate immune cells such as natural killer (NK) cells and macrophages,[13] and tumor necrosis factors (TNFs), which have been implicated in causing cell death.[14] The definition of a cytokine as a soluble factor, which is produced by one cell and acts on another cell, was based on the endocrine system of hormones.[15] Indeed, for certain aspects, cytokines can be considered "hormones" of immune and inflammatory responses. However, while hormones are mainly produced by a specific tissue or cell, cytokines are the products of most cells.[11] More importantly, cytokines have a far higher potency than hormones in molar basis. Indeed, even small concentrations of a cytokine can induce significant molecular changes in *in vitro* experiments.[16]

In MDD, distinct biomarkers of inflammation, including IL-1, IL-6, and TNF-α, have been measured. C-reactive protein (CRP) is also relevant in inflammation, as it is produced by hepatocytes in response to IL-6. Several meta-analyses have reported a significant increase in these immune molecules in both serum and plasma of patients with MDD.[17–19] More interestingly, some studies have demonstrated that such elevation in inflammatory cytokines is not only present in the periphery, but also in the central nervous system (CNS).[20] Postmortem studies have demonstrated an altered expression of genes of several inflammatory cytokines, including IL-1β, IL-6, and TNF-α in the prefrontal cortex of MDD patients[21] and of suicide victims when compared with normal control subjects.[22] In addition, such inflammatory abnormalities have also been reported in patients with CVD[23] or diabetes,[24] which are among the most prevalent comorbidities associated with MDD.[8] CVD patients show higher circulating pro-inflammatory cytokines, higher clinical markers of inflammation like CRP, and association with higher expression of inflammatory genes.[23,25] In particular, downregulation of the anti-inflammatory cytokine IL-10 and upregulation of the pro-inflammatory cytokines IL-6 and TNF- α have been reported in MDD patients with chronic heart failure.[26] Similarly, an increase in the concentration of IL-1β, TNF-α, and IL-6 has also been detected in patients with diabetes.[24] Indeed, these mediators have the potential to alter insulin sensitivity,[24] and ultimately predispose patients to the risk of developing several psychiatric diseases, including depression.[27]

It is still unknown, however, whether inflammatory changes are causally associated with the development of MDD or if they are instead a consequence of the illness. From an immunological point of view, for many years the brain was considered to be a privileged organ completely independent from the peripheral organs. Recent evidence has indeed demonstrated that, in response to insults, inflammatory cytokines are produced not only by immune cells, but also by cells in the CNS, including neurons.[28,29]

Several pathways through which peripheral blood factors, including inflammatory molecules, can be transmitted from the periphery to the brain of MDD patients have been investigated.[30] The blood-brain barrier (BBB) represents a strict control point, which regulates influx of several compounds from the periphery into the brain. The BBB is detected in all brain regions, with the exception of the circumventricular organs, and plays a fundamental role in neuronal functioning.[31] Previous studies have demonstrated that distinct cytokines, including IL-1β, IL-6, and TNF-α, known to be upregulated both in the periphery and in the CNS of MDD patients, are able to increase BBB permeability by direct actions on the endothelium.[32–34] For example, the production of IL-1β increased BBB permeability via reduction of the protein occludin, which regulates the morphological stability of the endothelium,[34] whereas IL-6 and TNF-α can cause changes in BBB permeability via induction of histamine and IFN-γ.[35] Although there is evidence that IFN-α crosses the BBB,[36,37] whether it influences BBB permeability is still not known. However, not only cytokines, but also other blood factors involved in MDD are known to affect the BBB. In one study, upregulation of several immunoglobulins (IgG, IgM, and IgA), detected in 41% of MDD patients, was associated with BBB damage.[38] Similarly, oxidation-dependent dysfunction of the brain endothelium may also participate in the occurrence of BBB alteration in MDD. Indeed, clinical and experimental studies in MDD suggest that increased oxidative molecules, including the vasodilator nitric oxide (NO), fibrin, and hemoglobin may contribute to endothelial dysfunction.[39,40]

Neurogenesis: A Potential Mechanism Mediating Immune-Related Depression

There is evidence that cytokines, together with other blood factors, penetrate the most permeable areas of the BBB to regulate different signaling in the brain.[41] Previous research has demonstrated the important role of blood factors on hippocampal neurogenesis, a process by which new neurons are generated from neural stem cells (NSCs).[42] In both humans and rodents, neurogenesis occurs in the subventricular zone (SVZ) of the lateral ventricles and the subgranular zone (SGZ) of the dentate gyrus (DG) in the hippocampus.[43] Current data have estimated that around 700 new neurons are added to the adult human hippocampus per day, suggesting that adult hippocampal neurogenesis has a critical role in mediating human brain functions, such as memory formation and cognition,[44,45] which are known to be impaired in patients with depression. Indeed, regulation of adult hippocampal neurogenesis is relevant to the development of depressive behavior and the ability to cope with stress.[46]

The role of neurogenesis in depression has been investigated for years, leading to the formulation of the "neurogenic theory of depression," proposed by Jacobs and colleagues in 2000.[47] This theory is based on three main prepositions. The first proposition of the neurogenic hypothesis is the hypothesis that hippocampal neurogenesis should be impaired in animal models of depression and in depressed patients. Recent evidence has shown that animal models of depression, using distinct inflammatory challenges, including cytokines[48,49] as well as chronic stressors, including chronic unpredictable mild stress[50] and social defeat stress,[51] reported an impaired hippocampal neurogenesis and concomitant development of depressive-like behaviors. In humans, however, findings are limited to postmortem brains. Three different studies provided an estimation of the total number of DG neural granule cells as a surrogate of hippocampal neurogenesis. The first study did not see any changes between controls and depressed patients[52]; however, the second and the third study detected a decrease (although not statistically significant in the second study) in NPCs in depressed patients,[53,54] therefore suggesting a potential alteration in hippocampal neurogenesis in depressed patients.[55]

The second proposition of the neurogenic hypothesis is that hippocampal neurogenesis is sufficient to induce the development of depression. Several animal studies have assessed whether inhibiting hippocampal neurogenesis can induce depressive-like behaviors. Most studies did not show any effects on baseline depression upon inhibition of hippocampal neurogenesis. However, one study showed increased anxiety-like behavior at baseline after hippocampal neurogenesis inhibition.[56] Another study found that only upon an acute stressor, mice with inhibited hippocampal neurogenesis experienced depressive-like behaviours,[57] potentially suggesting that hippocampal neurogenesis in animals is involved in cognitive functions linked to depression. In humans, evidence comes from cancer therapies, a surrogate of hippocampal neurogenesis inhibition,[58] in which patients show both altered memory and depressive symptoms.[58,59] Although indirect, this finding might indicate at least a partial involvement of hippocampal neurogenesis in the development of depression.

The third proposition of the neurogenic hypothesis is that hippocampal neurogenesis is required for antidepressant efficacy. Using an animal model, Santarelli and colleagues demonstrated that inhibiting hippocampal neurogenesis was sufficient to block the positive effects of fluoxetine.[60] Nonetheless, David and colleagues showed that some, but not all, of the antidepressant effect of fluoxetine was hippocampal neurogenesis–dependent.[61] With respect to human models, however, there is no evidence showing a hippocampal neurogenesis–dependent effect of antidepressant treatment. Finally, there are two studies (one in rodents and one in humans) showing that increasing hippocampal neurogenesis enables treatment of depression. The new P7C3 compound, known to selectively inhibit hippocampal neurogenesis, has now been used in rodents and has shown antidepressant-like effects.[62,63] Similarly, in humans, the new neurogenic NSI-189 compound, targeting hippocampal neurogenesis, has been used in a small clinical trial consisting of 22 MDD patients. Results were quite promising, showing a reduction in depressive symptoms after 28 days of treatment,[64] therefore providing further evidence for the involvement of hippocampal neurogenesis in treatment of depression.

Inflammatory Cytokines and Hippocampal Neurogenesis

Cytokines have been shown to play a central role in CNS functions.[65,66] Emerging evidence suggests that, during an inflammatory response, cytokines influence the neurogenic niche and regulate NPCs' proliferation and neurogenesis.[67–69] At present, only a few studies have looked at the role of cytokines on hippocampal neurogenesis, and they mainly focused on IL-1β, IL-6, IFN-α, and TNF-α.

In vivo and *in vitro* studies have shown that IL-1β reduces neuronal differentiation. Indeed, IL-1β alters neurogenesis in rat DG NPCs,[70] and in mice DG NPCs,[71] whereas co-treatment with an IL-1 receptor antagonist (ra) prevented the negative effect of IL-1β on cell proliferation.[48] Similarly, recombinant IL-1β decreases neurogenesis in human hippocampal NPCs[72–74] and in rat DG NPCs.[75] There are no *in vivo* studies looking at the effect of IL-6 on neurogenesis, while *in vitro* studies have shown contradictory findings. One study demonstrated that IL-6 promotes neurogenesis in rat DG NPCs.[76] Similarly, treatment with IL-6 increased neurogenesis in human hippocampal NPCs.[77] However, two other studies showed that IL-6 can decrease neurogenesis in rat DG NPCs[78] and in mice DG NPCs.[79] Only one *in vitro* study investigated the effect of IFN-α on neurogenesis, showing impaired proliferation in mice DG NPCs.[80] Similarly, two other *in vivo* studies have demonstrated that peripheral administration of IFN-α alters proliferation and neuronal differentiation in rodent DG NPCs.[80,81] Finally, *in vitro* studies have demonstrated the ability of TNF-α to reduce neurogenesis in rat DG NPCs[78] and in human hippocampal NPCs.[77] Comparably, *in vivo* evidence confirmed the effect of TNF-α on neurogenesis in mice DG NPCs.[72]

Overall, those findings demonstrate that inflammatory cytokines play both positive and negative roles in all processes integral to hippocampal neurogenesis, and thus in monitoring the normal functioning of the brain. Therefore, this strengthens the notion that inflammation, and particularly neuroinflammation, is relevant for the pathogenesis of neuropsychiatric and behavioral disorders, including depression and neurodegenerative diseases.

Moreover, IL-1, together with IL-18 and TNF-α, contributes to inhibition of synaptic plasticity and memory consolidation,[66,82] as can be seen in MDD patients or in experimental models of depression.[83] Similar abnormalities have also been reported in Alzheimer's and Parkinson's diseases, which are conditions characterized by progressive neurodegeneration as well as by an abnormal immune response, due to hyper-stimulation of microglia to produce inflammatory cytokines.[84] For example, in a mouse model of progressive Parkinson's disease, neurodegeneration is associated with an upregulation of IL-1β and TNF-α.[85] Similarly, brains of Alzheimer's disease patients show increased production of IL-1,[86] together with an increased depletion of the neural progenitor pool in distinct neurogenic areas.[87] Therefore, those results support the hypothesis that neuroinflammatory changes are pathological components of MDD as well as other neuropsychiatric conditions, such as Alzheimer's and Parkinson's diseases, potentially via regulation of neurogenic mechanisms (Figure 11.1).

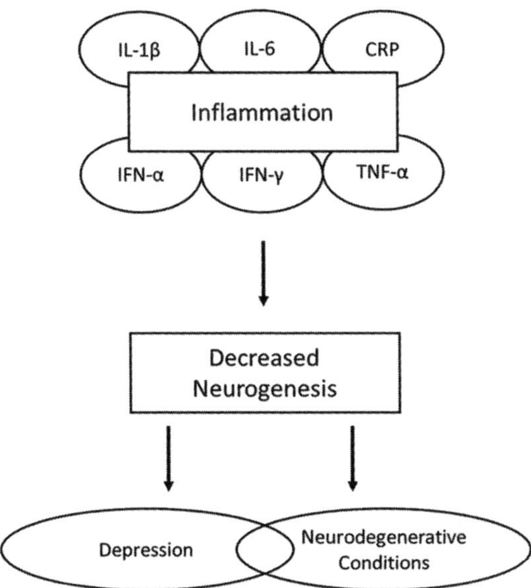

Figure 11.1 Possible involvement of inflammatory cytokines in the development of depression via regulation of neurogenic pathways.

THE EFFECT OF ANTI-INFLAMMATORY TREATMENT ON NEUROGENIC CHANGES IN THE CONTEXT OF DEPRESSION

Several studies have investigated the properties of anti-inflammatory agents as potential antidepressants. Non-steroidal anti-inflammatory drugs (NSAIDs),[88] in particular the selective cyclooxygenase 2 (COX-2) inhibitor celecoxib, and cytokine inhibitors,[89] such as TNF-α inhibitor minocycline or IL-6 antagonists (see, e.g., ClinicalTrials.gov identifier NCT02473289) are currently being tested in clinical trials for depression. However, due to small sample sizes, results in most clinical trials have been conflicting, particularly in NSAID studies. Some observational trials have associated NSAIDs with worse antidepressant treatment effects, especially in patients with no sub-chronic level of inflammation.[90,91] Although several adverse effects associated with anti-inflammatory treatment have been well described, evidence coming from small group studies reported positive outcomes, therefore suggesting that anti-inflammatory treatment are capable of exerting antidepressant effects.[92]

Research should therefore focus on identifying subgroups that may benefit more from anti-inflammatory intervention, such as patients with elevated inflammatory markers or a somatic comorbidity.[93] Preliminary evidence on the efficacy of specific anti-inflammatory agents, particularly celecoxib and minocycline, showed promising results, and therefore they should be further investigated in high-quality randomized clinical trials.[94]

Among the potential mechanisms through which these anti-inflammatory compounds may elicit antidepressant properties, neurogenesis might play an important role. Several studies have shown the ability of both NSAIDs[95] and cytokines inhibitors[96,97] to regulate neurogenic processes. Pharmacological inhibition of COX-2 with celecoxib prevented neuronal cell death in mouse hippocampus.[98] Similarly, minocycline reversed cognitive and depressive-like behaviors, and reductions in neurogenesis were induced by IFN-α[97] and ketamine in mice.[96]

CONCLUSION

Overall, evidence so far indicates a proof-of-concept concerning the use of anti-inflammatory compounds to treat depression. This fact provides supporting grounds for the proposed association between inflammation and MDD, potentially mediated by regulation of distinct neurogenic processes. Future research should therefore aim at identifying subgroups of depressed patients who might benefit from an anti-inflammatory treatment. Indeed, such studies could simultaneously measure, using available screening panels, the concentration of several inflammatory molecules both at baseline and after treatment in patients, and correlate changes with responses. Finally, future studies could also be designed in such a way that distinct molecular mechanisms underlying the antidepressant effect of anti-inflammatory agents could be investigated. This would ultimately allow a more effective future personalized treatment for patients with MDD.

REFERENCES

1. Bagby RM, Ryder AG, Schuller DR, Marshall MB. The Hamilton Depression Rating Scale: has the gold standard become a lead weight? *Am J Psychiatry.* 2004;161(12):2163–2177.
2. Beck AT, Ward CH, Mendelson M, Mock J, Erbaugh J. An inventory for measuring depression. *Arch Gen Psychiatry.* 1961;4:561–571.
3. Hasler G, Drevets WC, Manji HK, Charney DS. Discovering endophenotypes for major depression. *Neuropsychopharmacology.* 2004;29(10):1765–1781.
4. Krishnan V, Nestler EJ. The molecular neurobiology of depression. *Nature.* 2008;455(7215):894–902.
5. Richards D. Prevalence and clinical course of depression: a review. *Clin Psychol Rev.* 2011;31(7):1117–1125.
6. Gold PW, Machado-Vieira R, Pavlatou MG. Clinical and biochemical manifestations of depression: relation to the neurobiology of stress. *Neural Plast.* 2015;2015:581976.
7. Chesney E, Goodwin GM, Fazel S. Risks of all-cause and suicide mortality in mental disorders: a meta-review. *World Psychiatry.* 2014;13(2):153–160.
8. Hirschfeld RM. The comorbidity of major depression and anxiety disorders: recognition and management in primary care. *Prim Care Companion J Clin Psychiatry.* 2001;3(6):244–254.
9. Geerlings MI, den Heijer T, Koudstaal PJ, Hofman A, Breteler MM. History of depression, depressive symptoms, and medial temporal lobe atrophy and the risk of Alzheimer disease. *Neurology.* 2008;70(15):1258–1264.
10. Cherbuin N, Kim S, Anstey KJ. Dementia risk estimates associated with measures of depression: a systematic review and meta-analysis. *BMJ Open.* 2015;5(12):e008853.
11. Dinarello CA. Proinflammatory cytokines. *Chest.* 2000;118(2):503–508.
12. Brocker C, Thompson D, Matsumoto A, Nebert DW, Vasiliou V. Evolutionary divergence and functions of the human interleukin (IL) gene family. *Hum Genomics.* 2010;5(1):30–55.
13. Fensterl V, Sen GC. Interferons and viral infections. *Biofactors.* 2009;35(1):14–20.
14. Sun M, Fink PJ. A new class of reverse signaling costimulators belongs to the TNF family. *J Immunol.* 2007;179(7):4307–4312.
15. Turner MD, Nedjai B, Hurst T, Pennington DJ. Cytokines and chemokines: at the crossroads of cell signalling and inflammatory disease. *Biochim Biophys Acta.* 2014;1843(11):2563–2582.

16. Vilcek J, Feldmann M. Historical review: cytokines as therapeutics and targets of therapeutics. *Trends Pharmacol Sci.* 2004;25(4):201–209.
17. Dowlati Y, Herrmann N, Swardfager W, et al. A meta-analysis of cytokines in major depression. *Biol Psychiatry.* 2010;67(5):446–457.
18. Howren MB, Lamkin DM, Suls J. Associations of depression with C-reactive protein, IL-1, and IL-6: a meta-analysis. *Psychosom Med.* 2009;71(2):171–186.
19. Valkanova V, Ebmeier KP, Allan CL. CRP, IL-6 and depression: a systematic review and meta-analysis of longitudinal studies. *J Affect Disord.* 2013;150(3):736–744.
20. Levine J, Barak Y, Chengappa KN, Rapoport A, Rebey M, Barak V. Cerebrospinal cytokine levels in patients with acute depression. *Neuropsychobiology.* 1999;40(4):171–176.
21. Shelton RC, Claiborne J, Sidoryk-Wegrzynowicz M, et al. Altered expression of genes involved in inflammation and apoptosis in frontal cortex in major depression. *Mol Psychiatry.* 2011;16(7):751–762.
22. Pandey GN, Rizavi HS, Ren X, et al. Proinflammatory cytokines in the prefrontal cortex of teenage suicide victims. *J Psychiatric Res.* 2012;46(1):57–63.
23. Hansson GK. Inflammation, atherosclerosis, and coronary artery disease. *NEJM.* 2005;352(16):1685–1695.
24. Stuart MJ, Baune BT. Depression and type 2 diabetes: inflammatory mechanisms of a psychoneuroendocrine co-morbidity. *Neurosci Biobehav Rev.* 2012;36(1):658–676.
25. Libby P. Inflammation and cardiovascular disease mechanisms. *Am J Clin Nutr.* 2006;83(2):456S–460S.
26. Parissis JT, Adamopoulos S, Rigas A, et al. Comparison of circulating proinflammatory cytokines and soluble apoptosis mediators in patients with chronic heart failure with versus without symptoms of depression. *Am J Cardiol.* 2004;94(10):1326–1328.
27. Wagner J, Allen NA, Swalley LM, Melkus GD, Whittemore R. Depression, depression treatment, and insulin sensitivity in adults at risk for type 2 diabetes. *Diabetes Res Clin Pract.* 2009;86(2):96–103.
28. Tsakiri N, Kimber I, Rothwell NJ, Pinteaux E. Differential effects of interleukin-1 alpha and beta on interleukin-6 and chemokine synthesis in neurones. *Mol Cell Neurosci.* 2008;38(2):259–265.
29. Zunszain PA, Anacker C, Cattaneo A, et al. Interleukin-1beta: a new regulator of the kynurenine pathway affecting human hippocampal neurogenesis. *Neuropsychopharmacology.* 2012;37(4):939–949.
30. Louveau A, Harris TH, Kipnis J. Revisiting the mechanisms of CNS immune privilege. *Trends Immunol.* 2015;36(10):569–577.
31. Ballabh P, Braun A, Nedergaard M. The blood-brain barrier: an overview: structure, regulation, and clinical implications. *Neurobiol Dis.* 2004;16(1):1–13.
32. Schwaninger M, Sallmann S, Petersen N, et al. Bradykinin induces interleukin-6 expression in astrocytes through activation of nuclear factor-kappaB. *J Neurochem.* 1999;73(4):1461–1466.
33. Deli MA, Descamps L, Dehouck MP, et al. Exposure of tumor necrosis factor-alpha to luminal membrane of bovine brain capillary endothelial cells cocultured with astrocytes induces a delayed increase of permeability and cytoplasmic stress fiber formation of actin. *J Neurosci Res.* 1995;41(6):717–726.

34. Didier N, Romero IA, Creminon C, Wijkhuisen A, Grassi J, Mabondzo A. Secretion of interleukin-1beta by astrocytes mediates endothelin-1 and tumour necrosis factor-alpha effects on human brain microvascular endothelial cell permeability. *J Neurochem.* 2003;86(1):246–254.
35. Huber JD, Witt KA, Hom S, Egleton RD, Mark KS, Davis TP. Inflammatory pain alters blood-brain barrier permeability and tight junctional protein expression. *Am J Physiol Heart Circ Physiol.* 2001;280(3):H1241–H1248.
36. Billiau A, Heremans H, Ververken D, van Damme J, Carton H, de Somer P. Tissue distribution of human interferons after exogenous administration in rabbits, monkeys, and mice. *Archives Virol.* 1981;68(1):19–25.
37. Collins JM, Riccardi R, Trown P, O'Neill D, Poplack DG. Plasma and cerebrospinal fluid pharmacokinetics of recombinant interferon alpha A in monkeys: comparison of intravenous, intramuscular, and intraventricular delivery. *Cancer Drug Deliv.* 1985;2(4):247–253.
38. Bechter K, Reiber H, Herzog S, Fuchs D, Tumani H, Maxeiner HG. Cerebrospinal fluid analysis in affective and schizophrenic spectrum disorders: identification of subgroups with immune responses and blood-CSF barrier dysfunction. *J Psychiatric Res.* 2010;44(5):321–330.
39. Isingrini E, Belzung C, Freslon JL, Machet MC, Camus V. Fluoxetine effect on aortic nitric oxide–dependent vasorelaxation in the unpredictable chronic mild stress model of depression in mice. *Psychosom Med.* 2012;74(1):63–72.
40. Zhang XY, Zhou DF, Cao LY, Zhang PY, Wu GY, Shen YC. The effect of risperidone treatment on superoxide dismutase in schizophrenia. *J Clin Psychopharmacol.* 2003;23(2):128–131.
41. Reyes-Vazquez C, Prieto-Gomez B, Dafny N. Interferon modulates central nervous system function. *Brain Res.* 2012;1442:76–89.
42. Imayoshi I, Sakamoto M, Ohtsuka T, Kageyama R. Continuous neurogenesis in the adult brain. *Dev Growth Differ.* 2009;51(3):379–386.
43. Ernst A, Alkass K, Bernard S, et al. Neurogenesis in the striatum of the adult human brain. *Cell.* 2014;156(5):1072–1083.
44. Spalding KL, Bergmann O, Alkass K, et al. Dynamics of hippocampal neurogenesis in adult humans. *Cell.* 2013;153(6):1219–1227.
45. Deng W, Aimone JB, Gage FH. New neurons and new memories: how does adult hippocampal neurogenesis affect learning and memory? *Nat Rev Neurosci.* 2010;11(5):339–350.
46. Anacker C, Pariante CM. Stress and neurogenesis: can adult neurogenesis buffer stress responses and depressive behaviour? *Mol Psychiatry.* 2012;17(1):9–10.
47. Jacobs BL, van Praag H, Gage FH. Adult brain neurogenesis and psychiatry: a novel theory of depression. *Mol Psychiatry.* 2000;5(3):262–269.
48. Ryan SM, O'Keeffe GW, O'Connor C, Keeshan K, Nolan YM. Negative regulation of TLX by IL-1 beta correlates with an inhibition of adult hippocampal neural precursor cell proliferation. *Brain Behav Immun.* 2013;33:7–13.
49. Zhang K, Xu H, Cao L, Li K, Huang Q. Interleukin-1beta inhibits the differentiation of hippocampal neural precursor cells into serotonergic neurons. *Brain Res.* 2013;1490:193–201.

50. Lee KJ, Kim SJ, Kim SW, et al. Chronic mild stress decreases survival, but not proliferation, of new-born cells in adult rat hippocampus. *Exper Mol Med.* 2006;38(1):44–54.
51. Van Bokhoven P, Oomen CA, Hoogendijk WJ, Smit AB, Lucassen PJ, Spijker S. Reduction in hippocampal neurogenesis after social defeat is long-lasting and responsive to late antidepressant treatment. *Eur J Neurosci.* 2011;33(10):1833–1840.
52. Reif A, Fritzen S, Finger M, et al. Neural stem cell proliferation is decreased in schizophrenia, but not in depression. *Mol Psychiatry.* 2006;11(5):514–522.
53. Boldrini M, Underwood MD, Hen R, et al. Antidepressants increase neural progenitor cells in the human hippocampus. *Neuropsychopharmacology.* 2009;34(11):2376–2389.
54. Lucassen PJ, Stumpel MW, Wang Q, Aronica E. Decreased numbers of progenitor cells but no response to antidepressant drugs in the hippocampus of elderly depressed patients. *Neuropharmacology.* 2010;58(6):940–949.
55. Miller BR, Hen R. The current state of the neurogenic theory of depression and anxiety. *Curr Opin Neurobiol.* 2015;30:51–58.
56. Revest JM, Dupret D, Koehl M, et al. Adult hippocampal neurogenesis is involved in anxiety-related behaviors. *Mol Psychiatry.* 2009;14(10):959–967.
57. Snyder JS, Soumier A, Brewer M, Pickel J, Cameron HA. Adult hippocampal neurogenesis buffers stress responses and depressive behaviour. *Nature.* 2011;476(7361):458–461.
58. Pereira Dias G, Hollywood R, Bevilaqua MC, et al. Consequences of cancer treatments on adult hippocampal neurogenesis: implications for cognitive function and depressive symptoms. *Neuro Oncol.* 2014;16(4):476–492.
59. Lydiatt WM, Bessette D, Schmid KK, Sayles H, Burke WJ. Prevention of depression with escitalopram in patients undergoing treatment for head and neck cancer: randomized, double-blind, placebo-controlled clinical trial. *JAMA Otolaryngol Head Neck Surg.* 2013;139(7):678–686.
60. Santarelli L, Saxe M, Gross C, et al. Requirement of hippocampal neurogenesis for the behavioral effects of antidepressants. *Science.* 2003;301(5634):805–809.
61. David DJ, Samuels BA, Rainer Q, et al. Neurogenesis-dependent and -independent effects of fluoxetine in an animal model of anxiety/depression. *Neuron.* 2009;62(4):479–493.
62. Pieper AA, Xie S, Capota E, et al. Discovery of a proneurogenic, neuroprotective chemical. *Cell.* 2010;142(1):39–51.
63. Walker AK, Rivera PD, Wang Q, et al. The P7C3 class of neuroprotective compounds exerts antidepressant efficacy in mice by increasing hippocampal neurogenesis. *Mol Psychiatry.* 2015;20(4):500–508.
64. Fava M, Johe K, Ereshefsky L, et al. A Phase 1B, randomized, double blind, placebo controlled, multiple-dose escalation study of NSI-189 phosphate, a neurogenic compound, in depressed patients. *Mol Psychiatry.* 2016;21(10):1372–1380.
65. Shigemoto-Mogami Y, Hoshikawa K, Goldman JE, Sekino Y, Sato K. Microglia enhance neurogenesis and oligodendrogenesis in the early postnatal subventricular zone. *J Neurosci.* 2014;34(6):2231–2243.
66. Alboni S, Montanari C, Benatti C, et al. Interleukin 18 activates MAPKs and STAT3 but not NF-kappaB in hippocampal HT-22 cells. *Brain Behav Immun.* 2014;40:85–94.

67. Makhija K, Karunakaran S. The role of inflammatory cytokines on the aetiopathogenesis of depression. *Aust N Z J Psychiatry.* 2013;47(9):828–839.
68. Fuster-Matanzo A, Llorens-Martin M, Hernandez F, Avila J. Role of neuroinflammation in adult neurogenesis and Alzheimer disease: therapeutic approaches. *Mediat Inflammation.* 2013;2013:260925.
69. Borsini A, Zunszain PA, Thuret S, Pariante CM. The role of inflammatory cytokines as key modulators of neurogenesis. *Trends Neurosci.* 2015;38(3):145–157.
70. Koo JW, Duman RS. IL-1beta is an essential mediator of the antineurogenic and anhedonic effects of stress. *Proc Natl Acad Sci USA.* 2008;105(2):751–756.
71. Boehme M, Guenther M, Stahr A, et al. Impact of indomethacin on neuroinflammation and hippocampal neurogenesis in aged mice. *Neurosci Lett.* 2014;572:7–12.
72. Chen E, Xu D, Lan X, et al. A novel role of the STAT3 pathway in brain inflammation-induced human neural progenitor cell differentiation. *Curr Mol Med.* 2013;13(9):1474–1484.
73. Zunszain PA, Anacker C, Cattaneo A, et al. Interleukin-1beta: a new regulator of the kynurenine pathway affecting human hippocampal neurogenesis. *Neuropsychopharmacology.* 2011;37:939–949.
74. Borsini A, Alboni S, Horowitz MA, et al. Rescue of IL-1beta-induced reduction of human neurogenesis by omega-3 fatty acids and antidepressants. *Brain Behav Immun.* 2017;65:230–238.
75. Crampton SJ, Collins LM, Toulouse A, Nolan YM, O'Keeffe GW. Exposure of foetal neural progenitor cells to IL-1beta impairs their proliferation and alters their differentiation—a role for maternal inflammation? *J Neurochem.* 2012;120(6):964–973.
76. Oh J, McCloskey MA, Blong CC, Bendickson L, Nilsen-Hamilton M, Sakaguchi DS. Astrocyte-derived interleukin-6 promotes specific neuronal differentiation of neural progenitor cells from adult hippocampus. *J Neurosci Res.* 2010;88(13):2798–2809.
77. Johansson S, Price J, Modo M. Effect of inflammatory cytokines on major histocompatibility complex expression and differentiation of human neural stem/progenitor cells. *Stem Cells.* 2008;26(9):2444–2454.
78. Monje ML, Toda H, Palmer TD. Inflammatory blockade restores adult hippocampal neurogenesis. *Science.* 2003;302(5651):1760–1765.
79. Zonis S, Ljubimov VA, Mahgerefteh M, Pechnick RN, Wawrowsky K, Chesnokova V. p21(Cip) restrains hippocampal neurogenesis and protects neuronal progenitors from apoptosis during acute systemic inflammation. *Hippocampus.* 2013;23(12):1383–1394.
80. Zheng LS, Hitoshi S, Kaneko N, et al. Mechanisms for interferon-alpha-induced depression and neural stem cell dysfunction. *Stem Cell Rep.* 2014;3(1):73–84.
81. Kaneko N, Kudo K, Mabuchi T, et al. Suppression of cell proliferation by interferon-alpha through interleukin-1 production in adult rat dentate gyrus. *Neuropsychopharmacology.* 2006;31(12):2619–2626.
82. Pickering M, O'Connor JJ. Pro-inflammatory cytokines and their effects in the dentate gyrus. *Prog Brain Res.* 2007;163:339–354.
83. Zunszain PA, Hepgul N, Pariante CM. Inflammation and depression. *Curr Top Behav Neurosci.* 2013;14:135–151.

84. Das S, Basu A. Inflammation: a new candidate in modulating adult neurogenesis. *J Neurosci Res.* 2008;86(6):1199–1208.
85. Pisanu A, Lecca D, Mulas G, et al. Dynamic changes in pro- and anti-inflammatory cytokines in microglia after PPAR-gamma agonist neuroprotective treatment in the MPTPp mouse model of progressive Parkinson's disease. *Neurobiol Dis.* 2014;71C:280–291.
86. Griffin WST. What causes Alzheimer's. *Scientist.* 2011;25(9):36–40.
87. Taupin P. The therapeutic potential of adult neural stem cells. *Curr Opin Mol Ther.* 2006;8(3):225–231.
88. Muller N, Schwarz MJ, Dehning S, et al. The cyclooxygenase-2 inhibitor celecoxib has therapeutic effects in major depression: results of a double-blind, randomized, placebo controlled, add-on pilot study to reboxetine. *Mol Psychiatry.* 2006;11(7):680–684.
89. Dean OM, Kanchanatawan B, Ashton M, et al. Adjunctive minocycline treatment for major depressive disorder: a proof of concept trial. *Aust N Z J Psychiatry.* 2017;51(8):829–840.
90. Warner-Schmidt JL, Vanover KE, Chen EY, Marshall JJ, Greengard P. Antidepressant effects of selective serotonin reuptake inhibitors (SSRIs) are attenuated by antiinflammatory drugs in mice and humans. *Proc Natl Acad Sci USA.* 2011;108(22):9262–9267.
91. Gallagher PJ, Castro V, Fava M, et al. Antidepressant response in patients with major depression exposed to NSAIDs: a pharmacovigilance study. *Am J Psychiatry.* 2012;169(10):1065–1072.
92. Kohler O, Benros ME, Nordentoft M, et al. Effect of anti-inflammatory treatment on depression, depressive symptoms, and adverse effects: a systematic review and meta-analysis of randomized clinical trials. *JAMA Psychiatry.* 2014;71(12):1381–1391.
93. Zunszain PA. Improving the treatment for depressive symptoms and major depression with anti-inflammatory drugs. *Evid Based Ment Health.* 2015;18(4):116.
94. Na KS, Lee KJ, Lee JS, Cho YS, Jung HY. Efficacy of adjunctive celecoxib treatment for patients with major depressive disorder: a meta-analysis. *Prog Neuropsychopharmacol Biol Psychiatry.* 2014;48:79–85.
95. Ajmone-Cat MA, Cacci E, Minghetti L. Non-steroidal anti-inflammatory drugs and neurogenesis in the adult mammalian brain. *Curr Pharm Des.* 2008;14(14):1435–1442.
96. Lu Y, Giri PK, Lei S, et al. Pretreatment with minocycline restores neurogenesis in the subventricular zone and subgranular zone of the hippocampus after ketamine exposure in neonatal rats. *Neuroscience.* 2017;352:144–154.
97. Zheng LS, Kaneko N, Sawamoto K. Minocycline treatment ameliorates interferon-alpha- induced neurogenic defects and depression-like behaviors in mice. *Front Cell Neurosci.* 2015;9:5.
98. Jung KH, Chu K, Lee ST, et al. Cyclooxygenase-2 inhibitor, celecoxib, inhibits the altered hippocampal neurogenesis with attenuation of spontaneous recurrent seizures following pilocarpine-induced status epilepticus. *Neurobiol Dis.* 2006;23(2):237–246.

12

Immunology of Bipolar Disorder

IZABELA G. BARBOSA, MOISES E. BAUER, JAIR C. SOARES, AND ANTÔNIO L. TEIXEIRA ■

INTRODUCTION

The concept that circulating molecules might influence mood and behavior is not new. Ancient Egyptians were the first to describe mental symptoms, and they believed that these problems did not arise from affections in the brain. Later, Hippocrates (460–370 BC) proposed the "humoral theory" in which he believed mood, health, or illnesses were derived from the balance of circulating "humors," or basic fluids. According to the Hippocratic medicine principles, the human body is filled with four basic substances: blood, yellow bile, black bile, and phlegm. If they are in balance, the person is healthy, while any deficiency or excess of one or more of these fluids would cause different illnesses. The classic term "melancholia," an old correlate for severe depression, literally means "black bile": "when sadness and fear last for a long time. . . ." The idea that psychiatric symptoms were not related to brain dysfunction was sustained for centuries.

During the 19th century, knowledge about the biology of the brain increased significantly, and psychiatric disorders were conceptualized as "brain disorders." In the same period, the emergence of immunology was marked by the development of vaccines, including those for smallpox and rabies.[1] However, in the first half of the 20th century, there was a significant shift in the understanding of psychiatric disorders, notably anxiety and mood disorders, with an overemphasis on psychological explanations.

In the 1930s, the neuropathologist Hermann Lehmann-Facius (1899–1960) investigated whether psychiatric disorders could be caused by an autoimmune reaction with the production of antibodies against structures in the brain. This seems to be the first systematic study addressing the link between psychiatric disorders and the immune system.[1]

During the last two decades of the 20th century, there was another shift regarding the pathogenesis of psychiatric disorders, now toward biological explanations. In this context, the role of homeostatic systems, including the immune system, in the pathogenesis and physiopathology of these disorders has been proposed and investigated. Currently, BD is conceptualized as a multisystem condition, impairing behavior, cognition, and autonomic, endocrine, and immune functions. In this chapter, we will discuss the evidence of the involvement of immune system dysfunction in BD.

BIPOLAR DISORDER: AN OVERVIEW

Mood episodes like "melancholia" and mania have been described since Hippocrates. However, the definition of BD as a discrete entity was established only in the middle of the 19th century, by Jean-Pierre Falret. Falret proposed the term *folie circulaire* to describe a condition marked by alternating episodes of mania and "melancholia" punctuated by periods of relative well-being.[2] At the beginning of the 20th century, Emil Kraepelin introduced the term "manic-depressive insanity," emphasizing the need of observing the course of the illness marked by a variety of mood episodes over time, instead of relying only on a single presentation. More than a century later, the Kraepelinian concept based on recurrent episodes of depression and mania, independently of psychotic symptoms, remains the cornerstone for BD diagnosis.[2]

Epidemiological studies have shown that the lifetime worldwide prevalence of BD is estimated at 2.4%.[3] The first mood episodes usually emerge when the patient is around 20 years old, and the risk of new mood episodes remains until the person is around 70 years old.[4] The recurrence of mood episodes in patients with BD, even under pharmacotherapy, is common, and these patients present mood symptoms during at least half of their lifetime.[5] Between 25% and 40% of patients with BD type I may present "cycle acceleration" or "neuroprogression," i.e., a progressive course of illness characterized by increase in the frequency and severity of mood episodes, psychiatric and medical comorbidities, along with lower response to treatment.[6,7] The biological mechanisms underlying BD progression are unknown and may involve complex interactions among multiple genes, environmental factors, and physiological systems, including the immune system as ilustrate at Figure 12.1.

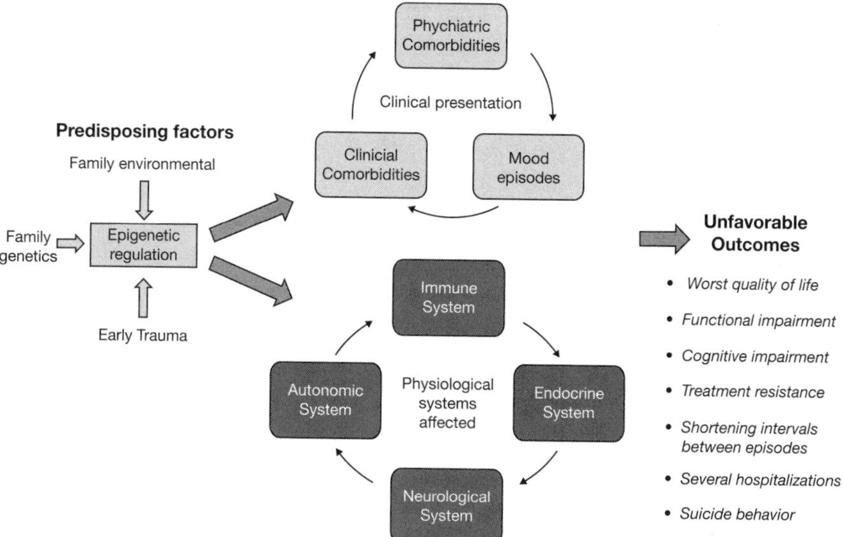

Figure 12.1 *Neurobiology of bipolar disorder.* The biological mechanisms underlying BD progression are unknown and may involve complex interactions among multiple genes, environmental factors, and physiological systems, including the immune system. The progressive course of illness is characterized by unfavorable outcomes, including the increase in the frequency and severity of mood episodes, psychiatric and medical comorbidities (e.g., functional and cognitive impairment), along with treatment resistance.

IMMUNE DYSFUNCTION IN BIPOLAR DISORDER

Immune-Related Genes

There is strong evidence indicating a genetic component in BD. The estimated heritability for BP is around 60%, and the remaining variance is attributed to non-shared environmental factors.[8] Approaches to mapping chromosome areas, candidate genes, rare variants, and copy number variations have not been able to identify a single alteration in BD.[9] Accordingly, BD has been understood as a polygenic disorder, and the genetic effects might contribute to clinically relevant phenotypes.[9,10]

In this scenario, great attention has been paid to the link of immune-related genes with clinical characteristics, comorbidities, and pharmacological response in BD. For instance, an association has been reported between pro-inflammatory gene expression and manic symptoms, especially increased speech and motor activity.[11] Nevertheless, genome-wide association studies (GWAS) failed to show significant relationships between immune-related genes and BD.[12–14]

Autoimmune Diseases and Autoantibodies

The association of autoimmune diseases and autoantibodies with BD has been reported in case series and epidemiological studies, suggesting a link between autoimmunity and BD. A cohort study including 3.57 million births and almost 10,000 patients with BD showed that a history of Guillain-Barré syndrome, Crohn's disease, and autoimmune hepatitis was associated with greater risk for BD.[15] The prevalence of BD among systemic lupus erythematosus patients was greater compared with controls,[16] while women with systemic lupus erythematosus may present a six-fold increase in the incidence of BD.[17] Patients with irritable bowel syndrome present higher incidence of BD compared with controls,[18] and multiple sclerosis patients present up to 30-fold increase in the incidence of BD.[19] The overall prevalence of autoimmune and allergic diseases was about 45% in BD patients,[20] almost two-fold higher than the estimated prevalence in the general population.[21] Asthma was associated with a higher risk of developing BD.[18]

Although some studies argued that patients with BD might exhibit a higher frequency of autoimmune thyroiditis,[22] a recent systematic review screening 340 studies did not confirm this association.[23] However, there is evidence of increased prevalence of circulating thyroid autoantibodies in BD,[23] independent of lithium exposure.[24] Patients with BD also present increased levels of several other autoantibodies in comparison with controls like antibodies against gliadin[25] and N-methyl-d-aspartate receptor (NMDAR).[26] Together, these findings suggest a link between autoimmunity and BD.

Infectious Diseases

Maternal exposure to infectious or inflammatory insults during pregnancy increases the risk of BD in the offspring. The infectious agents implicated in the pathogenesis of BD are the protozoan *Toxoplasma gondii*[27–29] and the viruses of the Herpesviridae family.[27,29,30] It has been hypothesized that antenatal period infection with agents with tropism for the central nervous system (CNS) might activate and/or disturb immune inflammatory cascades that directly affect neural connectivity and circuitry, impairing the normal development of the CNS. There is even experimental evidence that prenatal immune activation may negatively influence brain development and behavior across generations.[31]

Patients with BD also present increased prevalence of HIV, hepatitis B virus, and hepatitis C virus infections.[32] Nevertheless, this finding seems to be due to illness-related behavioral changes, such as increased exposure to risk situations, including substance use disorder, instead of a defective immune response. It remains to be established whether these chronic infection conditions influence the outcome of BD.

Immune Cell Phenotype and Function

Patients with BD and their offspring present a systemic activation of the mononuclear phagocytic system.[33,34] This has been demonstrated through state-related elevation of inflammatory monocyte gene expression,[35] increased proportion of monocytes (CD14+),[33,36] and higher interleukin (IL-)-6 production following lipopolysaccharide stimulation.[37] Moreover, patients with BD present a M1 signature (i.e., increased mRNA levels of IL-6 and CCL13 in peripheral blood mononuclear cells [PBMCs]) and a decrease of M2 markers (IL-10 and CCL11).[38] The M1 polarization of macrophages has been associated with pro-inflammatory responses in several disease models.

Studies have shown no differences in the number of B-cells (CD19+) and natural killer (NK) cells between patients with BD and controls.[39–42] Once the numbers of B cells and NK cells are similar in patients and controls, it seems that most of the lymphocyte dysfunction in BD is related to T cells.[43] The findings regarding T cells in patients with BD can be summarized as follows:

1. The frequency of cytotoxic T cells (CD8+) seems to be similar to controls',[41,44] although a significant expansion of CD8+CD28- T cells was demonstrated in parallel with shortened telomere length in PBMCs of patients with BD, suggesting premature cellular immunosenescence.[45]
2. Patients with BD do not present differences in the frequency of T helper cells (CD4+).[33,42]
3. Patients with BD present an increased proportion of activated T cells (CD4+CD25+).[33,46] Patients with BD also had increased activation of T cells after stress exposure *in vitro* in comparison with controls,[37,46] and T cells from patients are less sensitive to the suppressive glucocorticoid effects.[37,46]
4. There is reduction in the frequency of T regulatory (Treg) lymphocyte cells (CD4+CD25+Foxp3+),[41,46] reduction of IL-10 expressing Treg cells,[33] and increased expression of Toll-like receptor 2 (TLR-2) on Treg cells after stimulation.[47] Interestingly, a cohort study showed that BD offspring with low percentage of Tregs at adolescence had a higher risk to develop a mood disorder during adulthood.[48]

These findings strongly indicate an activated T cell profile, partly explaining the higher risk of autoimmune diseases in BD.

Studies evaluating the release of cytokines by stimulated T cells from patients with BD found increased production of pro-inflammatory molecules (IL-6, tumor necrosis factor [TNF], and interferon [IFN]-γ) in comparison with controls[41] with a tendency toward a Th1 profile.[33,41] It is worth mentioning that there are discordant results, with some studies showing decreased production of IL-2, IL-6, and IL-10,[49] while others report no significant differences.[37,50]

Th1 cells from the offspring of patients with BD present reduced IFN-γ production in comparison with controls.[48] In this same study, BD offspring had reduced percentage of Th1 and Th17 when compared with controls, but this alteration was noticed only when young adults were considered.[48] Accordingly, BD offspring might present a dynamic change of the immune profile from adolescence to adulthood.[48]

Circulating Levels of Immune Markers in Bipolar Disorder

BD has been associated with low-grade chronic inflammatory profiles. Recent meta-analyses confirmed increased circulating levels of the pro-inflammatory cytokines IL-1β, IL-6, IL-8, and TNF in patients with BD.[51,52] This pro-inflammatory profile is exacerbated during mood episodes, i.e., mania or depression.[53–55] For instance, in a six-week follow-up study, Uyanik and collaborators[56] reported higher levels of IL-6 and TNF levels in mania in comparison with their levels after treatment. In an eight-week follow-up study, Li and collaborators[57] found increased IL-23 and decreased transforming growth factor (TGF)-β plasma levels in patients with BD during mania episodes in comparison with their levels after effective treatment.

A few studies compared patients with BD and other psychiatric disorders. Patients with BD did not differ from patients with schizophrenia in peripheral levels of sTNFR1, IL-1ra, IL-6, IL-10, and IL-12,[58,59] or from patients with major depressive disorder (MDD), considering TNF, IL-6, and IL-12.[60,61] Goldsmith and collaborators[51] performed a meta-analysis including patients with schizophrenia, BD, and MDD.[51] IL-6, TNF, IL-1RA, and sIL-2R were significantly elevated in all three conditions during acute illness episodes.[51] Accordingly, it seems unlikely that a panel of immune/inflammatory-related markers will be able to discriminate BD in comparison with other major psychiatric disorders.

Patients with BD also present increased levels of:

1. Chemokines or chemotactic cytokines,[63,64]
2. Markers of endothelial cell activation, like intracellular adhesion molecules (ICAM) and vascular cell adhesion molecules (VCAM),[64,65]
3. Adipokines—mediators produced by the adipose tissue,[66]
4. Acute-phase proteins, such as C-reactive protein (CRP) levels, and[56,67]
5. Oxidative stress markers.[68,69]

The increase in oxidative stress and decrease in antioxidant enzymes in BD can determine lipid, protein, and DNA oxidative damage. Increased oxidative stress molecules produced by mitochondria may be potent activators of the inflammatory system, resulting in high levels of pro-inflammatory factors such as NF-κβ and IL-1β. Therefore, increased oxidative stress might be related to the activation of the pro-inflammatory profile in BD.

Evidence of Immune Dysfunction in the Central Nervous System in Bipolar Disorder

A small number of studies evaluated immune markers in the cerebrospinal fluid (CSF) of patients with BD. Söderlund and collaborators[70] found increased IL-1β levels in CSF of euthymic patients with BD. CSF IL-1β levels were higher in patients with BD during acute mood episodes in comparison with patients during euthymic episodes,[70] corroborating the observation of increased pro-inflammatory profile during acute mood episodes. Patients with BD also exhibited increased markers of monocyte and microglial activation (i.e., MCP1/CCL2, chitinase-3-like protein 1, and YKL-40), reinforcing the hypothesis of neuroinflammation in BD.[71] Interestingly, Zetterberg and collaborators[72] reported increased albumin ratio in the CSF of patients with BD, suggesting brain–blood barrier (BBB) impairment possibly driven by inflammatory mechanisms.[73]

A few studies have assessed immune markers in brain specimens of patients with BD, revealing increased inflammatory and decreased anti-inflammatory markers.[74,75] For instance, BD patients present increased protein and mRNA levels of IL-1β and its receptor (IL-1R) and NF-κB transcription factor subunits (p50 and p65).[74,75] Increased trans-membrane TNF protein level in the anterior cingulate area and decreased TNF receptor 2 protein levels in the dorsolateral prefrontal cortex were found in BD.[76] Patients with BD also exhibit increased IL-1β,

IL-6, and TNF levels in the frontal cortex.[77] Immune markers of mitochondrial production linked to activation of the NOD-like receptor family pyrin domain-containing 3 (NLRP3) a protein linked with apoptosis, are overexpressed in the prefrontal cortex of BD.[77]

Microglial markers (glial fibrillary acidic protein, inducible nitric oxide synthase, c-fos, and CD11b) are increased in postmortem studies involving patients with BD.[74,75] In line with this pathological finding, a PET study assessing [11C]-(R)-PK11195 expression found evidence of increased microglia activation in the right hippocampus of patients with BD.[78] There is also evidence of microglia activation in dorsolateral prefrontal cortex and anterior cingulate in BD,[74,75,79] brain regions known to be critical in mood and cognition modulation.[80]

The serotonin system is largely implicated in the pathophysiology of BD. Microglia may act as sensors and regulators of serotonin levels through the production of pro-inflammatory cytokines and their influence on the serotonin transporter (SERT) availability.[81] The SERT is a key regulator of central serotonergic activity, controlling the reuptake of serotonin. Chou and collaborators[82] found associations between plasma levels of the cytokines IL-1α, IL-1β, IL-6, and SERT availability in the midbrain, as assessed by single-photon-emission computed tomography, of patients with BD.

Microglia also play an important role in the kynurenine pathway, an alternative route of tryptophan metabolism that decreases serotonin-dependent neurotransmission. In microglia, kynurenine is degraded into 3-hydroxykynurenine (3-HK) by kynurenine 3-monooxidase, and later to quinolinic acid (QA). Both 3-HK and QA are neurotoxic. In astrocytes, kynurenine is degraded by kynurenine aminotransferase into kynurenic acid (KynA), which displays neuroprotective effect.[83] In comparison with controls, patients with BD have decreased KynA levels,[84] as well as a decreased 3HK/KYN ratio[84] and decreased KynA/QA ratio,[85] indicating a kynurenine pathway shift toward neurotoxicity.

In the long term, the persistent microglia activation in BD could determine neuronal death, leading to neurodegeneration.[86] Postmortem studies have shown loss of oligodendroglia, dendrites' atrophy, and decreased neuronal size and density in the brain of patients with BD, notably in the medial prefrontal cortex.[87]

Summarizing, a pro-inflammatory profile and microglial activation are present in the CNS of BD patients, and may influence brain structures. Neuroinflammation may exert neurodegenerative effects in BD. The possible microglia activation mechanisms and consequences in BD are shown in Figure 12.2.

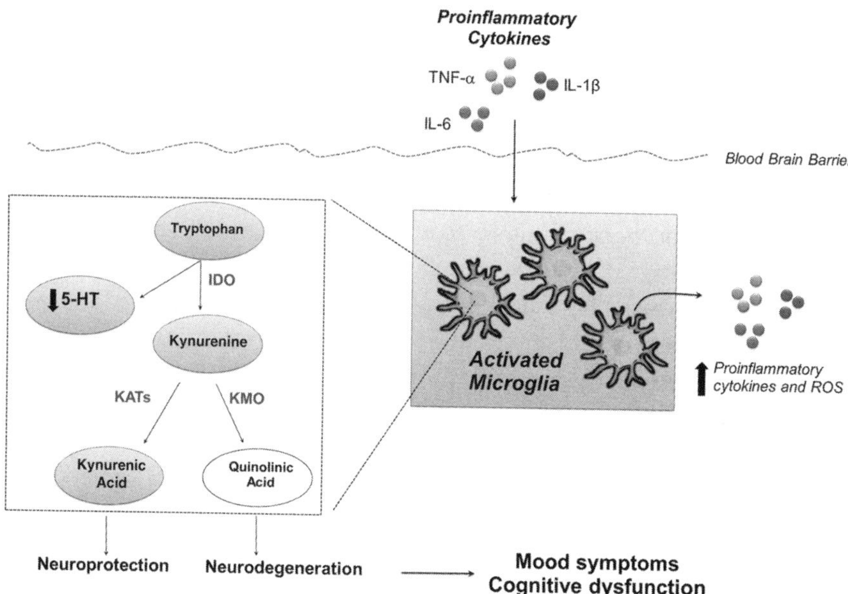

Figure 12.2 *Neuroinflammation is characterized by activated glial cells.* Circulating pro-inflammatory cytokines may gain access to the brain, activating microglia and astrocytes. The activated microglia may in turn contribute to neuroinflammation observed in BD by secreting pro-inflammatory cytokines and reactive oxygen species (ROS). Upon inflammatory cues, activated microglia cells enhance the activity of indoleamine 2,3-dioxygenase (IDO), converting tryptophan into the metabolites kynurenine, kynurenic acid, and quinolinic acid. This "kynurenine pathway" is known to deplete the brain stores of serotonin (5-HT) as well as to produce neuroprotective/neurodegeneration effects associated with core mood symptoms and cognitive dysfunction observed in BD.

PHARMACOLOGICAL STRATEGIES

The mechanisms underlying the therapeutic effect of mood stabilizers in BD are not fully understood. There is evidence suggesting that these drugs can modulate immune pathways. Lithium presents several anti-inflammatory properties, such as[88–91]:

1. Attenuates cyclooxygenase (COX)-2 expression;
2. Inhibits the synthesis of pro-inflammatory cytokines like IL-1β and TNF;
3. Inhibits nuclear factor (NF)-κB activity—which regulates the transcription of several immune-related genes—possibly through the inhibition of glycogen synthase kinase (GSK)-3β;

4. Attenuates microglia activation;
5. Inhibits the expression of TLR-4. *In vitro* studies showed that valproate decreases the production of the cytokines IL-1β, IL-2, IL-4, IL-6, IL-17, and TNF.[92,93]

However, a systematic review evaluating the evidence of the effect of single mood-stabilizing drugs on the circulating levels of cytokines found no effect of valproate.[91] Immune actions have also been ascribed to antipsychotics, and implicated in their metabolic effects, including weight gain and metabolic syndrome.[94]

The development of new therapeutic targets for the treatment of BD is highly needed, and the immune system is a promising target. Nery and collaborators[95] conducted the first study in this regard. They enrolled BD patients with depressive or mixed episodes for a double-blind, randomized, placebo-controlled add-on trial using the COX-2 selective inhibitor celecoxib. The treatment with celecoxib was associated with a faster improvement of depression.[95] Two double-blind, randomized, placebo-controlled add-on studies showed the efficacy of celecoxib to improve manic symptoms in patients with BD.[96,97] Minocycline is a second-generation tetracycline that exerts anti-inflammatory effects independent of its antibiotic mechanisms. An eight-week, open-label study with adjunctive minocycline reported reduction of depression severity in patients with BD.[98] A recent meta-analysis corroborated the role of N-acetylcysteine as an adjunctive strategy for the treatment of depressive symptoms in BD.[99]

It is worth noting that not all anti-inflammatory approaches will be eligible for the management of BD. For instance, potent anti-inflammatory strategies such as glucocorticoids and TNF-α antagonists (infliximab and Etanercept) have been implicated in the emergence of manic symptoms in subjects without previous history of mood disorders.[100] This clinical observation points to the complex relationship between BD and inflammation, which goes far beyond the described pro-inflammatory status represented by elevated circulating levels of inflammatory mediators.

CONCLUSION

BD is a complex multisystemic condition characterized not only by mood swings, but also by clinical and psychiatric comorbidities, cognitive impairment, and progressive changes in its phenomenology, poor response to the treatment, and compromised prognosis. Immune dysfunction has been pointed out as playing an important role in the pathophysiology of BD and its comorbidities. Immunomodulatory agents might be helpful to improve the treatment of a subset of BD patients.

REFERENCES

1. Steinberg H, Kirkby KC, Himmerich H. The historical development of immunoendocrine concepts of psychiatric disorders and their therapy. *Int J Mol Sci.* 2015;16(12):28841–28869.
2. Malhi GS, Bargh DM, Coulston CM, Das P, Berk M. Predicting bipolar disorder on the basis of phenomenology: implications for prevention and early intervention. *Bipolar Disord.* 2014;16(5):455–470.
3. Merikangas KR, Jin R, He JP, Kessler RC, et al. Prevalence and correlates of bipolar spectrum disorder in the World Mental Health Survey initiative. *Arch Gen Psychiatry.* 2011;68(3):241–251.
4. Angst J, Gamma A, Sellaro R, Lavori PW, Zhang H. Recurrence of bipolar disorders and major depression. A life-long perspective. *Eur Arch Psychiatry Clin Neurosci.* 2003;253(5):236–240.
5. Judd LL, Akiskal HS, Schettler PJ, et al. A prospective investigation of the natural history of the long-term weekly symptomatic status of bipolar II disorder. *Arch Gen Psychiatry.* 2003;60(3):261–269.
6. Martino DJ, Samamé C, Marengo E, Igoa A, Strejilevich SA. A critical overview of the clinical evidence supporting the concept of neuroprogression in bipolar disorder. *Psychiatry Res.* 2016;235:1–6.
7. Passos IC, Mwangi B, Vieta E, Berk M, Kapczinski F. Areas of controversy in neuroprogression in bipolar disorder. *Acta Psychiatr Scand.* 2016;134(2):91–103.
8. Song J, Bergen SE, Kuja-Halkola R, Larsson H, Landén M, Lichtenstein P. Bipolar disorder and its relation to major psychiatric disorders: a family-based study in the Swedish population. *Bipolar Disord.* 2015;17(2):184–193.
9. Goes FS. Genetics of bipolar disorder: recent update and future directions. *Psychiatr Clin North Am.* 2016;39(1):139–155.
10. Neale BM, Sklar P. Genetic analysis of schizophrenia and bipolar disorder reveals polygenicity but also suggests new directions for molecular interrogation. *Curr Opin Neurobiol.* 2015;30:131–138.
11. Harman BC, Riemersma-Van der Lek RF, Burger H, et al. Relationship between clinical features and inflammation-related monocyte gene expression in bipolar disorder—towards a better understanding of psychoimmunological interactions. *Bipolar Disord.* 2014;16(2):137–150.
12. Huang J, Perlis RH, Lee PH, et al. Cross-disorder genomewide analysis of schizophrenia, bipolar disorder, and depression. *Am J Psychiatry.* 2010;167(10):1254–1263.
13. Pedroso I, Lourdusamy A, Rietschel M, et al. Common genetic variants and gene-expression changes associated with bipolar disorder are over-represented in brain signaling pathway genes. *Biol Psychiatry.* 2012;72(4):311–317.
14. Chen DT, Jiang X, Akula N, et al. Genome-wide association study meta-analysis of European and Asian-ancestry samples identifies three novel loci associated with bipolar disorder. *Mol Psychiatry.* 2013;18(2):195–205.
15. Eaton WW, Pedersen MG, Nielsen PR, Mortensen PB. Autoimmune diseases, bipolar disorder, and non-affective psychosis. *Bipolar Disord.* 2010;12(6):638–646
16. Tiosano S, Nir Z, Gendelman O, et al. The association between systemic lupus erythematosus and bipolar disorder—a big data analysis. *Eur Psychiatry.* 2017;43:116–119.

17. Bachen EA, Chesney MA, Criswell LA. Prevalence of mood and anxiety disorders in women with systemic lupus erythematosus. *Arthritis Rheum.* 2009;61(6):822–829.
18. Liu CJ, Hu LY, Yeh CM, et al. Irritable brain caused by irritable bowel? A nationwide analysis for irritable bowel syndrome and risk of bipolar disorder. *PLoS One.* 2015;10(3):e0118209.
19. Edwards LJ, Constantinescu CS. A prospective study of conditions associated with multiple sclerosis in a cohort of 658 consecutive outpatients attending a multiple sclerosis clinic. *Mult Scler.* 2004;10(5):575–581.
20. Perugi G, Quaranta G, Belletti S, et al. General medical conditions in 347 bipolar disorder patients: clinical correlates of metabolic and autoimmune-allergic diseases. *J Affect Disord.* 2015;170:95–103.
21. Jacobson DL, Gange SJ, Rose NR, Graham NM. Epidemiology and estimated population burden of selected autoimmune diseases in the United States. *Clin Immunol Immunopathol.* 1997;84(3):223–243.
22. Degner D, Haust M, Meller J, Rüther E, Reulbach U. Association between autoimmune thyroiditis and depressive disorder in psychiatric outpatients. *Eur Arch Psychiatry Clin Neurosci.* 2015;265(1):67–72.
23. Barbuti M, Carvalho AF, Köhler CA, et al. Thyroid autoimmunity in bipolar disorder: A systematic review. *J Affect Disord.* 2017;221:97–106.
24. Kupka RW, Nolen WA, Post RM, et al. High rate of autoimmune thyroiditis in bipolar disorder: lack of association with lithium exposure. *Biol Psychiatry.* 2002;51(4):305–311.
25. Dickerson F, Stallings C, Origoni A, Vaughan C, Khushalani S, Yolken R. Markers of gluten sensitivity in acute mania: a longitudinal study. *Psychiatry Res.* 2012;196(1):68–71.
26. León-Caballero J, Pacchiarotti I, Murru A, et al. Bipolar disorder and antibodies against the N-methyl-d-aspartate receptor: A gate to the involvement of autoimmunity in the pathophysiology of bipolar illness. *Neurosci Biobehav Rev.* 2015;55:403–412.
27. Oliveira J, Oliveira-Maia AJ, Tamouza R, Brown AS, Leboyer M. Infectious and immunogenetic factors in bipolar disorder. *Acta Psychiatr Scand.* 2017;136(4):409–423.
28. de Barros JL, Barbosa IG, Salem H, et al. Is there any association between *Toxoplasma gondii* infection and bipolar disorder? A systematic review and meta-analysis. *J Affect Disord.* 2017;209:59–65.
29. Brown AS. The Kraepelinian dichotomy from the perspective of prenatal infectious and immunologic insults. *Schizophr Bull.* 2015;41(4):786–791.
30. Barichello T, Badawy M, Pitcher MR, et al. Exposure to perinatal infections and bipolar disorder: a systematic review. *Curr Mol Med.* 2016;16(2):106–118.
31. Weber-Stadlbauer U, Richetto J, Labouesse MA, Bohacek J, Mansuy IM, Meyer U. Transgenerational transmission and modification of pathological traits induced by prenatal immune activation. *Mol Psychiatry.* 2017;22(1):102–112.
32. Bauer-Staeb C, Jörgensen L, Lewis G, Dalman C, Osborn DPJ, Hayes JF. Prevalence and risk factors for HIV, hepatitis B, and hepatitis C in people with severe mental illness: a total population study of Sweden. *Lancet Psychiatry.* 2017;4(9):685–693.
33. Barbosa IG, Rocha NP, Assis F, et al. Monocyte and lymphocyte activation in bipolar disorder: a new piece in the puzzle of immune dysfunction in mood disorders. *Int J Neuropsychopharmacol.* 2014;18(1):1–7.

34. Padmos RC, Hillegers MH, Knijff EM, et al. A discriminating messenger RNA signature for bipolar disorder formed by an aberrant expression of inflammatory genes in monocytes. *Arch Gen Psychiatry*. 2008;65(4):395–407.
35. Becking K, Haarman BC, van der Lek RF, et al. Inflammatory monocyte gene expression: trait or state marker in bipolar disorder? *Int J Bipolar Disord*. 2015;3(1):20.
36. Knijff EM, Ruwhof C, de Wit HJ, et al. Monocyte-derived dendritic cells in bipolar disorder. *Biol Psychiatry*. 2006;59(4):317–326.
37. Knijff EM1, Breunis MN, Kupka RW, et al. An imbalance in the production of IL-1beta and IL-6 by monocytes of bipolar patients: restoration by lithium treatment. *Bipolar Disord*. 2007;9(7):743–753.
38. Brambilla P, Bellani M, Isola M, et al. Increased M1/decreased M2 signature and signs of Th1/Th2 shift in chronic patients with bipolar disorder, but not in those with schizophrenia. *Transl Psychiatry*. 2014;4:e406.
39. Torres KC, Souza BR, Miranda DM, et al. The leukocytes expressing DARPP-32 are reduced in patients with schizophrenia and bipolar disorder. *Prog Neuropsychopharmacol Biol Psychiatry*. 2009;33(2):214–219.
40. Karpiński P, Frydecka D, Sąsiadek MM, Misiak B. Reduced number of peripheral natural killer cells in schizophrenia but not in bipolar disorder. *Brain Behav Immun*. 2016;54:194–200.
41. do Prado CH, Rizzo LB, Wieck A, et al. Reduced regulatory T cells are associated with higher levels of Th1/TH17 cytokines and activated MAPK in type 1 bipolar disorder. *Psychoneuroendocrinology*. 2013;38(5):667–676.
42. Drexhage RC, Hoogenboezem TH, Versnel MA, Berghout A, Nolen WA, Drexhage HA. The activation of monocyte and T cell networks in patients with bipolar disorder. *Brain Behav Immun*. 2011;25(6):1206–1213.
43. Tsai SY, Chen KP, Yang YY, et al. Activation of indices of cell-mediated immunity in bipolar mania. *Biol Psychiatry*. 1999;45(8):989–994.
44. Rapaport MH. Immune parameters in euthymic bipolar patients and normal volunteers. *J Affect Disord*. 1994;32:149–156.
45. Rizzo LB, Do Prado CH, Grassi-Oliveira R, et al. Immunosenescence is associated with human cytomegalovirus and shortened telomeres in type I bipolar disorder. *Bipolar Disord*. 2013;15(8):832–838.
46. Wieck A, Grassi-Oliveira R, do Prado CH, et al. Differential neuroendocrine and immune responses to acute psychosocial stress in women with type 1 bipolar disorder. *Brain Behav Immun*. 2013;34:47–55.
47. Wieck A, Grassi-Oliveira R, do Prado CH, et al. Toll-like receptor expression and function in type I bipolar disorder. *Brain Behav Immun*. 2016;54:110–121.
48. Snijders G, Schiweck C, Mesman E, et al. A dynamic course of T cell defects in individuals at risk for mood disorders. *Brain Behav Immun*. 2016;58:11–17.
49. Boufidou F, Nikolaou C, Alevizos B, Liappas IA, Christodoulou GN. Cytokine production in bipolar affective disorder patients under lithium treatment. *J Affect Disord*. 2004;82(2):309–313.
50. Liu HC, Yang YY, Chou YM, Chen KP, Shen WW, Leu SJ. Immunologic variables in acute mania of bipolar disorder. *J Neuroimmunol*. 2004;150(1–2):116–122.
51. Goldsmith DR, Rapaport MH, Miller BJ. A meta-analysis of blood cytokine network alterations in psychiatric patients: comparisons between schizophrenia, bipolar disorder and depression. *Mol Psychiatry*. 2016;21(12):1696–1709.

52. Sayana P, Colpo GD, Simões LR, et al. A systematic review of evidence for the role of inflammatory biomarkers in bipolar patients. *J Psychiatr Res.* 2017;92:160–182.
53. Modabbernia A, Taslimi S, Brietzke E, Ashrafi M. Cytokine alterations in bipolar disorder: a meta-analysis of 30 studies. *Biol Psychiatry.* 2013;74(1):15–25.
54. Munkholm K, Braüner JV, Kessing LV, Vinberg M. Cytokines in bipolar disorder vs. healthy control subjects: a systematic review and meta-analysis. *J Psychiatr Res.* 2013;47(9):1119–1133.
55. Munkholm K, Vinberg M, Vedel Kessing L. Cytokines in bipolar disorder: a systematic review and meta-analysis. *J Affect Disord.* 2013;144(1–2):16–27.
56. Uyanik V, Tuglu C, Gorgulu Y, Kunduracilar H, Uyanik MS. Assessment of cytokine levels and hs-CRP in bipolar I disorder before and after treatment. *Psychiatry Res.* 2015;228(3):386–392.
57. Li H, Hong W, Zhang C, et al. IL-23 and TGF-β1 levels as potential predictive biomarkers in treatment of bipolar I disorder with acute manic episode. *J Affect Disord.* 2015;174:361–366.
58. Hope S, Melle I, Aukrust P, et al. Similar immune profile in bipolar disorder and schizophrenia: selective increase in soluble tumor necrosis factor receptor I and von Willebrand factor. *Bipolar Disord.* 2009;11(7):726–734.
59. Hope S, Dieset I, Agartz I, et al. Affective symptoms are associated with markers of inflammation and immune activation in bipolar disorders but not in schizophrenia. *J Psychiatr Res.* 2011;45(12):1608–1616.
60. Becking K, Boschloo L, Vogelzangs N, et al. The association between immune activation and manic symptoms in patients with a depressive disorder. *Transl Psychiatry.* 2013;3:e314.
61. Hung YJ, Hsieh CH, Chen YJ, et al. Insulin sensitivity, proinflammatory markers and adiponectin in young males with different subtypes of depressive disorder. *Clin Endocrinol (Oxf).* 2007;67(5):784–789.
62. Barbosa IG, Rocha NP, Bauer ME, et al. Chemokines in bipolar disorder: trait or state? *Eur Arch Psychiatry Clin Neurosci.* 2013;263(2):159–165.
63. Brietzke E, Kauer-Sant'Anna M, Teixeira AL, Kapczinski F. Abnormalities in serum chemokine levels in euthymic patients with bipolar disorder. *Brain Behav Immun.* 2009;23(8):1079–1082.
64. Reininghaus EZ, Lackner N, Birner A, et al. Extracellular matrix proteins matrix metallopeptidase 9 (MMP9) and soluble intercellular adhesion molecule 1 (sICAM-1) and correlations with clinical staging in euthymic bipolar disorder. *Bipolar Disord.* 2016;18(2):155–163.
65. Schaefer M, Sarkar S, Schwarz M, Friebe A. Soluble intracellular adhesion molecule-1 in patients with unipolar or bipolar affective disorders: results from a pilot trial. *Neuropsychobiology.* 2016;74(1):8–14.
66. Barbosa IG, Rocha NP, de Miranda AS, et al. Increased levels of adipokines in bipolar disorder. *J Psychiatr Res.* 2012;46(3):389–393.
67. Chang HH, Wang TY, Lee IH, et al. C-reactive protein: a differential biomarker for major depressive disorder and bipolar II disorder. *World J Biol Psychiatry.* 2017;18(1):63–70.
68. Tsai MC, Huang TL. Thiobarbituric acid reactive substances (TBARS) is a state biomarker of oxidative stress in bipolar patients in a manic phase. *J Affect Disord.* 2015;173:22–26.

69. Mansur RB, Rizzo LB, Santos CM, et al. Bipolar disorder course, impaired glucose metabolism and antioxidant enzymes activities: a preliminary report. *J Psychiatr Res.* 2016;80:38–44
70. Söderlund J, Olsson SK, Samuelsson M, et al. Elevation of cerebrospinal fluid interleukin-1ß in bipolar disorder. *J Psychiatry Neurosci.* 2011;36(2):114–118.
71. Jakobsson J, Bjerke M, Sahebi S, et al. Monocyte and microglial activation in patients with mood-stabilized bipolar disorder. *J Psychiatry Neurosci.* 2015;40(4):250–258.
72. Zetterberg H, Jakobsson J, Redsäter M, et al. Blood-cerebrospinal fluid barrier dysfunction in patients with bipolar disorder in relation to antipsychotic treatment. *Psychiatry Res.* 2014;217(3):143–146.
73. Patel JP, Frey BN. Disruption in the blood-brain barrier: the missing link between brain and body inflammation in bipolar disorder? *Neural Plast.* 2015;2015:708306.
74. Bezchlibnyk YB, Wang JF, McQueen GM, Young LT. Gene expression differences in bipolar disorder revealed by cDNA array analysis of post-mortem frontal cortex. *J Neurochem.* 2001;79(4):826–834.
75. Rao JS, Harry GJ, Rapoport SI, Kim HW. Increased excitotoxicity and neuroinflammatory markers in postmortem frontal cortex from bipolar disorder patients. *Mol Psychiatry.* 2010;15(4):384–392.
76. Dean B, Gibbons AS, Tawadros N, Brooks L, Everall IP, Scarr E. Different changes in cortical tumor necrosis factor-α-related pathways in schizophrenia and mood disorders. *Mol Psychiatry.* 2013;18(7):767–773.
77. Kim HK, Andreazza AC, Elmi N, Chen W, Young LT. Nod-like receptor pyrin containing 3 (NLRP3) in the post-mortem frontal cortex from patients with bipolar disorder: a potential mediator between mitochondria and immune-activation. *J Psychiatr Res.* 2016;72:43–50.
78. Haarman BC, Riemersma-Van der Lek RF, de Groot JC, et al. Neuroinflammation in bipolar disorder—a [(11)C]-(R)-PK11195 positron emission tomography study. *Brain Behav Immun.* 2014;40:219–225.
79. Haarman BC, Burger H, Doorduin J, et al. Volume, metabolites and neuroinflammation of the hippocampus in bipolar disorder—A combined magnetic resonance imaging and positron emission tomography study. *Brain Behav Immun.* 2016;56:21–33.
80. Townsend JD, Torrisi SJ, Lieberman MD, Sugar CA, Bookheimer SY, Altshuler LL. Frontal-amygdala connectivity alterations during emotion downregulation in bipolar I disorder. *Biol Psychiatry.* 2013;73(2):127–135.
81. Hsu JW, Lirng JF, Wang SJ, et al. Association of thalamic serotonin transporter and interleukin-10 in bipolar I disorder: a SPECT study. *Bipolar Disord.* 2014;16(3):241–248.
82. Chou YH, Hsieh WC, Chen LC, Lirng JF, Wang SJ. Association between the serotonin transporter and cytokines: implications for the pathophysiology of bipolar disorder. *J Affect Disord.* 2016;191:29–35.
83. Dantzer R, O'Connor JC, Lawson MA, Kelley KW. Inflammation-associated depression: from serotonin to kynurenine. *Psychoneuroendocrinology.* 2011;36(3):426–436.
84. Birner A, Platzer M, Bengesser SA, et al. Increased breakdown of kynurenine towards its neurotoxic branch in bipolar disorder. *PLoS One.* 2017;12(2):e0172699.

85. Savitz J, Dantzer R, Wurfel BE, et al. Neuroprotective kynurenine metabolite indices are abnormally reduced and positively associated with hippocampal and amygdalar volume in bipolar disorder. *Psychoneuroendocrinology*. 2015;52:20411.
86. Streit WJ. Microglial senescence: does the brain's immune system have an expiration date? *Trends Neurosci*. 2006;29(9):506–510.
87. Savitz JB, Price JL, Drevets WC. Neuropathological and neuromorphometric abnormalities in bipolar disorder: view from the medial prefrontal cortical network. *Neurosci Biobehav Rev*. 2014;42:132–147.
88. Nassar A, Azab AN. Effects of lithium on inflammation. *ACS Chem Neurosci*. 2014;5(6):451–458.
89. Valvassori SS, Tonin PT, Varela RB, et al. Lithium modulates the production of peripheral and cerebral cytokines in an animal model of mania induced by dextroamphetamine. *Bipolar Disord*. 2015;17(5):507–517.
90. Li N, Zhang X, Dong H, Zhang S, Sun J, Qian Y. Lithium ameliorates LPS-induced astrocytes activation partly via inhibition of Toll-like receptor 4 expression. *Cell Physiol Biochem*. 2016;38(2):714–725.
91. van den Ameele S, van Diermen L, Staels W, et al. The effect of mood-stabilizing drugs on cytokine levels in bipolar disorder: a systematic review. *J Affect Disord*. 2016;203:364–373.
92. Himmerich H, Bartsch S, Hamer H, et al. Modulation of cytokine production by drugs with antiepileptic or mood stabilizer properties in anti-CD3- and anti-Cd40-stimulated blood *in vitro*. *Oxid Med Cell Longev*. 2014;2014:806162.
93. Himmerich H, Bartsch S, Hamer H, et al. Impact of mood stabilizers and antiepileptic drugs on cytokine production in-vitro. *J Psychiatr Res*. 2013;47(11):1751–1759.
94. Fonseka TM, Müller DJ, Kennedy SH. Inflammatory cytokines and antipsychotic-induced weight gain: review and clinical implications. *Mol Neuropsychiatry*. 2016;2(1):1–14.
95. Nery FG, Monkul ES, Hatch JP, et al. Celecoxib as an adjunct in the treatment of depressive or mixed episodes of bipolar disorder: a double-blind, randomized, placebo-controlled study. *Hum Psychopharmacol*. 2008;23(2):87–94.
96. Mousavi SY, Khezri R, Karkhaneh-Yousefi MA, et al. A randomized, double-blind placebo-controlled trial on effectiveness and safety of celecoxib adjunctive therapy in adolescents with acute bipolar mania. *J Child Adolesc Psychopharmacol*. 2017;27(6):494–500.
97. Arabzadeh S, Ameli N, Zeinoddini A, et al. Celecoxib adjunctive therapy for acute bipolar mania: a randomized, double-blind, placebo-controlled trial. *Bipolar Disord*. 2015;17(6):606–614.
98. Soczynska JK, Kennedy SH, Alsuwaidan M, et al. A pilot, open-label, 8-week study evaluating the efficacy, safety and tolerability of adjunctive minocycline for the treatment of bipolar I/II depression. *Bipolar Disord*. 2017;19(3):198–213.
99. Rosenblat JD, Kakar R, Berk M, et al. Anti-inflammatory agents in the treatment of bipolar depression: a systematic review and meta-analysis. *Bipolar Disord*. 2016;18(2):89–101.
100. Austin M, Tan YC. Mania associated with infliximab. *Aust N Z J Psychiatry*. 2012;46(7):684–685.

13

Immunology of Post-Traumatic Stress Disorder (PTSD)

ANDREA WIECK, IZABELA G. BARBOSA, ANTONIO L. TEIXEIRA, AND MOISES E. BAUER ■

INTRODUCTION

The Greek philosopher Herodotus (c. 484–425 BC) told the history of a soldier from Atenas, who, despite of not suffering any injury during the Battle of Marathon, was stricken permanently blind after seeing his friend dead. Throughout history, descriptions of several intense emotional reactions related to war trauma can be found. However, only in the 17th century, following the Great Fire of London (1666), one of the first reports of post-traumatic symptoms was found in Samuel Pepys's diary. Six months after Pepys survived the fire, he reported the following: "It is strange to think how to this very day I cannot sleep a night without great terrors of fire; and this very night could not sleep till almost 2:00 in the morning through thoughts of fire (...) so great was our fear that it was enough to take away our senses."[1] These symptoms were later designated as "traumatic neurosis" by Herman Oppenheim (1889), a neurologist who proposed that functional problems would result from subtle changes in the central nervous system. Following the formal proposal of the Post-Traumatic Stress Disorder (PTSD) diagnosis by the American Psychiatry Association (APA) in the 1980s, the search for these "subtle changes" related to the disorder was intensified.

PTSD is characterized by constant re-experiencing of the original traumatic event, with disturbing thoughts, feelings, or dreams related to the events, as well as mood changes and exaggerated reactivity to stress. The increase of "fight-or-flight" responses may lead to chronic dysregulations of stress-responsive mechanisms (i.e., the allostatic load), including the hypothalamic-pituitary-adrenal (HPA) axis, sympathetic nervous system (SNS), and immune system.[2] The chronic low-grade inflammation, with increased inflammatory biomarkers in the blood, is the most common immune imbalance reported in PTSD. Similarly, there is extensive evidence indicating the role of inflammation in the pathophysiology of mood disorders, including major depressive disorder (MDD) and bipolar disorder (BD).[3] Considering the close relationship between PTSD and MDD, it has been proposed that a considerable part of the physiological mechanisms involved in MDD might be relevant to PTSD as well. Accordingly, shared biomarkers could be detected in both conditions. However, in contrast with the significant literature on immune alterations in mood disorders, the field of PTSD immunity is still incipient.

PTSD patients have increased risk of developing several medical comorbidities associated with inflammatory and/or autoimmune pathogenesis.[4] For example, cardiovascular diseases have been associated with PTSD, and PTSD was considered a strong predictor of heart disease.[5–9] In addition, PTSD greatly increases the risk for developing rheumatoid arthritis,[9–11] inflammatory bowel disease, multiple sclerosis, lupus, psoriasis, thyroid diseases,[5,12] fibromyalgia, and metabolic syndrome.[11] Therefore, the development of inflammatory comorbidities adds further detrimental effects to patients with PTSD. We review here the cellular and molecular features of the chronic low-grade inflammation that characterizes PTSD and how they relate to dysregulated stress-responses.

CHRONIC LOW-GRADE INFLAMMATION

An increasing body of evidence indicates that PTSD is associated with chronic peripheral inflammation. Several pro-inflammatory cytokines, such as interleukin-6 (IL-6), tumor necrosis factor–alpha (TNF-α) and its soluble receptors, interferon-gamma (IFN-γ), interleukin-1 beta (IL1-β), and C-reactive protein (CRP) were found to be increased in PTSD patients.[13–17] In support, a recent meta-analysis found increased levels of several pro-inflammatory cytokines in PTSD.[18] Illness duration was found to be positively associated with IL-1β levels, and the severity of the illness with IL-6 levels.[18]

Increased IL-6 levels immediately following traumatic events were also found associated with PTSD development six months after the trauma.[19] In parallel, MDD patients with PTSD symptoms presented high levels of TNF-α soluble receptors (sTNFR1 and sTNFR2) in comparison with a healthy control group.[20] Specifically, sTNFR1 levels were correlated to clinical severity, supporting the role for chronic inflammation in the PTSD development.[20] Interestingly, high plasma levels of CRP in Marines before deployment were associated with PTSD development following combat.[21] Also, the Marines who developed PTSD following combat exposure presented altered gene expression patterns (i.e., interferon and monocyte-related) collected before combat compared to those who did not develop PTSD following combat.[22]

The causative role between inflammation and PTSD development has not been established yet, nor the mechanisms through which the immune system influences PTSD. Gill and colleagues (2011) showed that women who recovered from PTSD had lower concentrations of pro-inflammatory markers (i.e., IL-6 and CRP) when compared to women who did not recover from PTSD, and these levels were comparable to those observed for non-traumatized healthy individuals.[23] Although this finding did not establish any causality, it strongly supports a major role for inflammation in PTSD development and symptomatology.

The underlying mechanisms of the immune imbalance in PTSD also remain largely unknown. One possible explanation relies on functional impairment of the HPA axis, especially the observation of hypocortisolemia. The immune system is critically regulated by glucocorticoids (GCs), and all immune cells express glucocorticoid receptors (GRs) to varying degrees. Because of the reduced cortisol levels, increased inflammatory response could be observed due to lack of proper immune control.

Another possible explanation involves epigenetic changes. Trauma exposure could lead to long-lasting epigenetic changes, resulting in a chronic inflammation profile.[24,25] In fact, gene transcription of proteins related to neurotransmitter regulation, such as serotoninergic and GABAergic receptors, as well as genes involved in immune and endocrine responses, were found altered in patients with PTSD.[26] Also, genes related to inflammatory response and cellular activation, such as calcium channel blockers, leptin regulatory genes, and Toll-like receptor genes, were found to be demethylated. Differential methylation of immune-regulating genes was found to be associated with increased inflammation in PTSD. Hypermethylation of inflammatory initiator genes and demethylation of inflammatory regulatory genes were also found altered in patients with PTSD, indicating a propensity to an enhanced inflammatory response.[24,25,27,28]

This chronic low-grade inflammation could contribute to depressive symptoms, usually very common in patients with PTSD. Peripheral cytokines may reach the brain via humoral and neural routes, driving CNS inflammation and modulating key brain areas involved with mood regulation. The humoral route includes the passage of cytokines through leaky regions of the blood–brain barrier (BBB), such as the circumventricular organs. Cytokines may also signal peripheral information to the brain via binding to receptors expressed in peripheral nerve fibers (e.g., the vagus nerve)—the "neural route." In the brain, cytokines change the production, metabolism, and transport of neurotransmitters that synergistically affect mood, including dopamine, glutamate, and serotonin.[29] For example, cytokines stimulate indoleamine 2,3-dioxygenase (IDO), an enzyme that increases tryptophan metabolism and hence reduces serotonin levels—because tryptophan is the primary precursor of serotonin. Furthermore, IDO simultaneously enhances kynurenine levels, a tryptophan metabolite, which can be metabolized by two cellular pathways: (1) microglia, generating 3-hydroxykynurenine (3-HK) and quinolinic acid (QA); and (2) astrocytes, producing kynurenic acid (KA). 3-HK is an oxidative stressor, whereas QA is an N-methyl-D-aspartate (NMDA) receptor agonist, increasing glutamate release and blocking glutamate reuptake by astrocytes.[3] Increased levels of QA were found in microglia of suicide patients with MDD.[30] QA is also associated with lipid peroxidation and oxidative stress. In combination, these activities can lead to excitotoxicity and neurodegeneration. Furthermore, the binding of glutamate to extrasynaptic NMDA receptors can lead to decreased production of brain-derived neurotrophic factor (BDNF), impairing neuroplasticity and contributing to impaired cognition.[31] In contrast to QA, the KA can reduce glutamate and dopamine release, both of which can contribute to cognitive dysfunction.[3] To date, no studies have investigated these molecules in PTSD.

ACTIVATION OF CELLULAR IMMUNITY AND PREMATURE SENESCENCE

PTSD was associated with increased cellular immune responses.[4,32,33] including increased counts in peripheral blood mononuclear cells (PBMC), particularly natural killer (NK) and T cells.[4,34,35] Also, increased activated T cells (CD2+HLADR+), B cells (CD20+CD23+), and NK cells (CD16+CD71+) were observed in women with PTSD.[36] Sommerhof and colleagues (2009) also showed increased percentages of central memory T cells (CD45RA-CCR7+) as well as effector memory cells (CD45RA-CCR7-) in patients with PTSD, specifically those exposed to sexual abuse during childhood.[35] In contrast, some studies reported reduced numbers of both T helper (CD4+) and T cytotoxic (CD8+) cells.[34,37] Patients with PTSD showed approximately 50% fewer regulatory (FOXP3+) T cells when compared to healthy controls.[35] Interestingly, psychotherapeutic treatment, i.e., twelve sessions of narrative exposure therapy, decreased PTSD symptoms and increased the proportion of regulatory T cells.[38]

Stress, including psychological stress, is known to play a key role in the acceleration of immunological aging (i.e., immunosenescence). A recent study showed that past-year history of PTSD was associated with increased aged T cell phenotypes in adults living in an urban center. The changes were particularly noted in the CD8+ subset and included a higher ratio of late-differentiated effector to naïve T cells, a higher percentage of KLRG1+ cells, and a higher percentage of CD57+ cells.[39] There is also compelling evidence linking psychosocial stress to accelerated telomere erosion, an important marker of cellular aging. It has been shown that adult war veterans with current PTSD had shorter PBMC telomere lengths than their age-matched controls.[40]

The molecular mechanisms implicated in the cellular immunity changes in PTSD encompass differential expression of genes related to canonical pathways of immune cell development. One relevant pathway altered in PTSD is the T helper (Th) differentiation, biasing T cells to either pro-inflammatory or anti-inflammatory phenotypes. According to Bam and colleagues (2016), three important genes in this pathway are upregulated in PTSD: STAT4, TBX21, and HLA-DQA1; the STAT4 and TBX21 are important genes associated with T-cell fate.[28] Chemokines and their receptors (CCLA, CCL5, CXCL1, CXCL2, CXCL3, CXCL6, CXCL8, CXCR1, and CXCR8) were also upregulated in PTSD patients.[28] One possible explanation relies on epigenetics, since several CpG (cytosine and guanine nucleotides) sites in the promoter region of investigated genes were found to be methylated. Conversely, upregulated genes were found to be targets of miRNAs, which, in turn, were downregulated. The miRNAs are important regulators of gene expression, since their interaction with target miRNAs leads to its destabilization and, consequently, no protein synthesis.[28]

NEUROENDOCRINE ALTERATIONS IN STRESS-RESPONSIVE SYSTEMS

The neuroendocrine research in PTSD has focused on two central systems involved in stress physiology: the HPA axis and the SNS. The HPA axis is the major neuroendocrine system involved in stress responses. After stressful stimuli, cortisol levels rapidly rise in the circulation. Following cessation of stress, cortisol returns to its baseline level via negative feedback mechanisms in the brain. Because virtually all cells in the body express GRs, chronic GC elevations are associated with several systemic effects, including metabolic and immune changes.

Considering that the adaptive response to stress involves HPA axis activation, it was expected that PTSD patients would show chronic elevated cortisol levels. However, early studies indicated that PTSD was associated with low cortisol levels at baseline (hypocortisolemia).[41,42] Indeed, some studies have observed an association between PTSD and low cortisol levels (plasma, saliva, or urine levels) when compared to healthy controls.[43] PTSD is also associated with low diurnal cortisol levels and increased cortisol production in the night.[4] In contrast, Elzinga and colleagues (2003) observed increased salivary cortisol levels in women with PTSD exposed to personalized trauma scripts.[44] Anticipatory and recovery cortisol levels were also increased in those patients. Cortisol levels during and after tests were strongly associated to PTSD symptom severity, especially arousal and re-experiencing.[44] These data suggest that, rather than a static pattern of hypo- or hypercortisolemia, the recurrent re-experience of the traumatic event, or exposure to new stressful situations (traumatic or not) may determine the cortisol levels. Accordingly, reduced cortisol levels (hypocortisolemia) may reflect a compensation for the hypercortisolemia occurring during new stressor exposures or re-experiences. As discussed, immune alterations observed in PTSD have been related to impaired neuroendocrine function and regulation, especially in the HPA axis (see Figure 13.1).

It has been speculated that hypocortisolemia would be associated with enhanced central and systemic sensitivity to GCs. A meta-analysis reported increased negative feedback sensitivity of GRs in traumatized patients with and without PTSD compared to non-traumatized individuals.[45] Patients with PTSD had increased sensitivity to cortisol, as demonstrated by exacerbated immunosuppression in the dexamethasone suppression test,[46] and increased *in vitro* GC sensitivity in PBMCs.[47,48] Cellular sensitivity to GCs can be estimated by examining the effects of dexamethasone in suppressing cytokine production *in vitro*. One possible explanation is the increased expression of glucocorticoid receptor α (GRα) in patients with PTSD.[47] However, the paradox of concurrent enhanced sensitivity to GCs and chronic low-grade inflammation remains to be understood. With better GC-related control of immune responses, one should observe less inflammation, not more, as is usually seen in patients with PTSD.

Epigenetic changes may explain the HPA axis abnormalities. Previous studies reported an experience-dependent programming of the HPA axis in brain neurons via epigenetic alterations. Murgatroyd and colleagues (2010) reported that neuron gene expression can be altered via methylation in response to social experiences, producing long-lasting changes in gene expression via epigenetic mechanisms.[49] Sensory and cognitive areas of the brain would register social or physical experiences responsible for activating signaling pathways in the epigenetics machinery.[49] Methylation patterns in the promoter region of genes involved in HPA functioning, such as the GR, arginine vasopressin (AVP), and BDNF, have been shown to be altered in brains from animal models of maternal care and perinatal stress.[50–54] Following these findings in animal models, McGowan and colleagues (2009) found increased methylation in the promoter region of GR in brains from suicide victims with a history of childhood trauma when compared to suicide victims without history of childhood trauma. As a result, increased levels of GR were also observed in the hypermethylated brains.[55] Also, altered methylation patterns in the promoter region of GR were observed in mixed lymphocyte populations from individuals with history of childhood trauma.[56–58] Altogether, these data suggest that stress and trauma exposure can

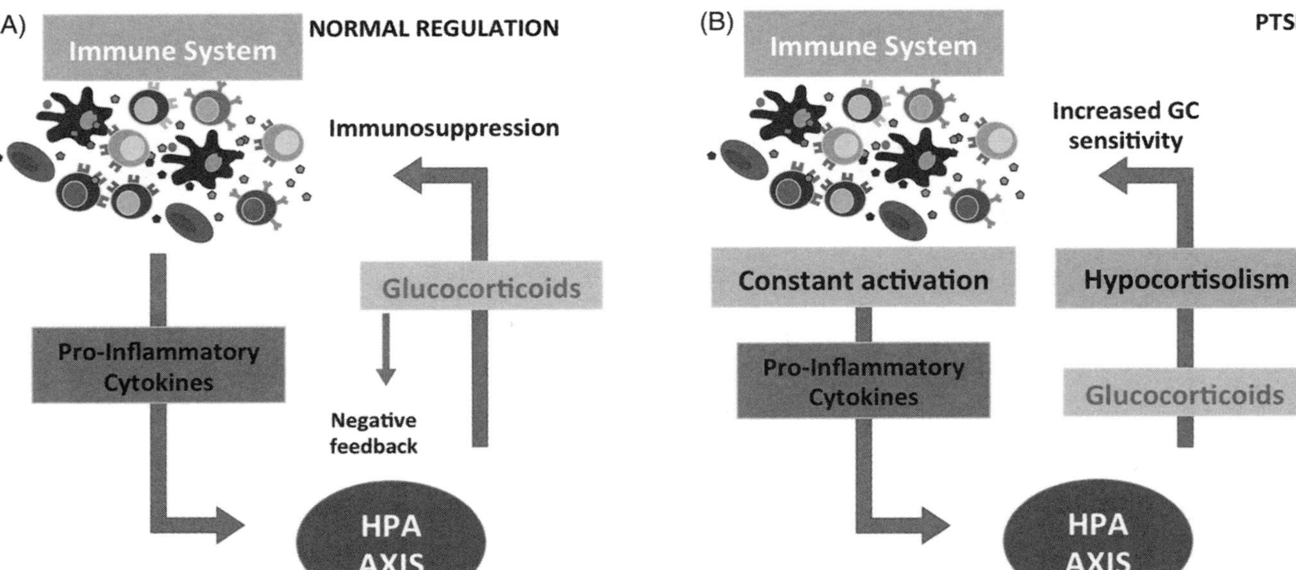

Figure 13.1 *Bidirectional communication between immune system and HPA axis.* **(A)** In a normal, homeostatic situation, inflammatory response (i.e., infection) activates the HPA axis, which, in turn, increases cortisol secretion controlling immune responses. **(B)** Persistent re-experience of traumatic events in PTSD patients is a situation of chronic stress exposure. In this way, communication between the immune system and HPA axis is impaired, resulting in hypocortisolism, increased cellular glucocorticoid resistance, and consequent immune system activation, leading to inflammation.

lead to long-lasting epigenetic alterations in genes involved with stress reactivity and HPA axis functioning.

PTSD has also been associated with anatomical changes in brain areas involved with stress processing. A reduced hippocampal volume in PTSD could be due to chronic exposure to high levels of cortisol during trauma and subsequent stressful situations.[59] The hippocampus expresses the highest GR densities and is thus highly sensitive to GC fluctuations. The hippocampus plays an important role not only in HPA axis functioning, but also in memory and learning. Patients with PTSD frequently exhibit memory impairment. On the other hand, reduced hippocampal volume may lead to increased amygdala reactivity and, consequently, overreactions during stressful situations, as observed in patients with PTSD.[60] The direction of this association is still debatable: whether reduced hippocampus would increase PTSD risk, or PTSD is responsible for inducing hippocampal shrinkage and its related cognitive and behavioral changes.

There is evidence linking PTSD to SNS activation, as observed by increased heart rate and blood pressure, as well as overreaction to stressful situations. Catecholamines released in those situations are responsible for the "fight-or-flight" response. It has been shown that PTSD patients had increased, centrally and peripherally, catecholamine levels.[4] One study suggested that increased heart rate combined with norepinephrine release at the time of the trauma are good predictors for subsequent PTSD development.[61]

CONCLUSIONS AND PERSPECTIVES

The physiological stress response includes adaptive neuroimmune mechanisms engaged to restore body's homeostasis during stressful situations. During a stressful situation, immune and stress-reactive systems (HPA and SNS axes) are activated, coordinating the individual's physiological responses aimed to enhance survival. However, an intense activation of these stress-reactive systems during traumatic events may produce long-lasting epigenetic, neuroanatomical, and neuroendocrine changes. In this context, low-grade inflammation could emerge as a maladaptive or dysregulated physiological response mounted to stress. It has also been proposed that dysregulation of the stress axes following the onset of the PTSD contributes to symptoms by having a direct effect on brain regions critical for emotion regulation, such as the prefrontal cortex, insula, amygdala, and hippocampus.[62]

Increased pro-inflammatory mediators (e.g., cytokines) in the circulation may further activate the HPA axis and modify the brain milieu, leading to depressive symptoms and cognitive impairment. In healthy individuals, the stress-related HPA axis activation is beneficial as it modulates immune response in a way that it is not detrimental. Once the stressful situation ends, both the HPA axis and immune responses return to basal levels of functioning. However, chronic stress is associated with allostatic load and dysregulations of the main stress-responsive systems, including the SNS, HPA axis, and immune system.

Persistent re-experience of traumatic events is one of the cardinal symptoms of the PTSD. Physiologically, this symptom is perceived by the subject as a chronic stress, possibly resulting in maladaptive neuroendocrine and immune responses. It is debatable whether these dysregulated responses cause and/or are caused by PTSD. However, such alterations are not a simple epiphenomenon, but rather are intimately implicated in the pathophysiology of PTSD. New therapeutic strategies could emerge based on this knowledge. Indeed, several anti-inflammatory strategies have been tested with great success for the treatment of mood disorders (i.e., MDD and BD), namely the polyunsaturated fatty acids (PUFAs), cyclooxygenase (COX) inhibitors, anti-TNF-α, and minocycline.[63,64] A recent preclinical study indicated the potential use of minocycline in preventing inflammatory and behavioral changes in an animal model of PTSD.[65] Further clinical investigations of the use of minocycline, as well as other anti-inflammatory strategies, in PTSD are needed.

REFERENCES

1. Daly, R. J. (1983). Samuel Pepys and post-traumatic stress disorder. *Br J Psychiatry, 143*, 64–68.
2. Olff, M., & van Zuiden, M. (2017). Neuroendocrine and neuroimmune markers in PTSD: pre-, peri- and post-trauma glucocorticoid and inflammatory dysregulation. *Curr Opin Psychol, 14*, 132–137.
3. Miller, A. H., & Raison, C. L. (2016). The role of inflammation in depression: from evolutionary imperative to modern treatment target. *Nat Rev Immunol, 16*(1), 22–34.
4. Pace, T. W., & Heim, C. M. (2011). A short review on the psychoneuroimmunology of post-traumatic stress disorder: from risk factors to medical comorbidities. *Brain Behav Immun, 25*(1), 6–13.
5. Boscarino, J. A. (2004). Post-traumatic stress disorder and physical illness: results from clinical and epidemiologic studies. *Ann N Y Acad Sci, 1032*, 141–153.
6. Boscarino, J. A. (2008). A prospective study of PTSD and early-age heart disease mortality among Vietnam veterans: implications for surveillance and prevention. *Psychosom Med, 70*(6), 668–676.
7. Kibler, J. L. (2009). Post-traumatic stress and cardiovascular disease risk. *J Trauma Dissoc, 10*(2), 135–150.
8. Kibler, J. L., Joshi, K., & Ma, M. (2009). Hypertension in relation to post-traumatic stress disorder and depression in the US National Comorbidity Survey. *Behav Med, 34*(4), 125–132.

9. Boscarino, J. A., Forsberg, C. W., & Goldberg, J. (2010). A twin study of the association between PTSD symptoms and rheumatoid arthritis. *Psychosom Med, 72*(5), 481–486.
10. Qureshi, S. U., Pyne, J. M., Magruder, K. M., Schulz, P. E., & Kunik, M. E. (2009). The link between post-traumatic stress disorder and physical comorbidities: a systematic review. *Psychiatr Q, 80*(2), 87–97.
11. Neigh, G. N., & Ali, F. F. (2016). Co-morbidity of PTSD and immune system dysfunction: opportunities for treatment. *Curr Opin Pharmacol, 29*, 104–110.
12. O'Donovan, A., Cohen, B. E., Seal, K. H., et al. (2015). Elevated risk for autoimmune disorders in Iraq and Afghanistan veterans with post-traumatic stress disorder. *Biol Psychiatry, 77*(4), 365–374.
13. Bauer, M. E., Wieck, A., Lopes, R. P., Teixeira, A. L., & Grassi-Oliveira, R. (2010). Interplay between neuroimmunoendocrine systems during post-traumatic stress disorder: a minireview. *Neuroimmunomodulation, 17*(3), 192–195.
14. Pervanidou, P. (2008). Biology of post-traumatic stress disorder in childhood and adolescence. *J Neuroendocrinol, 20*(5), 632–638.
15. Lindqvist, D., Wolkowitz, O. M., Mellon, S., et al. (2014). Proinflammatory milieu in combat-related PTSD is independent of depression and early life stress. *Brain Behav Immun, 42*, 81–88.
16. Michopoulos, V., & Jovanovic, T. (2015). Chronic inflammation: a new therapeutic target for post-traumatic stress disorder? *Lancet Psychiatry, 2*(11), 954–955.
17. Michopoulos, V., Norrholm, S. D., & Jovanovic, T. (2015). Diagnostic biomarkers for post-traumatic stress disorder: promising horizons from translational neuroscience research. *Biol Psychiatry, 78*(5), 344–353.
18. Passos, I. C., Vasconcelos-Moreno, M. P., Costa, L. G., et al. (2015). Inflammatory markers in post-traumatic stress disorder: a systematic review, meta-analysis, and meta-regression. *Lancet Psychiatry, 2*(11), 1002–1012.
19. Pervanidou, P., Kolaitis, G., Charitaki, S., et al. (2007). Elevated morning serum interleukin (IL)-6 or evening salivary cortisol concentrations predict post-traumatic stress disorder in children and adolescents six months after a motor vehicle accident. *Psychoneuroendocrinology, 32*(8–10), 991–999.
20. Grassi-Oliveira, R., Brietzke, E., Pezzi, J. C., Lopes, R. P., Teixeira, A. L., & Bauer, M. E. (2009). Increased soluble tumor necrosis factor-alpha receptors in patients with major depressive disorder. *Psychiatry Clin Neurosci, 63*(2), 202–208.
21. Eraly, S. A., Nievergelt, C. M., Maihofer, A. X., et al. (2014). Assessment of plasma C-reactive protein as a biomarker of post-traumatic stress disorder risk. *JAMA Psychiatry, 71*(4), 423–431.
22. Breen, M. S., Maihofer, A. X., Glatt, S. J., et al. (2015). Gene networks specific for innate immunity define post-traumatic stress disorder. *Mol Psychiatry, 20*(12), 1538–1545.
23. Gill, J. M., Saligan, L., Lee, H., Rotolo, S., & Szanton, S. (2013). Women in recovery from PTSD have similar inflammation and quality of life as non-traumatized controls. *J Psychosom Res, 74*(4), 301–306.
24. Smith, A. K., Conneely, K. N., Kilaru, V., et al. (2011). Differential immune system DNA methylation and cytokine regulation in post-traumatic stress disorder. *Am J Med Genet B Neuropsychiatr Genet, 156B*(6), 700–708.

25. Uddin, M., Aiello, A. E., Wildman, D. E., et al. (2010). Epigenetic and immune function profiles associated with post-traumatic stress disorder. *Proc Natl Acad Sci U S A, 107*(20), 9470–9475.
26. Segman, R. H., Shefi, N., Goltser-Dubner, T., Friedman, N., Kaminski, N., & Shalev, A. Y. (2005). Peripheral blood mononuclear cell gene expression profiles identify emergent post-traumatic stress disorder among trauma survivors. *Mol Psychiatry, 10*(5), 500–513, 425.
27. Bam, M., Yang, X., Zhou, J., et al. (2016). Evidence for epigenetic regulation of pro-inflammatory cytokines, interleukin-12 and interferon gamma, in peripheral blood mononuclear cells from PTSD patients. *J Neuroimmune Pharmacol, 11*(1), 168–181.
28. Bam, M., Yang, X., Zumbrun, E. E., et al. (2016). Dysregulated immune system networks in war veterans with PTSD is an outcome of altered miRNA expression and DNA methylation. *Sci Rep, 6*, 31209.
29. Miller, A. H., Maletic, V., & Raison, C. L. (2009). Inflammation and its discontents: the role of cytokines in the pathophysiology of major depression. *Biol Psychiatry, 65*(9), 732–741.
30. Steiner, J., Walter, M., Gos, T., et al. (2011). Severe depression is associated with increased microglial quinolinic acid in subregions of the anterior cingulate gyrus: evidence for an immune-modulated glutamatergic neurotransmission? *J Neuroinflammation, 8*, 94.
31. Hardingham, G. E., Fukunaga, Y., & Bading, H. (2002). Extrasynaptic NMDARs oppose synaptic NMDARs by triggering CREB shut-off and cell death pathways. *Nat Neurosci, 5*(5), 405–414.
32. Spivak, B., Shohat, B., Mester, R., et al. (1997). Elevated levels of serum interleukin-1 beta in combat-related post-traumatic stress disorder. *Biol Psychiatry, 42*(5), 345–348.
33. Maes, M., Lin, A. H., Delmeire, L., et al. (1999). Elevated serum interleukin-6 (IL-6) and IL-6 receptor concentrations in post-traumatic stress disorder following accidental man-made traumatic events. *Biol Psychiatry, 45*(7), 833–839.
34. Kawamura, N., Kim, Y., & Asukai, N. (2001). Suppression of cellular immunity in men with a past history of post-traumatic stress disorder. *Am J Psychiatry, 158*(3), 484–486.
35. Sommershof, A., Aichinger, H., Engler, H., et al. (2009). Substantial reduction of naive and regulatory T cells following traumatic stress. *Brain Behav Immun, 23*(8), 1117–1124.
36. Sabioncello, A., Kocijan-Hercigonja, D., Rabatic, S., et al. (2000). Immune, endocrine, and psychological responses in civilians displaced by war. *Psychosom Med, 62*(4), 502–508.
37. Glover, D. A., Steele, A. C., Stuber, M. L., & Fahey, J. L. (2005). Preliminary evidence for lymphocyte distribution differences at rest and after acute psychological stress in PTSD-symptomatic women. *Brain Behav Immun, 19*(3), 243–251.
38. Morath, J., Gola, H., Sommershof, A., et al. (2014). The effect of trauma-focused therapy on the altered T cell distribution in individuals with PTSD: evidence from a randomized controlled trial. *J Psychiatr Res, 54*, 1–10.
39. Aiello, A. E., Dowd, J. B., Jayabalasingham, B., et al. (2016). PTSD is associated with an increase in aged T cell phenotypes in adults living in Detroit. *Psychoneuroendocrinology, 67*, 133–141.

40. Jergovic, M., Tomicevic, M., Vidovic, A., et al. (2014). Telomere shortening and immune activity in war veterans with post-traumatic stress disorder. *Prog Neuropsychopharmacol Biol Psychiatry, 54*, 275–283.
41. Yehuda, R., Giller, E. L., Southwick, S. M., Lowy, M. T., & Mason, J. W. (1991). Hypothalamic-pituitary-adrenal dysfunction in post-traumatic stress disorder. *Biol Psychiatry, 30*(10), 1031–1048.
42. Olff, M., Guzelcan, Y., de Vries, G. J., Assies, J., & Gersons, B. P. (2006). HPA- and HPT-axis alterations in chronic post-traumatic stress disorder. *Psychoneuroendocrinology, 31*(10), 1220–1230.
43. Ironson, G., Cruess, D., & Kumar, M. (2007). Immune and neuroendocrine alterations in post-traumatic stress disorder. In R. Ader (Ed.), *Psychoneuroimmunology* (4th ed., pp. 531–547). San Diego, CA: Academic Press.
44. Elzinga, B. M., Schmahl, C. G., Vermetten, E., van Dyck, R., & Bremner, J. D. (2003). Higher cortisol levels following exposure to traumatic reminders in abuse-related PTSD. *Neuropsychopharmacology, 28*(9), 1656–1665.
45. Morris, M. C., Compas, B. E., & Garber, J. (2012). Relations among post-traumatic stress disorder, comorbid major depression, and HPA function: a systematic review and meta-analysis. *Clin Psychol Rev, 32*(4), 301–315.
46. McFarlane, A. C., Barton, C. A., Yehuda, R., & Wittert, G. (2010). Cortisol response to acute trauma and risk of post-traumatic stress disorder. *Psychoneuroendocrinology, 36*(5), 720–727.
47. Pitts, K. P., Joksimovic, L., Steudte-Schmiedgen, S., Rohleder, N., & Wolf, J. M. (2016). Determinants of altered intracellular endocrine-immune interplay in Bosnian war refugees suffering from PTSD. *Biol Psychol, 118*, 1–7.
48. Rohleder, N., Joksimovic, L., Wolf, J. M., & Kirschbaum, C. (2004). Hypocortisolism and increased glucocorticoid sensitivity of pro-Inflammatory cytokine production in Bosnian war refugees with post-traumatic stress disorder. *Biol Psychiatry, 55*(7), 745–751.
49. Murgatroyd, C., Wu, Y., Bockmuhl, Y., & Spengler, D. (2010). Genes learn from stress: how infantile trauma programs us for depression. *Epigenetics, 5*(3), 194–199.
50. Blaze, J., Scheuing, L., & Roth, T. L. (2013). Differential methylation of genes in the medial prefrontal cortex of developing and adult rats following exposure to maltreatment or nurturing care during infancy. *Dev Neurosci, 35*(4), 306–316.
51. Murgatroyd, C., Patchev, A. V., Wu, Y., et al. (2009). Dynamic DNA methylation programs persistent adverse effects of early-life stress. *Nat Neurosci, 12*(12), 1559–1566.
52. Roth, T. L., Lubin, F. D., Funk, A. J., & Sweatt, J. D. (2009). Lasting epigenetic influence of early-life adversity on the BDNF gene. *Biol Psychiatry, 65*(9), 760–769.
53. Weaver, I. C., Cervoni, N., Champagne, F. A., et al. (2004). Epigenetic programming by maternal behavior. *Nat Neurosci, 7*(8), 847–854.
54. Zhang, T. Y., Hellstrom, I. C., Bagot, R. C., Wen, X., Diorio, J., & Meaney, M. J. (2010). Maternal care and DNA methylation of a glutamic acid decarboxylase 1 promoter in rat hippocampus. *J Neurosci, 30*(39), 13130–13137.
55. McGowan, P. O., Sasaki, A., D'Alessio, A. C., et al. (2009). Epigenetic regulation of the glucocorticoid receptor in human brain associates with childhood abuse. *Nat Neurosci, 12*(3), 342–348.

56. Klengel, T., Mehta, D., Anacker, C., et al. (2013). Allele-specific FKBP5 DNA demethylation mediates gene–childhood trauma interactions. *Nat Neurosci, 16*(1), 33–41.
57. Tyrka, A. R., Price, L. H., Marsit, C., Walters, O. C., & Carpenter, L. L. (2012). Childhood adversity and epigenetic modulation of the leukocyte glucocorticoid receptor: preliminary findings in healthy adults. *PLoS One, 7*(1), e30148.
58. Yang, X., Ewald, E. R., Huo, Y., et al. (2012). Glucocorticoid-induced loss of DNA methylation in non-neuronal cells and potential involvement of DNMT1 in epigenetic regulation of Fkbp5. *Biochem Biophys Res Commun, 420*(3), 570–575.
59. Felmingham, K., Williams, L. M., Whitford, T. J., et al. (2009). Duration of post-traumatic stress disorder predicts hippocampal grey matter loss. *Neuroreport, 20*(16), 1402–1406.
60. Jones, T., & Moller, M. D. (2011). Implications of hypothalamic-pituitary-adrenal axis functioning in post-traumatic stress disorder. *J Am Psychiatr Nurses Assoc, 17*(6), 393–403.
61. Yehuda, R., McFarlane, A. C., & Shalev, A. Y. (1998). Predicting the development of post-traumatic stress disorder from the acute response to a traumatic event. *Biol Psychiatry, 44*(12), 1305–1313.
62. Michopoulos, V., Vester, A., & Neigh, G. (2016). Post-traumatic stress disorder: A metabolic disorder in disguise? *Exp Neurol, 284*(Pt B), 220–229.
63. Nery, F. G., Monkul, E. S., Hatch, J. P., et al. (2008). Celecoxib as an adjunct in the treatment of depressive or mixed episodes of bipolar disorder: a double-blind, randomized, placebo-controlled study. *Hum Psychopharmacol, 23*(2), 87–94.
64. Fond, G., Hamdani, N., Kapczinski, F., et al. (2014). Effectiveness and tolerance of anti-inflammatory drugs' add-on therapy in major mental disorders: a systematic qualitative review. *Acta Psychiatr Scand, 129*(3), 163–179.
65. Levkovitz, Y., Fenchel, D., Kaplan, Z., Zohar, J., & Cohen, H. (2015). Early post-stressor intervention with minocycline, a second-generation tetracycline, attenuates post-traumatic stress response in an animal model of PTSD. *Eur Neuropsychopharmacol, 25*(1), 124–132.

14

Immunology of Eating Disorders

ADALIENE VERSIANI MATOS FERREIRA,
LAÍS BHERING MARTINS, NAYARA MUSSI MONTEZE,
GENEVIÈVE MARCELIN, AND KARINE CLÉMENT ■

INTRODUCTION

Eating disorders are characterized by the dysregulation in eating behavior leading to extreme increase or decrease in food intake that, in turn, causes change in body weight, adiposity, and physical health.[1,2] Anorexia nervosa (AN), bulimia nervosa (BN), and binge eating disorder (BED) are the three major eating disorders. These disorders usually begin at adolescence or young adulthood.[3] The prevalence of eating disorders in the United States is around 4.4%, and about 1.5 million people in Europe are affected by these conditions.[3–5] Eating disorders occur primarily in women, and approximately 13% of young women have related symptoms at some point of their lives.[3,4]

It is well known that food and nutrients can influence the immune response and modulate the protection against environmental factors such as microorganisms and other noxious insults.[6,7] For instance, the intake of macronutrients and specific vitamins and minerals is important for adequate production of immune cells.[7] Malnutrition, especially during childhood, is associated with atrophy of the bone marrow and thymus (primary sites of lymphocyte maturation), decrease of hematopoiesis, alteration of cytokine production, impairment of the growth of enterocytes, increasing intestinal permeability, and the risk of infections.[8] Excessive intake of nutrients or bioactive components can influence the immune response as well, by impairing the antioxidant/oxidant balance in the immune cells. Moreover, the excess of some micronutrients may deplete others, affecting the immune function. For instance, excessive consumption of

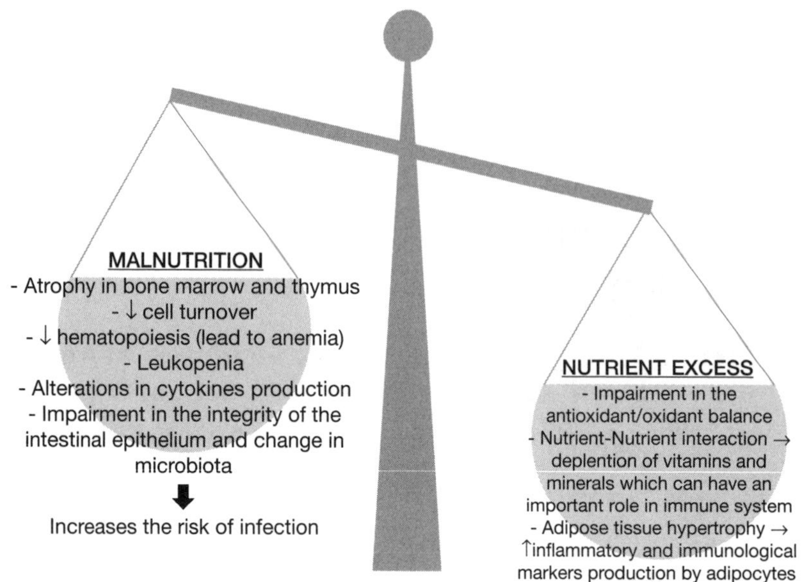

Figure 14.1 Immune consequences of malnutrition and overeating.

zinc results in lower bioavailability of copper, which is an essential co-factor for oxidation-reduction reactions.

Prolonged and excessive energy consumption may lead to adipose tissue hypertrophy and increase in the release of inflammatory molecules by adipocytes[7,9] (Figure 14.1). Obesity-related chronic low-grade inflammation also impairs insulin sensitivity and contributes to the development of metabolic (e.g., diabetes) and cardiovascular diseases.[10]

In this chapter, we review the current understanding of the immune imbalance that occurs in eating disorders.

ANOREXIA NERVOSA

AN is an eating disorder characterized by three main features: (1) significant low weight; (2) intense fear of gaining weight or of becoming fat, or persistent behavior that interferes with weight gain, even when the person has a very low weight; and (3) body image disturbance.[1] The lifetime prevalence of AN is around 3.6%.[11] AN occurs most commonly in adolescents and young women, with a 10:1 female-to-male ratio, and it is the third most prevalent chronic condition in adolescent girls.[11–13] Fewer than half of patients with AN reach full recovery, and 20% remain chronically ill, with a mortality rate around 5%.[14] These patients usually develop serious health complications such as osteoporosis, infertility, cardiovascular diseases, and thyroid and gastrointestinal diseases, which contribute to the high mortality rate in these patients.[13]

Peculiar immune abnormalities occur in parallel to malnutrition in AN.[15] Patients with AN usually develop leukopenia with lymphocytosis. Despite that, at least in the early phases of the AN, these patients seem to be resistant to common viral infections compared to patients with typical malnutrition, suggesting a qualitative change in lymphocyte populations.[16,17] Previous studies have shown an increase in CD4+ T lymphocyte proportion in patients with AN.[18,19] Satio et al. (2007) compared total lymphocyte count, lymphocyte subsets, and nutritional markers (body mass index [BMI], serum insulin-like growth factor-1 [IGF-I] and serum zinc) between 33 AN patients and 10 healthy controls. CD4+ proportion was increased, while total lymphocyte count, BMI, IGF-I, and zinc were decreased in AN.[19] Zinc is an important micronutrient involved in innate and adaptive immune responses.[20] Previous studies also observed lower urine levels of zinc in patients with AN, indicating deficiency of this nutrient.[21,22] Other nutrient deficiencies can occur due to inappropriate eating behaviors. For instance, vomiting, long periods of fasting, and laxative and diuretic misuse can cause hypokalemia, hypocalcemia, hypomagnesemia, and hypophosphatemia, with a significant impairment in metabolic homeostasis and possibly in immune response.[23]

A typical characteristic of the AN patient is the loss of the adipose tissue mass. As the adipose tissue acts as an endocrine organ producing several cytokines and adipokines that influence immune responses,[24] it is expected that the severe weight loss related to AN will lead to changes in the inflammatory and immunological status of these patients.[25] Low levels of leptin and high levels of adiponectin have consistently been observed in AN patients.[16,26,27] It is worth noting that this scenario is quite different from the one that happens in obesity, reflecting its direct association with the amount of adipose tissue mass.[28,29]

Adipokines play an important role in the regulation of immune pathways. Leptin deficiency alters inflammatory markers, macrophage function, and T helper cell response.[28] Furthermore, leptin is directly involved with the central regulation of the appetite.[26,28] Adiponectin is also involved with macrophage function, regulating macrophage proliferation, plasticity, and polarization.[29] Adiponectin modulates inflammation through downregulation of pro-inflammatory cytokines such as tumor necrosis factor (TNF) and interleukin-6 (IL-6), and upregulation of anti-inflammatory molecules such as IL-10 and arginase-1.[29] Despite the changes in the levels of leptin and adiponectin in AN and their role in the immune response, studies investigating the immune profile of AN patients have reported controversial results. According to a meta-analysis of cross-sectional and longitudinal studies including 22 studies with 924 participants, AN patients have higher levels of TNF, IL-1-β, IL-6, and soluble TNF receptor II, and lower

C-reactive protein (CRP) levels and soluble IL-6 receptors compared to heathy individuals.[30] Instead of a widespread increase of inflammatory mediators in AN, some mediators are even reduced.

It is well known that cytokines are potent cell-signaling molecules that have direct and indirect effects on the CNS, modulating several functions, including hunger and satiety.[31] TNF is a pro-inflammatory cytokine that reduces insulin sensitivity,[32] increases the production of other inflammatory cytokines such as IL-6 and IL-1β,[33] and stimulates lipolysis and weight loss.[34] Chronic elevation of TNF as seen in AN acts in anorexigenic pathways through the activation of the hypothalamus-pituitary-adrenocortical (HPA) axis, determining catecholamine release, glucocorticoid secretion, upregulation of anorexigenic neuropeptides (cholecystokinin, CCK; proopiomelanocortin, POMC; cocaine- and amphetamine-regulated transcript, CART), and downregulation of the orexigenic neuropeptides (beta-endorphin; neuropeptide Y, NPY; and vasoactive intestinal peptide, VIP).[31,35] Interestingly, a patient with AN treated with TNF antagonists (infliximab for four months followed by adalimumab for six months) had significant weight gain and improvement of psychopathological parameters.[36,37] In this context, a better understanding of the metabolic and immunological pathways involved in the AN physiopathology may provide interesting therapeutic venues in the future.

BULIMIA NERVOSA

BN was first described as a "chronic phase of anorexia nervosa" by Gerald Russell, a British psychiatrist.[38] BN was later recognized as a distinct condition with its own characteristics.[39] The estimated lifetime prevalence of BN is between 0.9% and 3%, being more prevalent in Hispanic/Latino population.[40] Three features are essential for BN diagnosis: (1) recurrent episodes of binge eating; (2) recurrent, inappropriate compensatory behaviors to prevent weight gain; and (3) self-evaluation that is unduly influenced by body shape and weight. The most common compensatory behavior is self-induced vomiting, but other purging behaviors may occur, such as misuse of laxatives and diuretics, fasting, or excessive exercise.[1] Compensatory behaviors are responsible for a variety of medical complications in patients with BN, including cutaneous manifestations (e.g., alopecia, xerosis, hypertrichosis, and nail fragility), and dental and gastrointestinal problems.[41] Laboratory abnormalities may also occur, especially in low-weight bulimics and those who have severe purging behaviors.[42] For instance, electrolyte imbalance can occur due to repeated episodes of vomiting.[41]

Nagata et al. (2006) reported that BN patients had their lymphocyte proliferative response, evaluated by the mitogenic response of T lymphocytes to phytohemagglutinin A (PHA), negatively correlated with anxiety symptoms. No difference in specific subpopulations of lymphocytes was found in BN patients when compared to controls, unlike what is observed in AN.[43,44]

The results regarding peripheral levels of cytokines are still controversial. Nakai et al. (2000) observed significantly higher levels of TNF in BN patients when compared with controls.[45] However, other studies did not find any difference in the serum levels of cytokines like IL-1, IL-2, IL-6, and TNF.[43,46,47] BN involves several cyclical stages of starvation and overeating that may influence the study of immune parameters. Indeed, feeding and fasting states are important determinants of cytokine and immune cell profiles.[42,43] Another confounding factor is psychiatric comorbidity. More than 60% of patients with BN have depressive and/or anxiety disorders[42] that may also influence the immune function.

In sum, due to the limited availability of studies and lack of adequate control of confounding factors, it is not possible to draw definite conclusions on immune dysfunction in BN.

BINGE-EATING DISORDER

BED is characterized by recurrent episodes of overeating in a short period of time without a sense of control, often very quickly and to the point of discomfort. Unlike AN and BN, BED is not associated with a compensatory strategy to reduce the body weight gain.[1] BED seems to be the most prevalent eating disorder, affecting more women than men.[48] In the adult population, the prevalence varies from 2% to 5%, rising up to 50% among obese adults seeking weight reduction.[1] Although BED is distinct from obesity from a phenomenological standpoint, there is a robust link between these two conditions.

Overnutrition can influence the immune response directly or indirectly with a subsequent high risk of infections and certain types of cancer.[49] BED patients show depressed immune function, possibly due to factors associated with body weight gain, such as the higher amounts of fat and micronutrient intake and the expansion of the adipose tissue.[49]

Elevated amounts of fats and determined essential micronutrients such as zinc and iron are likely to impair immunity.[50,51] High fat intake alters the profile of lymphocytes and cell membrane composition, specifically the lymphoid phospholipid composition, affecting the activity of several membrane receptors. Regarding micronutrients, men who ingested 150 mg of elemental zinc twice a day for six weeks showed lower lymphocyte proliferation as well as lower chemotaxis and phagocytosis, compared with baseline parameters.[52] Zinc supplementation was also associated with a reduction of T cell functions and apoptosis of B cells.[53]

Obesity is characterized by excess body fat and abnormal cytokine production, associated with a chronic low-grade inflammation.[24,54] Adipose tissue is an important organ that regulates many physiological processes, mainly due to some molecules called adipocytokines that are produced by this tissue. Adiponectin, leptin, TNF, IL-6, IL-1, and CC-chemokine ligand 2 (CCL2; also known as MCP-1) are the main cytokines produced by adipocytes.[24,55] The adipose tissue expansion increases the synthesis of acute-phase reactants, such as the C-reactive protein (CRP), and the activation of pro-inflammatory signaling pathways. Furthermore, during adipose expansion, the number of macrophages, CD8+ T cells, and CD4+ T cells expressing inflammatory cytokines and chemokines such as TNF, CCL2, IL-6, IFN-γ, and IL-17 also increase substantially.[54] Adipocytes contribute to roughly one-third of the IL-6 levels in the circulation of patients who are obese, while macrophages in adipose tissue seem to be the main source of TNF.[24]

Adipocyte-derived CCL2 has been identified as a potential factor contributing to macrophage infiltration into adipose tissue. Once macrophages are in the adipose tissue, they can perpetuate a vicious cycle of macrophage recruitment and production of pro-inflammatory cytokines. As a consequence, this "adipocytokine–cytokine cocktail" might favor a pro-inflammatory milieu,[24,54,55] expressed, for instance, as a chronic low-grade inflammation seen in obese patients, which, in turn, is associated with lower insulin sensitivity and metabolic disease development.[10]

It is important to notice that recent studies have also shown that inflammation during obesity might have beneficial effects. Inflammation would have an essential role in maintaining a healthy adipose tissue expansion through tissue remodeling, increased energy expenditure, and suppressed food intake. Altogether, these events would favor the development of insulin sensitivity.[56–61] Studies in rodents and humans have shown that the blockade of pro-inflammatory cytokine signaling by using pharmacological agents or knockout mice promoted increases in body weight and adiposity.[58,60,62]

Immune function parameters in people with BED still need to be investigated. Due to the significant overlap between BED and obesity, a careful control for obesity-related confounding factors will be necessary.

CONCLUSION

The bidirectional influence of the immune system on the metabolism, and vice versa, has been acknowledged. This field is still in its infancy, and further understanding of specific immune pathways that are involved in the control of eating and fasting behaviors needs to be unveiled. In the future, with the advance of immunotherapeutic approaches, other strategies may help the treatment of inflammation-related diseases and eating disorders.

REFERENCES

1. American Psychiatric Association. *Diagnostic and Statistical Manual of Mental Disorders, Fifth Edition*. Arlington, VT: American Psychiatric Association; 2013.
2. Traviss-Turner GD, West RM, Hill AJ. Guided self-help for eating disorders: a systematic review and metaregression. *Eur Eat Disord Rev*. 2017;25(3):148–164.
3. Hudson JI, Hiripi E, Pope HG, Jr., Kessler RC. The prevalence and correlates of eating disorders in the National Comorbidity Survey Replication. *Biol Psychiatry*. 2007;61(3):348–358.
4. Stice E, Marti CN, Rohde P. Prevalence, incidence, impairment, and course of the proposed DSM-5 eating disorder diagnoses in an 8-year prospective community study of young women. *J Abnorm Psychol*. 2013;122(2):445–457.
5. Gustavsson A, Svensson M, Jacobi F, Allgulander C, Alonso J, Beghi E, et al. Cost of disorders of the brain in Europe 2010. *Eur Neuropsychopharmacol*. 2011;21(10):718–779.
6. Calder PC, Kew S. The immune system: a target for functional foods? *Br J Nutr*. 2002;88(Suppl 2):S165–S177.
7. Alpert P. The role of vitamins and minerals in the immune system. *Home Health Care Manag Pract*. 2017;29(3):1–4.
8. França TGD, Ishikawa LLW, Zorzella-Pezavento SFG, F. C, da Cunha MLRS, Sartori A. Impact of malnutrition on immunity and infection. *J Venom Anim Toxins Incl Trop Dis* 2009;15(3):375.
9. Rogge MM. The case for an immunologic cause of obesity. *Biol Res Nurs*. 2002;4(1):43–53.
10. Shoelson SE, Herrero L, Naaz A. Obesity, inflammation, and insulin resistance. *Gastroenterology*. 2007;132(6):2169–2180.
11. Mustelin L, Silen Y, Raevuori A, Hoek HW, Kaprio J, Keski-Rahkonen A. The DSM-5 diagnostic criteria for anorexia nervosa may change its population prevalence and prognostic value. *J Psychiatr Res*. 2016;77:85–91.
12. World Health Organization. *International Classification of Diseases (10th revision)*. Geneva: WHO; 1990.
13. Meczekalski B, Podfigurna-Stopa A, Katulski K. Long-term consequences of anorexia nervosa. *Maturitas*. 2013;75(3):215–220.
14. Steinhausen HC. The outcome of anorexia nervosa in the 20th century. *Am J Psychiatry*. 2002;159(8):1284–1293.
15. Marcos A. Eating disorders: a situation of malnutrition with peculiar changes in the immune system. *Eur J Clin Nutr*. 2000;54(Suppl 1):S61–S64.
16. Omodei D, Pucino V, Labruna G, Procaccini C, Galgani M, Perna F, et al. Immune-metabolic profiling of anorexic patients reveals an anti-oxidant and anti-inflammatory phenotype. *Metabolism*. 2015;64(3):396–405.

17. Nova E, Marcos A. Immunocompetence to assess nutritional status in eating disorders. *Expert Rev Clin Immunol.* 2006;2(3):433–444.
18. Nagata T, Kiriike N, Tobitani W, Kawarada Y, Matsunaga H, Yamagami S. Lymphocyte subset, lymphocyte proliferative response, and soluble interleukin-2 receptor in anorexic patients. *Biol Psychiatry.* 1999;45(4):471–474.
19. Saito H, Nomura K, Hotta M, Takano K. Malnutrition induces dissociated changes in lymphocyte count and subset proportion in patients with anorexia nervosa. *Int J Eat Disord.* 2007;40(6):575–579.
20. Hojyo S, Fukada T. Roles of zinc signaling in the immune system. *J Immunol Res.* 2016;2016:6762343.
21. Katz RL, Keen CL, Litt IF, Hurley LS, Kellams-Harrison KM, Glader LJ. Zinc deficiency in anorexia nervosa. *J Adolesc Health Care.* 1987;8(5):400–406.
22. McClain CJ, Stuart MA, Vivian B, McClain M, Talwalker R, Snelling L, et al. Zinc status before and after zinc supplementation of eating disorder patients. *J Am Coll Nutr.* 1992;11(6):694–700.
23. Winston AP. The clinical biochemistry of anorexia nervosa. *Ann Clin Biochem.* 2012;49(s):132–143.
24. Tilg H, Moschen AR. Adipocytokines: mediators linking adipose tissue, inflammation and immunity. *Nat Rev Immunol.* 2006;6(10):772–783.
25. Pisetsky DS, Trace SE, Brownley KA, Hamer RM, Zucker NL, Roux-Lombard P, et al. The expression of cytokines and chemokines in the blood of patients with severe weight loss from anorexia nervosa: an exploratory study. *Cytokine.* 2014;69(1):110–115.
26. Baranowska-Bik A, Baranowska B, Martynska L, Litwiniuk A, Kalisz M, Kochanowski J, et al. Adipokine profile in patients with anorexia nervosa. *Endokrynol Pol.* 2017;68(4):422–429.
27. Dostalova I, Smitka K, Papezova H, Kvasnickova H, Nedvidkova J. Increased insulin sensitivity in patients with anorexia nervosa: the role of adipocytokines. *Physiol Res.* 2007;56(5):587–594.
28. Bluher S, Shah S, Mantzoros CS. Leptin deficiency: clinical implications and opportunities for therapeutic interventions. *J Invest Med.* 2009;57(7):784–788.
29. Luo Y, Liu M. Adiponectin: a versatile player of innate immunity. *J Mol Cell Biol.* 2016;8(2):120–128.
30. Solmi M, Veronese N, Favaro A, Santonastaso P, Manzato E, Sergi G, et al. Inflammatory cytokines and anorexia nervosa: a meta-analysis of cross-sectional and longitudinal studies. *Psychoneuroendocrinology.* 2015;51:237–252.
31. Corcos M, Guilbaud O, Paterniti S, Moussa M, Chambry J, Chaouat G, et al. Involvement of cytokines in eating disorders: a critical review of the human literature. *Psychoneuroendocrinology.* 2003;28(3):229–249.
32. Stephens JM, Pekala PH. Transcriptional repression of the GLUT4 and C/EBP genes in 3T3-L1 adipocytes by tumor necrosis factor-alpha. *J Biol Chem.* 1991;266(32):21839–845.
33. McArdle MA, Finucane OM, Connaughton RM, McMorrow AM, Roche HM. Mechanisms of obesity-induced inflammation and insulin resistance: insights into the emerging role of nutritional strategies. *Front Endocrinol (Lausanne).* 2013;4:52.

34. Ryden M, Arvidsson E, Blomqvist L, Perbeck L, Dicker A, Arner P. Targets for TNF-alpha-induced lipolysis in human adipocytes. *Biochem Biophys Res Commun.* 2004;318(1):168–175.
35. Holden RJ, Pakula IS. Tumor necrosis factor-alpha: is there a continuum of liability between stress, anxiety states and anorexia nervosa? *Med Hypotheses.* 1999;52(2):155–162.
36. Solmi M, Santonastaso P, Caccaro R, Favaro A. A case of anorexia nervosa with comorbid Crohn's disease: beneficial effects of anti-TNF-alpha therapy? *Int J Eat Disord.* 2013;46(6):639–641.
37. Barber J, Sheeran T, Mulherin D. Anti-tumour necrosis factor treatment in a patient with anorexia nervosa and juvenile idiopathic arthritis. *Ann Rheum Dis.* 2003;62(5):490–491.
38. Russell G. Bulimia nervosa: an ominous variant of anorexia nervosa. *Psychol Med.* 1979;9(3):429–448.
39. Russell G. Thoughts on the 25th anniversary of bulimia nervosa. *Eur Eat Disord Rev.* 2004;12(3):139–152.
40. Castillo M, Weiselberg E. Bulimia nervosa/purging disorder. *Curr Probl Pediatr Adolesc Health Care.* 2017;47(4):85–94.
41. Mehler PS, Rylander M. Bulimia nervosa—medical complications. *J Eat Disord.* 2015;3:12.
42. Rushing JM, Jones LE, Carney CP. Bulimia nervosa: a primary care review. *Prim Care Companion J Clin Psychiatry.* 2003;5(5):217–224.
43. Nagata T, Yamada H, Iketani T, Kiriike N. Relationship between plasma concentrations of cytokines, ratio of CD4 and CD8, lymphocyte proliferative responses, and depressive and anxiety state in bulimia nervosa. *J Psychosom Res.* 2006;60(1):99–103.
44. Pirke K, Nerl C, Krieg J, Fichter M. Immunological findings in anorexia and bulimia nervosa. *Int J Eat Disord.* 1990;11(2):185–189.
45. Nakai Y, Hamagaki S, Takagi R, Taniguchi A, Kurimoto F. Plasma concentrations of tumor necrosis factor-alpha (TNF-alpha) and soluble TNF receptors in patients with bulimia nervosa. *Clin Endocrinol (Oxf).* 2000;53(3):383–388.
46. Nagata T, Yamada H, Iketani T, Kiriike N. Relationship between plasma concentrations of cytokines, ratio of CD4 and CD8, lymphocyte proliferative responses, and depressive and anxiety state in bulimia nervosa. *J Psychosom Res.* 2006;60:99–103.
47. Monteleone P, Maes M, Fabrazzo M, Tortorella A, Lin A, Bosmans E, et al. Immunoendocrine findings in patients with eating disorders. *Neuropsychobiology.* 1999;40(3):115–120.
48. Cuesto G, Everaerts C, Leon LG, Acebes A. Molecular bases of anorexia nervosa, bulimia nervosa and binge eating disorder: shedding light on the darkness. *J Neurogenet.* 2017:1–22.
49. Samartína S, Chandraa R. Obesity, overnutrition and the immune system. *Nutr Res.* 2001;21(1–2):243–262.
50. Kumari BS, Chandra RK. Overnutrition and immune responses. *Nutr Res.* 1993;13:3–18.
51. Chandra RK, McBean LD. Zinc and immunity. *Nutrition.* 1994;10(1):79–80.

52. Chandra RK. Excessive intake of zinc impairs immune responses. *JAMA*. 1984;252(11):1443–1446.
53. Ibs KH, Rink L. Zinc-altered immune function. *J Nutr*. 2003;133(5 Suppl 1):1452S–1456S.
54. Kau AL, Ahern PP, Griffin NW, Goodman AL, Gordon JI. Human nutrition, the gut microbiome and the immune system. *Nature*. 2011;474(7351):327–336.
55. Monteleone P, Di Lieto A, Tortorella A, Longobardi N, Maj M. Circulating leptin in patients with anorexia nervosa, bulimia nervosa or binge-eating disorder: relationship to body weight, eating patterns, psychopathology and endocrine changes. *Psychiatry Res*. 2000;94(2):121–129.
56. Ye J, McGuinness OP. Inflammation during obesity is not all bad: evidence from animal and human studies. *Am J Physiol Endocrinol Metab*. 2013;304(5):E466–E477.
57. Pang C, Gao Z, Yin J, Zhang J, Jia W, Ye J. Macrophage infiltration into adipose tissue may promote angiogenesis for adipose tissue remodeling in obesity. *Am J Physiol Endocrinol Metab*. 2008;295(2):E313–E322.
58. Wallenius V, Wallenius K, Ahren B, Rudling M, Carlsten H, Dickson SL, et al. Interleukin-6-deficient mice develop mature-onset obesity. *Nat Med*. 2002;8(1):75–79.
59. Oliveira MC, Menezes-Garcia Z, Henriques MC, Soriani FM, Pinho V, Faria AM, et al. Acute and sustained inflammation and metabolic dysfunction induced by high refined carbohydrate-containing diet in mice. *Obesity (Silver Spring)*. 2013;21(9):E396–E406.
60. Menezes-Garcia Z, Oliveira MC, Lima RL, Soriani FM, Cisalpino D, Botion LM, et al. Lack of platelet-activating factor receptor protects mice against diet-induced adipose inflammation and insulin-resistance despite fat pad expansion. *Obesity (Silver Spring)*. 2014;22(3):663–672.
61. Kotas ME, Medzhitov R. Homeostasis, inflammation, and disease susceptibility. *Cell*. 2015;160(5):816–827.
62. Briot K, Garnero P, Le Henanff A, Dougados M, Roux C. Body weight, body composition, and bone turnover changes in patients with spondyloarthropathy receiving anti-tumour necrosis factor (alpha) treatment. *Ann Rheum Dis*. 2005;64(8):1137–1140.

15

Immunology of Late-Life Psychiatric Disorders and Dementias

NATALIA P. ROCHA, ERICA LEANDRO VIEIRA, BRENO SATLER DINIZ, AND ANTONIO L. TEIXEIRA ■

INTRODUCTION

Population aging is a worldwide phenomenon. Virtually every country is experiencing an increase in the proportion of elderly people due to fertility rate decline and life expectancy rise. Therefore, the population aging is considered one of the most significant social challenges in the 21st century, increasing the demand for care, services and technologies to prevent and treat chronic conditions associated with aging.[1]

Psychiatric disorders are major contributors to the loss of independence of older adults. Neurocognitive disorders are the most common psychiatric disorders in late life. The prevalence of dementia is estimated to be 5–10% in the elderly population (defined as 60+ or 65+ years in developing and developed countries, respectively).[2] Around 47 million people have dementia worldwide, and there are nearly 10 million new cases every year. In 2015, dementia was the seventh leading cause of death. There is a huge economic impact associated with dementia, with current costs estimated at $818 billion per year.[3] Alzheimer's disease (AD) is the most common type of dementia (62% of cases), followed by vascular dementia (VaD, 17%), mixed dementia (i.e., AD and VaD, 10%), dementia with Lewy bodies (DLB, 4%), and frontotemporal dementia (FTD, 2%).[4]

Information on the prevalence of other psychiatric disorders in older adults is limited. Data from a large-scale study conducted in the United States from 1980–1984 revealed a 12.3% prevalence of any psychiatric disorder among the non-institutionalized population aged 65 years or older. As expected, the prevalence of psychiatric disorders is higher in clinical and institutional samples, ranging from 68–94%.[5] A more recent and comprehensive study investigating DSM-IV psychiatric disorders in older adults revealed a 6.8% prevalence of any mood disorder among adults aged 55 years and older in the United States. The most prevalent mood disorder was major depression (5.6%). A higher proportion of older adults reported any anxiety disorders (11.4%), specific phobias being the most common (5.8%). The prevalence of any substance use disorders among older adults was 3.8%.[6] Although most psychiatric disorders occur less frequently in older populations when compared with adults of other age ranges, it is worth mentioning that many elderly subjects present with symptoms that do not meet the formal diagnosis criteria for a particular psychiatric disorder but are clinically significant, suggesting that current definitions may be less applicable for older adults.[5]

Psychiatric disorders seem to increase the risk of dementia. Schizophrenia has been associated with increased risk of dementia, especially when diagnosed in older age. For instance, patients with schizophrenia with late- or very-late first contact with the psychiatric system are at two to three times higher risk of subsequently receiving a diagnosis of dementia compared to patients with musculoskeletal diseases or the general population.[7]

Depression also has a complex and probably bidirectional association with dementia. Patients with mild cognitive impairment (MCI) or dementia present with increased frequency of depressive disorders when compared with non–cognitive impaired people of the same age group. Episodes of significant depressive symptoms, especially when recurrent, were associated with increased risk for dementia.[8] The coexistence of depression and MCI in older people is associated with an increased risk of dementia, as 20–60% of them develop AD within a few years after the onset of depression.[9] Based on the assumption that the underlying neuropathological condition that causes MCI or dementia is also responsible for depressive symptoms, it has been hypothesized that depression in late life is an early sign of neurocognitive disorders rather than their cause. In this scenario, late-life depression (LLD), MCI, and dementia could constitute a clinical continuum.[10]

AGING-RELATED BIOLOGICAL CHANGES

Aging is a complex process associated with restructuring and/or deterioration of physiological functions across all body systems,[11] especially the central nervous system (CNS), the endocrine, and the immune systems (Figure 15.1). The CNS is one the most affected by aging due to its limited capacity of neuronal regeneration. Age-associated changes in the CNS result, for instance, in slowing in motor and cognitive processes.[12] However, these changes and the consequent functional impairment vary from person to person, based on different genetic backgrounds and lifestyles.

Regarding the endocrine system, the levels of most hormones, including estrogen, testosterone, growth hormone, and insulin-like growth factor-I, decrease with aging. Accordingly, the aging process is often characterized by reduced protein synthesis, decrease in lean body mass and bone mass, increased fat mass and insulin resistance, higher cardiovascular disease risk, fatigue, decreased libido, and erectile dysfunction.[13]

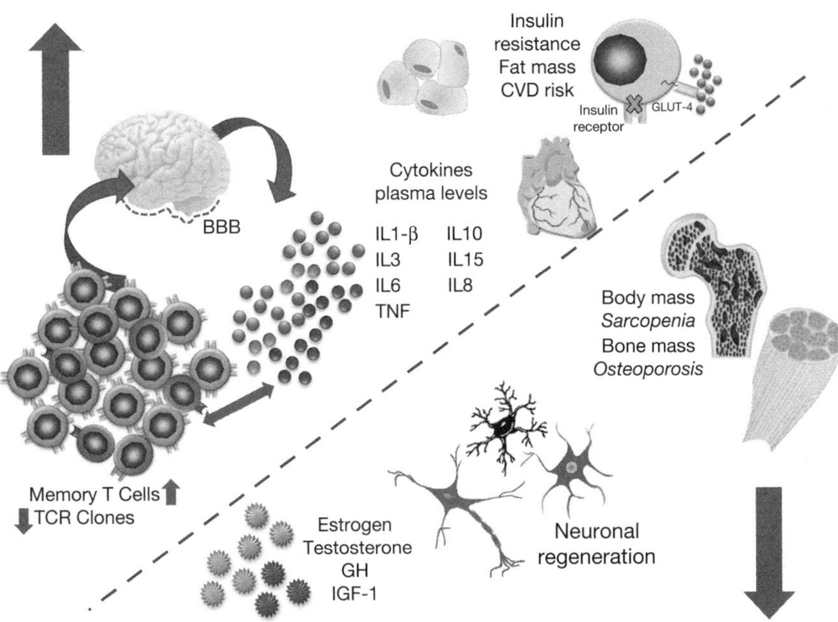

Figure 15.1 *Most important age-associated changes involving the central nervous system (CNS), the endocrine, and the immune systems.* Several functions are increased (upper part of the figure) or decreased (lower part) with aging. While the capacity of neuronal regeneration; speed of motor and cognitive processes; levels of estrogen, testosterone, growth hormone (GH), and insulin-like growth factor-I (IGF-I); body mass (sarcopenia), and bone mass (osteoporosis) decrease: fat mass, insulin resistance, cardiovascular disease (CVD) risk, fatigue, levels of inflammatory cytokines (*inflammaging*), and number of memory T cells increase. Interestingly, the increase of memory T cells is associated with the reduction of TCR clones and an "exhaustion" of the naïve T cells, a scenario that predisposes to recurrent infections and autoimmunity. There is also impairment of the blood–brain barrier (BBB).

Along with endocrine changes, the immune system undergoes profound alterations with age. Both the innate and adaptive branches of the immune system modify with aging, determining greater susceptibility to infectious diseases and reduced response to vaccination in elderly subjects.[14] The age-related changes in the immune system are collectively known as *immunosenescence*. The hallmark of this phenomenon is the alteration in T cell subsets with shrinkage of T cell repertoires. There is an accumulation of memory T cells, or T cell clones, and an exhaustion of the naïve T cell pool. In addition, T cell proliferation and cooperation capacities are reduced.[11] As a consequence, advanced age is marked by a decrease in the capacity to cope with stressors. Paradoxically, aging is associated with increased susceptibility to autoimmune diseases.[15] For instance, the incidence of rheumatoid arthritis significantly increases with aging, going from 7.3/100,000 in the 18–34-year age group to 107.3/100,000 in the 75–84-year age group.[16] The multifaceted immunosenescence profile also includes a constitutive low-grade inflammation that may contribute to the development and/or progression of chronic diseases, including cardiovascular and neurodegenerative diseases, and age-specific ailments such as frailty.[17] This low-grade chronic inflammation observed in aging is a phenomenon known as *inflammaging*.

INFLAMMAGING: AGING-RELATED LOW-GRADE INFLAMMATION

The term "inflammaging" was introduced by Franceschi and colleagues almost two decades ago to describe the progressive increase in the pro-inflammatory status observed with aging. This phenomenon would be determined by long-standing antigenic load and stress throughout life.[18] The age-related pro-inflammatory status, or inflammaging, is characterized by low-grade, systemic, chronic, and subclinical inflammation,[18] being associated with the pathophysiology of most age-related diseases. Atherosclerosis, AD, osteoporosis, type 2 diabetes mellitus, and sarcopenia/frailty syndrome are age-related conditions that exhibit very distinct clinical phenotypes but share inflammatory mechanisms in their pathogenesis.[19] Other common conditions among elderly people such as depression are also associated with elevated levels of inflammatory markers, highlighting the pivotal role played by inflammation in age-related diseases.

The aforementioned low-grade inflammation is evidenced by a two- to four-fold increase in the circulating levels of inflammatory mediators in the elderly, such as cytokines and acute phase proteins.[20] A range of studies has described increases in the circulating levels of different molecules, including interleukin

(IL)-3, IL-6, IL-8, IL-10, IL-15, and tumor necrosis factor (TNF).[11] IL-6 is the cytokine most consistently associated with aging. IL-6 levels are low or undetectable in most young people, increasing stepwise with age.[11,21]

A wide range of factors has been implicated in the development of this low-grade inflammation, including an increased amount of fat tissue, decreased production of sex steroid hormones, lifestyle habits, and chronic diseases. A very important factor seems to be the chronic load of antigenic stress, which affects the immune system throughout life with the progressive activation of immune cells and reduction of naïve cell repertoire.[18]

INFLAMMAGING AND AGE-ASSOCIATED DISEASES

The increase in lifespan is associated with higher incidence and prevalence of diseases such as cardiovascular diseases, sarcopenia, cognitive impairment, and dementia. The imbalance in the production and release of cytokines and the maintenance of a pro-inflammatory state might contribute to the pathogenesis and/or progression of these diseases. Therefore, inflammaging seems to be associated with increased morbidity and mortality in the elderly.

It is still controversial whether aging-related inflammation plays a causal or a counter-regulatory role, and how systemic low-grade inflammation affects peripheral tissues and the CNS.[22] It is worth emphasizing that the changes associated with inflammaging are part of the physiological aging processes. However, some individuals are capable of coping better with these changes than others. Genetic predisposition, lifestyle-related factors, and interaction among them may explain why individuals are more or less prone to develop age-related diseases.

The aging-related increase in inflammatory mediators has been linked not only to morbidity, but also to increased mortality in the elderly population. Higher circulating levels of IL-6 and C-reactive protein (CRP) were associated with mortality in non-disabled participants in the Iowa 65+ Rural Health Study.[23] In addition to IL-6 and CRP, low-grade increase in the levels of circulating TNF-α and soluble IL-2 receptor (sIL2R), as well as low levels of albumin, are strong predictors of mortality risk in longitudinal studies. These findings were independent of preexisting morbidity and other traditional risk factors for death, suggesting that inflammatory mediators may have direct biological effects [for a review, see [22]]. A prospective study with 80-year-old people found that baseline serum TNF-α levels were associated with mortality in men, but not in women. Low-grade increase in serum IL-6 was a strong predictor of all-cause mortality in both sexes during the following period of six years.[24] Detectable serum levels of TNF-α were found to be predictors of earlier mortality in elderly institutionalized patients.[25] A study with centenarians found that TNF-α, but not IL-6 or IL-8, was an independent prognostic marker for mortality. CRP had an effect that disappeared when TNF-α was included in this analysis.[21] Whether different inflammatory mediators have distinct effects in elderly population remains to be determined.

LATE-LIFE DEPRESSION AND IMMUNE CHANGES

Cytokines have been extensively implicated in the pathophysiology of sickness behavior. The term "sickness behavior" refers to a series of cytokine-induced behavioral changes originally described in the context of infectious diseases. These symptoms include lethargy, fatigue, psychomotor slowing, reduced appetite, sleep alterations, and inattention, among others, resembling symptoms typical of a depressive episode.[26] The recognition of the similarities between sickness behavior and depression prompted the investigation of inflammatory/immune mechanisms in the pathophysiology of depression.

Major depressive episodes are common in the elderly population, with one-year prevalence rates ranging from 5–10%.[27] The underlying biological mechanisms are not fully understood, but depression in the elderly involves significant changes in the immune-inflammatory responses.

Depressive disorders have been associated with innate immunity hyperactivity and increases in circulating levels of inflammatory mediators—mainly cytokines such as IL-6, IL-1β, and TNF-α, but also CRP—the same features observed in inflammaging.[26,28,29] Given the increase in cytokine concentrations with age, the elevated frequency of depression in elderly populations is not surprising.[19] Studies consistently reported higher circulating levels of inflammatory mediators such as soluble TNF receptor (sTNFR)2[29] and IL-1β[28] in antidepressant-free patients with late-life depression, even when compared with non-depressed elderly subjects, corroborating the hypothesis that inflammaging might be involved in the development of late-life depression as well.

An important facet of late-life depression is its relationship with neurocognitive disorders and an increased risk of dementia.[30,31] Using proteomic approaches, studies have shown that biomarkers and biological pathways related to immune-inflammatory control are related to cognitive impairment in late-life depression.[32–34] Changes in the same biomarkers and biological pathways are also reported in patients with AD (see further in this chapter). Altogether, these findings suggest that abnormalities in immune-inflammatory response are a putative biological link between late-life depression and increased risk of dementia in older adults.

DEMENTIA AND IMMUNE CHANGES

Cognitive decline is one of the most disabling age-related disorders due to the progressive loss of independence and functioning. Several studies have been conducted in order to evaluate the association between chronic elevation of inflammatory cytokines and cognitive functions (for a review, see [35]). Cross-sectional studies have found that increased circulating levels of CRP,[36] pro-inflammatory cytokines such as IL-1β,[37] and TNF-α[38] and its receptors[39] are associated with cognitive impairment and dementia, especially AD. Increased levels of inflammatory mediators have also been prospectively associated with cognitive decline and AD. For instance, in the community-based Framingham Study cohort, elderly subjects presenting with increased production of IL-1β or TNF-α by peripheral blood mononuclear cells were at higher risk of developing AD.[40] Higher serum levels of IL-6 and CRP were associated with poorer cognitive performance at baseline and a greater risk of cognitive impairment after two years of follow-up.[41] The same group showed in a five-year prospective observational study that participants with metabolic syndrome and high inflammation, defined as a combination of CRP and IL-6 levels, were more likely than those without metabolic syndrome to develop cognitive impairment. Interestingly, subjects presenting with metabolic syndrome but low inflammation did not exhibit an increased likelihood of cognitive impairment, suggesting an important role for inflammation in metabolic syndrome–related cognitive impairment.[42]

The elevation of inflammatory markers associated with MCI (which can be regarded as a prodromal phase of AD) in comparison with elderly subjects without cognitive impairment suggests that inflammation may be an early marker for the neurodegenerative cascades associated with AD. Moreover, there is a consistent association between depression and cognitive impairment in the elderly population. It is well known that the presence of depressive symptoms is a risk factor for the progression to dementia in subjects with MCI.[30] Inflammation is one of main mechanisms shared by both depression and dementia.[33] A recent meta-analysis found that peripheral levels of IL-1β, IL-2, IL-6, IL-18, sTNFR1, sTNFR2, homocysteine, high-sensitivity CRP (hsCRP), interferon (IFN)-γ, CXCL-10, epidermal growth factor (EGF), vascular cell adhesion molecule-1 (VCAM-1), α1-antichymotrypsin, and transferrin are increased in patients with AD compared with healthy controls, emphasizing the role of peripheral inflammation in AD.[43]

The pathological hallmarks of AD are abnormal amyloid beta (Aβ) accumulation in the intercellular space and the presence of neurofibrillary tangles composed of aggregates of hyperphosphorylated tau protein inside neurons. Although the formation of amyloid plaques and neurofibrillary tangles is regarded as the main contributor to the neurodegeneration in AD, interventions targeting these specific elements have been limited in modulating the clinical trajectory of the disease or modifying its natural history. While Aβ and tau remain strong candidates for AD therapy, additional mechanisms may be involved in the development and/or progression of the disease.

Mounting evidence suggests that immune changes/inflammatory processes are involved in AD pathology from the early stages of disease, when disease-modifying therapies might be more effective.[43] Postmortem studies have found increased expression of inflammatory mediators in the brain of patients with AD. The first studies showed the presence of complement proteins[44] and the expression of major histocompatibility complex (MHC) class II molecules on microglia surrounding amyloid plaques[45] (reviewed in [46]). These findings were corroborated by further studies that found increased expression of cytokines, chemokines, and complement proteins in AD brains. Not only immunohistochemistry-based studies but also western-blot, enzyme-linked immunosorbent assay, and mRNA assessments confirmed the increase in inflammatory/immune mediators in the brains of people who suffered from AD (reviewed in [47]). More recently, studies using positron emission tomography (PET) techniques have demonstrated increased *in vivo* microglial activation in patients with AD, even in early phases of cognitive impairment.[48] The inflammation-related changes in the CNS of patients were also confirmed by studies evaluating cerebrospinal fluid (CSF) samples. The levels of several inflammatory mediators, including the cytokines TNF-α, IL-6, IL-8, and the granulocyte-macrophage colony-stimulating factor (GM-CSF) were found to be increased in the CSF of patients with AD in comparison with controls.[48]

Immunophenotyping studies have also provided evidence of peripheral immune alterations in AD. Peripheral blood from patients with AD had less circulating B (CD19+) lymphocytes than controls. The number of human leukocyte antigen–antigen D related (HLA-DR)+ CD19+ B lymphocytes was also reduced in AD. HLA-DR is an MHC class II molecule expressed by antigen presenting (such as CD14+ monocytes/macrophages and CD19+ B cells) to activate T cells. In addition, it was observed that there was a lower percentage of late-stage effector (CCR7-RA+CD4+) and a higher percentage of effector memory (CCR7-RO+CD4+) T lymphocytes in AD samples in comparison with controls. Patients with AD also displayed reduced numbers of naïve CD8+ T cells (CCR7+RA+CD8+) and increased numbers of late-stage effector CD8+ cells (CCR7-RA+CD8+).[49] Altogether, these results indicate that patients with AD exhibit changes in the naïve/memory subtypes of CD4+ and CD8+ T cells.

Based on the evidence that AD is associated with immune/inflammatory changes, studies have investigated whether anti-inflammatory- and immune-based strategies are effective in the treatment of AD. Epidemiological and observational studies have described a protective effect of the chronic use of nonsteroidal anti-inflammatory drugs (NSAIDs) in AD patients, reducing or delaying the development of the disease. Nevertheless, the efficacy of NSAIDs in the treatment of AD was not proven in randomized trials.[48] One hypothesis to explain this apparent discrepancy is that NSAIDs would work only in the prodromal or very early stages of the disease, not exerting clinically meaningful effects once the disease is established due to amyloid and tau burden.

Immunotherapy has been regarded as a major breakthrough in the development of disease-modifying treatments for AD. The immunization with $A\beta_{42}$ led to disappearance of amyloid plaques in a transgenic mouse model of AD. However, the trials using the first-generation vaccines in patients with mild to moderate AD had to be terminated due to severe adverse effects, including life-threatening meningoencephalitis. Nevertheless, immunotherapy still holds the potential to be further explored. Next-generation vaccines targeting more specific epitopes and inducing a more controlled immune response are currently in development.[50]

Immune/inflammatory mechanisms seem to be associated with neurodegeneration in general rather than being exclusively related to AD pathology. Although the great majority of studies in this regard has been conducted in AD patients, there is also evidence of inflammatory/immune changes in other neurodegenerative diseases, such as DLB and FTD.

DLB is the second-most-common neurodegenerative dementia. The hallmark of DLB is the presence of Lewy bodies (intracytoplasmic, spherical, eosinophilic, neuronal inclusion bodies composed mainly of abnormal deposits of α-synuclein) in the brainstem and cortex. DLB is often misdiagnosed, mainly due to the overlap of its symptoms with those of AD and Parkinson's disease (PD). The onset and the time frame of symptoms' development are crucial for the clinical diagnosis of DLB. While AD onset is insidious and progresses gradually, DLB often presents as a more rapid clinical syndrome, with relatively fast cognitive and motor decline. In addition, DLB should be diagnosed when the patient presents with dementia before or within the first year of onset of Parkinsonian motor signs.[51] Postmortem studies have provided evidence of neuroinflammation in DLB brains. Overexpression of IL-1α, TNF-α, and inducible nitric oxide synthase (iNOS) was shown in amygdala, hippocampus, entorhinal, and insular cortices of DLB brains in comparison with controls.[52] In addition, patients with pure DLB (i.e., presenting Lewy bodies but no AD pathology) had a significantly greater number of microglial cells than non-demented control individuals, but fewer than patients with either pure AD, or DLB combined with AD pathology. Interestingly, there was a positive correlation between the number of activated microglia and the Lewy body burden in different brain regions.[53] Microglia activation in DLB has also been demonstrated by *in vivo* studies. One PET study reported increased microglia activation in the *substantia nigra*, putamen, and several cortical regions in early-stage patients with DLB.[54] The microglial involvement in DLB pathophysiology was confirmed by postmortem data showing that both complement proteins and microglia are associated with degenerating neurons with Lewy bodies.[55] Two immunohistochemistry studies, however, failed to demonstrate increased microglial activation in postmortem samples of DLB brains.[56,57] Different methods for microglia quantification and limited statistical power due to small sample sizes might explain these divergent results.

The contribution of inflammation to the pathological process in DLB has been further confirmed by data obtained from CSF and peripheral blood samples. One study found upregulation of proteins involved in acute phase reactants (APRS)/immune response in the CSF of patients with DLB in comparison with controls. Complement C3, complement C4a, transthyretin, pigment epithelium-derived factor, and prothrombin were two- to seven-fold upregulated, while inter-alpha-trypsin inhibitor heavy chain (ITIH4) showed 60- to 88-fold upregulation in DLB samples.[58] In addition, higher serum IL-6 and TNF-α levels were associated with worse cognitive impairment and higher severity of neuropsychiatric symptoms, respectively.[59] Conversely, one study reported decreased CSF levels of IL-6 in DLB samples in comparison with controls and AD.[60]

Neuroinflammation might also be involved in the pathophysiology of FTD. FTD is clinically characterized by marked behavior and/or language changes. Patients with the semantic variant of primary progressive aphasia (semantic dementia), a subtype of FTD, present with increased rates of autoimmune diseases. Non-thyroid-spectrum autoimmune diseases are twice as common in FTD patients as in patients with AD and non-demented controls. In addition, plasma levels of TNF-α were significantly elevated in patients with semantic dementia in comparison with controls (highlighted in [61]).

As previously described in AD and DLB, PET studies with translocator protein (TSPO) ligands have demonstrated increased microglia activation in FTD patients.[62] Peripheral blood from patients with FTD presented a decreased number of circulating B and T lymphocytes.[49] Patients with FTD exhibited changes in the expression of molecules associated with co-stimulatory signaling on peripheral blood mononuclear cells. Cytotoxic T-Lymphocyte Associated Protein (CTLA)-4 expression on CD4+ T cells from FTD was reduced when compared with AD and controls.[63] CSF levels of YKL-40, a biomarker of inflammation, were found to be elevated in FTD patients as well as in samples from AD and MCI patients in comparison with controls.[64] The levels of the chemokines CCL2/monocyte chemoattractant protein (MCP)-1 and CXCL8/IL-8 were increased in the CSF of patients with FTD in comparison with controls,[65,66] as well as the cytokines TNF-α, transforming growth factor (TGF)-β,[67] and IL-15.[68]

Inflammatory processes have been associated not only with degenerative dementias, but also with vascular dementia. Damage to the blood vessels in the brain has been attributed to atherosclerosis, which is considered an inflammatory condition.[19] The combination of high levels of CRP and IL6 was associated with an increased risk of vascular dementia.[69] Here again, the causality direction has not been fully elucidated. There is still an ongoing debate about whether the increase in circulating levels of inflammatory mediators represents a causative agent implicated in cognitive decline, or the result from inflammatory processes in the CNS determined by the vascular insult. It is also possible that both hypotheses are true.[22]

CONCLUSION

Aging is a highly complex process associated with progressive changes of several physiological systems and functions. The immune system is one the most affected systems. Elderly people exhibit a low-grade chronic inflammatory state known as *inflammaging*. Interacting with genetic background and lifestyle habits, inflammaging seems to contribute to the onset and progression of age-related conditions. Accordingly, immune/inflammatory changes have been described in dementias and other age-related psychiatric disorders such as late-life depression. It remains to be elucidated whether inflammation plays a primary (causative) or secondary (responsive) role in the physiopathology of these conditions.

REFERENCES

1. United Nations, Department of Economic and Social Affairs, Population Division. *World Population Ageing 2015*. 2015; http://www.un.org/en/development/desa/population/publications/pdf/ageing/WPA2015_Report.pdf; Accessed 10/12/2017.
2. Hugo J, Ganguli M. Dementia and cognitive impairment: epidemiology, diagnosis, and treatment. *Clin Geriatr Med*. 2014;30(3):421–442.
3. WHO. Fact sheet: dementia. 2017; http://www.who.int/mental_health/neurology/dementia/en/. Accessed 10/12/2017.
4. Prince MKM, Guerchet M, McCrone P, et al. *Dementia UK: update*. 2014; alzheimers.org.uk/dementiauk. Accessed 10/12/2017.
5. Hybels CF, Blazer DG. Epidemiology of late-life mental disorders. *Clin Geriatr Med*. 2003;19(4):663–696, v.
6. Reynolds K, Pietrzak RH, El-Gabalawy R, Mackenzie CS, Sareen J. Prevalence of psychiatric disorders in U.S. older adults: findings from a nationally representative survey. *World Psychiatry*. 2015;14(1):74–81.

7. Korner A, Lopez AG, Lauritzen L, Andersen PK, Kessing LV. Late and very-late first-contact schizophrenia and the risk of dementia—a nationwide register based study. *Int J Geriatr Psychiatry.* 2009;24(1):61–67.
8. Dotson VM, Beydoun MA, Zonderman AB. Recurrent depressive symptoms and the incidence of dementia and mild cognitive impairment. *Neurology.* 2010;75(1):27–34.
9. Polyakova M, Sonnabend N, Sander C, et al. Prevalence of minor depression in elderly persons with and without mild cognitive impairment: a systematic review. *J Affect Disord.* 2014;152–154:28–38.
10. Panza F, Frisardi V, Capurso C, et al. Late-life depression, mild cognitive impairment, and dementia: possible continuum? *Am J Geriatr Psychiatry.* 2010;18(2):98–116.
11. Hunt KJ, Walsh BM, Voegeli D, Roberts HC. Inflammation in aging, part 1: physiology and immunological mechanisms. *Biol Res Nurs.* 2010;11(3):245–252.
12. Yankner BA, Lu T, Loerch P. The aging brain. *Annu Rev Pathol.* 2008;3:41–66.
13. Chahal HS, Drake WM. The endocrine system and ageing. *J Pathol.* 2007;211(2):173–180.
14. Frasca D, Blomberg BB. Inflammaging decreases adaptive and innate immune responses in mice and humans. *Biogerontology.* 2016;17(1):7–19.
15. Goronzy JJ, Weyand CM. Immune aging and autoimmunity. *Cell Mol Life Sci.* 2012;69(10):1615–1623.
16. Doran MF, Pond GR, Crowson CS, O'Fallon WM, Gabriel SE. Trends in incidence and mortality in rheumatoid arthritis in Rochester, Minnesota, over a forty-year period. *Arthritis Rheum.* 2002;46(3):625–631.
17. Goronzy JJ, Weyand CM. Understanding immunosenescence to improve responses to vaccines. *Nat Immunol.* 2013;14(5):428–436.
18. Franceschi C, Bonafe M, Valensin S, et al. Inflamm-aging. An evolutionary perspective on immunosenescence. *Ann N Y Acad Sci.* 2000;908:244–254.
19. Hunt KJ, Walsh BM, Voegeli D, Roberts HC. Inflammation in aging, part 2: implications for the health of older people and recommendations for nursing practice. *Biol Res Nurs.* 2010;11(3):253–260.
20. Vasto S, Candore G, Balistreri CR, et al. Inflammatory networks in ageing, age-related diseases and longevity. *Mech Ageing Dev.* 2007;128(1):83–91.
21. Bruunsgaard H, Andersen-Ranberg K, Hjelmborg J, Pedersen BK, Jeune B. Elevated levels of tumor necrosis factor alpha and mortality in centenarians. *Am J Med.* 2003;115(4):278–283.
22. Krabbe KS, Pedersen M, Bruunsgaard H. Inflammatory mediators in the elderly. *Exp Gerontol.* 2004;39(5):687–699.
23. Harris TB, Ferrucci L, Tracy RP, et al. Associations of elevated interleukin-6 and C-reactive protein levels with mortality in the elderly. *Am J Med.* 1999;106(5):506–512.
24. Bruunsgaard H, Ladelund S, Pedersen AN, Schroll M, Jorgensen T, Pedersen BK. Predicting death from tumour necrosis factor-alpha and interleukin-6 in 80-year-old people. *Clin Exp Immunol.* 2003;132(1):24–31.
25. Mooradian AD, Reed RL, Osterweil D, Scuderi P. Detectable serum levels of tumor necrosis factor alpha may predict early mortality in elderly institutionalized patients. *J Am Geriatr Soc.* 1991;39(9):891–894.

26. Dantzer R, O'Connor JC, Freund GG, Johnson RW, Kelley KW. From inflammation to sickness and depression: when the immune system subjugates the brain. *Nat Rev Neurosci.* 2008;9(1):46–56.
27. Byers AL, Yaffe K, Covinsky KE, Friedman MB, Bruce ML. High occurrence of mood and anxiety disorders among older adults: The National Comorbidity Survey Replication. *Arch Gen Psychiatry.* 2010;67(5):489–496.
28. Diniz BS, Teixeira AL, Talib L, Gattaz WF, Forlenza OV. Interleukin-1beta serum levels is increased in antidepressant-free elderly depressed patients. *Am J Geriatr Psychiatry.* 2010;18(2):172–176.
29. Diniz BS, Teixeira AL, Talib LL, Mendonca VA, Gattaz WF, Forlenza OV. Increased soluble TNF receptor 2 in antidepressant-free patients with late-life depression. *J Psychiatr Res.* 2010;44(14):917–920.
30. Mourao RJ, Mansur G, Malloy-Diniz LF, Castro Costa E, Diniz BS. Depressive symptoms increase the risk of progression to dementia in subjects with mild cognitive impairment: systematic review and meta-analysis. *Int J Geriatr Psychiatry.* 2016;31(8):905–911.
31. Diniz BS, Butters MA, Albert SM, Dew MA, Reynolds CF, 3rd. Late-life depression and risk of vascular dementia and Alzheimer's disease: systematic review and meta-analysis of community-based cohort studies. *Br J Psychiatry.* 2013;202(5):329–335.
32. Diniz BS, Sibille E, Ding Y, et al. Plasma biosignature and brain pathology related to persistent cognitive impairment in late-life depression. *Mol Psychiatry.* 2015;20(5):594–601.
33. Mendes-Silva AP, Pereira KS, Tolentino-Araujo GT, et al. Shared biologic pathways between Alzheimer disease and major depression: a systematic review of microRNA expression studies. *Am J Geriatr Psychiatry.* 2016;24(10):903–912.
34. Diniz BS, Reynolds CF, 3rd, Sibille E, et al. Enhanced molecular aging in late-life depression: the senescent-associated secretory phenotype. *Am J Geriatr Psychiatry.* 2017;25(1):64–72.
35. Lim A, Krajina K, Marsland AL. Peripheral inflammation and cognitive aging. *Mod Trends Pharmacopsychiatry.* 2013;28:175–187.
36. Noble JM, Manly JJ, Schupf N, Tang MX, Mayeux R, Luchsinger JA. Association of C-reactive protein with cognitive impairment. *Arch Neurol.* 2010;67(1):87–92.
37. Forlenza OV, Diniz BS, Talib LL, et al. Increased serum IL-1beta level in Alzheimer's disease and mild cognitive impairment. *Dement Geriatr Cogn Disord.* 2009;28(6):507–512.
38. Holmes C, Cunningham C, Zotova E, et al. Systemic inflammation and disease progression in Alzheimer disease. *Neurology.* 2009;73(10):768–774.
39. Diniz BS, Teixeira AL, Ojopi EB, et al. Higher serum sTNFR1 level predicts conversion from mild cognitive impairment to Alzheimer's disease. *J Alzheimers Dis.* 2010;22(4):1305–1311.
40. Tan ZS, Beiser AS, Vasan RS, et al. Inflammatory markers and the risk of Alzheimer disease: the Framingham Study. *Neurology.* 2007;68(22):1902–1908.
41. Yaffe K, Lindquist K, Penninx BW, et al. Inflammatory markers and cognition in well-functioning African-American and white elders. *Neurology.* 2003;61(1):76–80.
42. Yaffe K, Kanaya A, Lindquist K, et al. The metabolic syndrome, inflammation, and risk of cognitive decline. *JAMA.* 2004;292(18):2237–2242.

43. Lai KSP, Liu CS, Rau A, et al. Peripheral inflammatory markers in Alzheimer's disease: a systematic review and meta-analysis of 175 studies. *J Neurol Neurosurg Psychiatry.* 2017;88(10):876–882.
44. Eikelenboom P, Stam FC. Immunoglobulins and complement factors in senile plaques. An immunoperoxidase study. *Acta Neuropathol.* 1982;57(2–3):239–242.
45. McGeer PL, Itagaki S, Boyes BE, McGeer EG. Reactive microglia are positive for HLA-DR in the substantia nigra of Parkinson's and Alzheimer's disease brains. *Neurology.* 1988;38(8):1285–1291.
46. Wyss-Coray T. Inflammation in Alzheimer disease: driving force, bystander or beneficial response? *Nat Med.* 2006;12(9):1005–1015.
47. Akiyama H, Barger S, Barnum S, et al. Inflammation and Alzheimer's disease. *Neurobiol Aging.* 2000;21(3):383–421.
48. Calsolaro V, Edison P. Neuroinflammation in Alzheimer's disease: current evidence and future directions. *Alzheimers Dement.* 2016;12(6):719–732.
49. Busse M, Michler E, von Hoff F, et al. Alterations in the peripheral immune system in dementia. *J Alzheimers Dis.* 2017;58(4):1303–1313.
50. Kulshreshtha A, Piplani P. Current pharmacotherapy and putative disease-modifying therapy for Alzheimer's disease. *Neurol Sci.* 2016;37(9):1403–1435.
51. McKeith IG, Galasko D, Kosaka K, et al. Consensus guidelines for the clinical and pathologic diagnosis of dementia with Lewy bodies (DLB): report of the consortium on DLB international workshop. *Neurology.* 1996;47(5):1113–1124.
52. Katsuse O, Iseki E, Kosaka K. Immunohistochemical study of the expression of cytokines and nitric oxide synthases in brains of patients with dementia with Lewy bodies. *Neuropathology.* 2003;23(1):9–15.
53. Mackenzie IR. Activated microglia in dementia with Lewy bodies. *Neurology.* 2000;55(1):132–134.
54. Iannaccone S, Cerami C, Alessio M, et al. In vivo microglia activation in very early dementia with Lewy bodies, comparison with Parkinson's disease. *Parkinsonism Relat Disord.* 2013;19(1):47–52.
55. Togo T, Iseki E, Marui W, Akiyama H, Ueda K, Kosaka K. Glial involvement in the degeneration process of Lewy body-bearing neurons and the degradation process of Lewy bodies in brains of dementia with Lewy bodies. *J Neurol Sci.* 2001;184(1):71–75.
56. Streit WJ, Xue QS. Microglia in dementia with Lewy bodies. *Brain Behav Immun.* 2016;55:191–201.
57. Shepherd CE, Thiel E, McCann H, Harding AJ, Halliday GM. Cortical inflammation in Alzheimer disease but not dementia with Lewy bodies. *Arch Neurol.* 2000;57(6):817–822.
58. Dieks JK, Gawinecka J, Asif AR, et al. Low-abundant cerebrospinal fluid proteome alterations in dementia with Lewy bodies. *J Alzheimers Dis.* 2013;34(2):387–397.
59. Clough Z, Jeyapaul P, Zotova E, Holmes C. Proinflammatory cytokines and the clinical features of dementia with Lewy bodies. *Alzheimer Dis Assoc Disord.* 2015;29(1):97–99.
60. Wennstrom M, Hall S, Nagga K, Londos E, Minthon L, Hansson O. Cerebrospinal fluid levels of IL-6 are decreased and correlate with cognitive status in DLB patients. *Alzheimers Res Ther.* 2015;7(1):63.

61. Piguet O. Neurodegenerative disease: frontotemporal dementia—time to target inflammation? *Nat Rev Neurol.* 2013;9(6):304–305.
62. Zhang J. Mapping neuroinflammation in frontotemporal dementia with molecular PET imaging. *J Neuroinflammation.* 2015;12:108.
63. Santos RR, Torres KC, Lima GS, et al. Reduced frequency of T lymphocytes expressing CTLA-4 in frontotemporal dementia compared to Alzheimer's disease. *Prog Neuropsychopharmacol Biol Psychiatry.* 2014;48:1–5.
64. Alcolea D, Carmona-Iragui M, Suarez-Calvet M, et al. Relationship between beta-secretase, inflammation and core cerebrospinal fluid biomarkers for Alzheimer's disease. *J Alzheimers Dis.* 2014;42(1):157–167.
65. Galimberti D, Venturelli E, Villa C, et al. MCP-1 A-2518G polymorphism: effect on susceptibility for frontotemporal lobar degeneration and on cerebrospinal fluid MCP-1 levels. *J Alzheimers Dis.* 2009;17(1):125–133.
66. Galimberti D, Schoonenboom N, Scheltens P, et al. Intrathecal chemokine levels in Alzheimer disease and frontotemporal lobar degeneration. *Neurology.* 2006;66(1):146–147.
67. Sjogren M, Folkesson S, Blennow K, Tarkowski E. Increased intrathecal inflammatory activity in frontotemporal dementia: pathophysiological implications. *J Neurol Neurosurg Psychiatry.* 2004;75(8):1107–1111.
68. Rentzos M, Zoga M, Paraskevas GP, et al. IL-15 is elevated in cerebrospinal fluid of patients with Alzheimer's disease and frontotemporal dementia. *J Geriatr Psychiatry Neurol.* 2006;19(2):114–117.
69. Ravaglia G, Forti P, Maioli F, et al. Blood inflammatory markers and risk of dementia: the Conselice Study of Brain Aging. *Neurobiol Aging.* 2007;28(12):1810–1820.

16

Immune-Based Biomarkers and Therapies in Psychiatry

Future Prospects

GAURAV SINGHAL AND BERNHARD T. BAUNE ■

INTRODUCTION

There is currently an international crisis concerning psychiatric disorders, as it is estimated that one in four of the global population has mental illness.[1] People with mental illness are more likely to develop metabolic disorders, such as type 2 diabetes mellitus, stroke, and myocardial infarction, thereby increasing the mortality rate ratio by two to three times and reducing life expectancy by at least 10 years.[2] While these statistics can be attributed to developing-country health systems, sociocultural issues, and natural disasters,[3] it may also indicate that the prevailing models to diagnose and treat psychiatric disorders are inadequate in spite of extensive investigations at the molecular level during the last two decades.

More recently, the role of immune system in the brain is being investigated to elucidate the multiple mechanistic pathways responsible for the onset and chronic occurrence of psychiatric disorders. Over- or under-expression, as well as morphological alterations of glial cells[4–6]; increases in the levels of the pro-inflammatory cytokines TNF-α, IL-1β, and IL-6 in the brain resulting in neuroinflammation[7–9]; and development of antigenic proteins such as tau[10,11] and beta-amyloid (Aβ)[12] are some of the immune-related factors associated with the psychiatric disorders and related mood changes, and cognitive and memory deficits.

In this chapter, we will discuss the immune-based biomarkers, treatment approaches, and future research directions for some of the major neuropsychiatric disorders, i.e., depression, bipolar disorder (BD), schizophrenia, Alzheimer's disease (AD), and Parkinson's disease (PD).

MOOD DISORDERS (DEPRESSION, BIPOLAR DISORDER)

Depression is the leading debilitating mental disorder, affecting more than 300 million people globally.[13] Clinical depression or major depressive disorder (MDD) is characterized by a distinct change of mood accompanied by sadness, irritability, loss of interest in all activities and events, as well as psychophysiological changes.[14] Similarly, people with bipolar disorder (BD) have manic, or both manic and depressive, episodes, separated by periods of normal mood. BD affects 60 million people in the world.[15] Several factors, ranging from genetic to environmental factors,[16] hyperactivity of hypothalamic-pituitary-adrenal axis,[17,18] serotonergic dysfunctions,[19,20] impaired growth factor (e.g., BDNF) metabolism,[21,22] as well as immune imbalance[4,5] have been implicated in the causation of depressive-like behavior and BD. Factors like adverse neurobiological changes such as chronic neuroinflammation with resultant neurodegeneration, and loss of neuroplasticity, neuroprotection, and cellular resilience, directly result in the development of mood disorders.[23,24]

Immune-Based Biomarkers of Mood Disorders

Glial cells play a significant role in the disruption of neuroplasticity and exacerbation of mood disorders during psychological stress.[4,5] These immune cells, which are essential for the maintenance of neuroimmune homeostasis, express various pro-inflammatory cytokines (e.g., TNF-α, IL-1β, IL-6, IFN-γ), chemokine receptors (CXCR1, CXCR3, CCR3, CCR4, CCR5, CCR6, CXCR2, CXCR4, CXCR5), class I and II major histocompatibility complex (MHC) antigens, and toxic molecules (e.g., superoxide anions, nitric oxide),[25,26] as well as activate intracellular multiprotein complexes called inflammasomes in the presence of pathogenic and/or infectious stimuli,[27] exacerbating neurodegenerative changes and leading to depression-like behavior.[28] There is increased expression of pro-inflammatory cytokines, such as TNF-α, IL-1β, and IL-6, in the brain, resulting in neuroinflammation and associated mood changes and cognitive and memory deficits.[7,8] The over-expressed TNF-α has been shown to cause hippocampal degeneration and microglial apoptosis, both regarded as neuropathological underpinnings of depression.[29,30] Also, TNF-α-induced apoptotic cascades resulting in neuronal and glial loss have been reported during BD.[31] The pro-inflammatory cytokines IL-1 and TNF-α secrete adhesion molecules that attach to the endothelium of blood vessels in the brain, facilitating migration of leukocytes from blood to the brain tissues.[32] Moreover, IFN-α promotes expression of pro-inflammatory surface markers MHC II, CD86, and CD54 (M1 polarisation), causing neuroinflammation, and thereby, depression and BD.[33,34]

An excess of glucocorticoids, which affects microglial morphology and phenotype, has been shown to result in depression.[35,36] Similarly, a decrease in the levels of serum BDNF[22] and impaired expression of cytokines in the brain[34] have been linked to BD. Several morphological and functional changes of astrocytes, such as decreased astrocyte cell number and Glial fibrillary acidic protein (GFAP) protein in the hippocampus, and reduced cytokine and chemokine expression on astrocytes, have been reported in patients with MDD.[37]

A number of immune and non-immune markers in the serum, such as pro-inflammatory cytokines, chemokines, and cortisol could help the diagnosis and/or stratification of mood disorders. These proteins could also be serially measured in serum for therapeutic reasons. For research purposes, postmortem brain immunohistochemical analysis could be performed to estimate the percentages of GFAP and ionized calcium-binding adapter molecule 1 (IBA1) (indicators of astrocytes and microglia expression, respectively). Likewise, the level of neurotrophins in the brain could be measured using ELISA on serum and cerebrospinal fluid, and FACS (Fluoroscence Activated Cell Sorting) could be used to estimate the proportion of T cells subsets (CD4+ and CD8+ cells) in blood, as well as in the cervical lymph nodes that drains the brain.

Immune-Based Treatment for Mood Disorders

Since neuroinflammation is a pathological factor for mood disorders, therapeutic use of non-steroidal anti-inflammatory drugs NSAIDs has been recommended.[38,39] However, a recent critical review of the use of NSAIDs in MDD found inconsistent findings and substantial methodological heterogeneity in studies, and concluded that the efficacy of NSAIDs for depressive symptoms appears negligible.[40] As discussed before, long-term use of NSAIDs can also result in side effects such as gastric ulceration and damage to kidneys.[41,42] Conversely, omega-3 polyunsaturated fatty acids (PUFAs) have been shown to improve the short-term course of illness in patients with both depression[43] and bipolar disorder.[44]

Antidepressants are the first line of treatment for MDD, and they target molecular pathways associated with the pathogenesis of depression, including selective serotonin reuptake inhibitors (SSRIs), serotonin and norepinephrine reuptake inhibitors (SNRIs), and norepinephrine reuptake inhibitors (NRIs), which tend to correct monoaminergic dysfunctions responsible for MDD, thereby reducing depression[45] and improving cognition.[46] Other types of antidepressants are atypical antidepressants, tricyclic antidepressants (TCAs), and monoamine oxidase inhibitors (MAOIs) that regulate levels of monoamines in the brain through independent mechanisms. However, the efficacy of currently available monoamine antidepressants is far from perfect in treating depression. This calls for non-pharmacological approaches such as physical exercise, which has been shown to induce neurogenesis and improve immune functions in the brain, thereby reducing depressive-like behaviour,[47] to be used in combination with antidepressants for treating depression. Moreover, with each bout of physical exercise, there is a decrease in the levels of pro-inflammatory cytokines: in particular, TNF-α, both in serum and CSF.[48]

Future Directions for Immune-Based Research in Mood Disorders

Taken together, it is clear that while several pharmacological and immunomodulatory approaches have been proposed, tested, and clinically applied for treating mood disorders, there is no proven way of treating them entirely. Also, the immunomodulatory role of physical exercise in the brain needs to be investigated further before devising treatment and prevention strategies based on it. More importantly, better alternatives to NSAIDs are required; for example, drugs that induce anti-inflammatory effects in the brain with minimal or no side effects. Nonetheless, psychological treatments can be useful as an adjunctive treatment alongside antidepressant medication for depression at this stage. Similarly, mood-stabilisers, such as lithium carbonate and sodium valproate, and psychosocial support are effective treatments of the acute phase of BD and help prevent relapse.[49]

PSYCHOTIC DISORDERS (SCHIZOPHRENIA)

Schizophrenia is a debilitating mental disorder affecting 21 million people worldwide.[50] Schizophrenia is listed in the DSM-5 as displaying the heterogeneous symptoms of hallucinations, delusions, and disorganised speech and behaviour.[51] Theories to explain the mechanisms of schizophrenia include glutamatergic dysfunction,[52] dopaminergic hypothesis,[53] immune function impairment,[54,55] environmental factors and gene interaction,[56] and N methyl-D-aspartate (NMDA) receptor hypofunction.[57] The geographic location also proves to contribute to schizophrenia, as prevalence is higher amongst those in urban regions than in rural; more time spent living in urban areas increases the risk of schizophrenia.[56] There is growing evidence now of the immune involvement during schizophrenia.

Immune-Based Biomarkers of Schizophrenia

Neuroinflammation with resultant neurodegeneration is a mechanism implicated in schizophrenia.[58,59] Over-expression of Th1 pro-inflammatory cytokines in serum and CSF,[60–62] as well as microglial dysfunction,[58] have been reported in schizophrenic patients. Elevated levels of peripheral cytokines during schizophrenia could be related to the bidirectional communication between the central nervous system (CNS) and the systemic circulation through blood–brain barrier (BBB). Moreover, lymphocytes' response to antigens is also delayed during schizophrenia,[63] suggesting diminished cellular immunity. Indeed, the soluble intercellular adhesion molecule-1 (sICAM-1), a marker for cellular immunity activation, shows significantly reduced expression on macrophages and lymphocytes in schizophrenia.[64] A clinical study reported a higher frequency of CD3+ and CD4+ cells and a higher CD4/CD8 ratio during the acute state of schizophrenia than in healthy controls.[65] This suggests that the relative impairment of both humoral and cell-mediated immunity may be either a causative factor or a result of the psychotic state.

Recent research suggests a role for the complement classical pathway components in the pathogenesis of schizophrenia. It has been reported that the mean values of the hemolytic activities of the C1, C3, and C4 complement components were significantly higher, while C2 complement component was significantly lower in the serum of the schizophrenic patients.[66] This could explain the neurodevelopmental changes, such as impaired neurogenesis[67] and synaptic dysfunction,[68] found in the brain of schizophrenic patients, taking into account the role played by the complement cascade in the remodeling of synaptic connections, as well as differentiation and maturation of neuronal progenitor cells in the developing brain.[69,70] Likewise, the levels of mitogen-activated protein kinases (MAPKs), which are specific protein kinases (serine-threonine specific) and regulate immunomodulatory functions of the brain,[71] are elevated in the cerebellar vermis of schizophrenic patients, resulting in abnormalities of structure, function, and signal transduction in the brain.[72]

Immune markers, such as Th1 pro-inflammatory cytokines, sICAM-1, and complement components, could be measured in the blood of patients for diagnostic and/or prognostic purposes in schizophrenia.

Immune-Based Treatment for Schizophrenia

Antipsychotics, such as chlorpromazine and other phenothiazines, modulate dopaminergic pathways and have been the drug of choice for schizophrenia for a long time now. Research has shown that antipsychotics also inhibit human natural killer (NK) cell activity and antibody-dependent cell-mediated cytotoxicity in a dose-dependent manner.[55] In addition, chlorpromazine has a lowering effect on the levels of circulating TNF-α and IFN-γ,[55] suggesting an anti-inflammatory effect of antipsychotics. However, fewer than 50% of schizophrenic patients respond to antipsychotic therapy, which has been attributed to the pathophysiological heterogeneity among patients.[73] Moreover, chlorpromazine and other first-generation antipsychotics, when used in the long run, can result in tardive dyskinesia.[74] The risk of tardive dyskinesia is now reduced with second-generation antipsychotics, though. All pathophysiological aspects and symptoms need to be considered to determine an effective treatment, including psychosocial support, for schizophrenia. Effective prevention and management strategies for people with schizophrenia include facilitation of assisted living, supported housing, and supported employment.

Future Directions for Immune-Based Research in Schizophrenia

Although neuroinflammation seems to be an important pathogenic factor for schizophrenia, not much research has been conducted in this direction yet. NSAIDs, such as celecoxib and aspirin, have been tested and reported to reduce symptom severity in schizophrenia,[75] and can, therefore, be used as adjunctives to antipsychotics. However, more investigation is required before NSAIDs are clinically applied for treating schizophrenia.

Omega-3 PUFA[76] and non-pharmacological approaches such as physical exercise[77] have also been investigated for treating schizophrenia, but with less success. Further research into anti-inflammatory strategies used alone and/or in combination is required for devising better prevention and treatment measures for schizophrenia.

NEURODEGENERATIVE DISEASES (ALZHEIMER'S DISEASE [AD], PARKINSON'S DISEASE [PD])

Neurodegenerative diseases, such as AD and PD, are characterized by progressive cellular and molecular deficits in the brain.[11,78] These two neurodegenerative diseases put a heavy burden on the health systems of both developed and developing countries. Once established, these conditions are non-treatable; therefore, exploring therapeutic approaches to control them at the initial stages is of great importance.

For AD, the primary pathological mechanisms involve extracellular deposition of amyloid beta (Aβ) in the brain,[79] decline in the synthesis of the excitatory neurotransmitter acetylcholine,[80] formation of abnormal tau protein leading to the disintegration of microtubules during cell division,[11] glial cells–induced oxidative and inflammatory stress,[81] as well as aging and genetic predisposition.[82] PD, on the other hand, is characterized by the depletion of dopamine due to the death of dopaminergic neurons in the pars compacta of the substantia nigra, causing impairment of motor functions with accompanying symptoms of brakykinesia, resting tremor, rigidity, postural instability, fatigue, sleep abnormalities, depression, and dementia.[83] Some of the molecular changes that lead to dopaminergic neuron loss include inhibition of the mitochondrial complex I, causing aggregation of protein α-synuclein and hence impairments in protein handling and detoxification.[84] Detoxification is necessary to remove excess calcium, otherwise dopaminergic neurons show excess cytotoxicity and undergo death and degeneration during PD.[85]

Immune-Based Biomarkers of Neurodegenerative Diseases and Alzheimer's Dementia

While suggested pathological mechanisms for neurodegenerative diseases are many, a common link between them is the development of the inflammatory cascade in the brain. Insoluble Aβ deposits and neurofibrillary tangles formed in the brain have been shown to sensitize neurons, providing stimuli for neuroinflammation during the AD.[90] Studies have shown over-expression of pro-inflammatory IL-1β and IL-18 cytokines in the microglia, astrocytes, as well as neurons, which are co-localized with both Aβ plaques and tau protein in the brain of AD patients.[91,92] Elevated levels of TNF-α in serum[93] and IL-1β in CSF[94] have been reported in AD patients. In addition, researchers have observed upregulation of the complement system, such as C1q and C9, and their activation products C4d and C3d,[95] in the human brain during AD and other neurodegenerative diseases.[95] The complement system consists of distinct plasma proteins that act as opsonins and initiate a series of inflammatory responses.[96] Similarly, high levels of CRP increase the paracellular permeability of the BBB, induce reactive gliosis, and impair CNS function.[97] High levels of CRP in the hippocampus has also been linked to AD.[98]

In PD, a significant increase in microglia number has been reported in the substantia nigra pars compacta, being activated by the aggregated α-synuclein and leading to persistent and progressive nigral degeneration.[99] Over-expression of microglia could be neuroinflammatory and cytotoxic, as they secrete several neurotoxic products, such as proteinases, cytokines, and reactive oxygen and nitrogen intermediates.[100] A decrease in the number of reactive astrocytes, as observed during PD,[6] is also a known outcome of neuroinflammation, which may have been caused by cytotoxic T cells, antibodies, and/or cytokines.[101]

Interestingly, healthy elderly individuals with no cognitive deficit have also shown significant amyloid deposition in the brain.[102] This suggests that development of inflammatory cascade seems to be a requisite for the onset of neurodegenerative diseases. Likewise, even though the burden of tau proteins is higher in the brain of AD patients,[10] tau phosphorylation is essential to protect neurons against oxidative stress.[103] Moreover, inflammatory cascade in the brain in itself can augment deposition of Aβ[12] and phosphorylation of tau proteins,[104] initiating a never-ending vicious cycle.

Presence of insoluble Aβ deposits and neurofibrillary tangles, glial changes, and neurodegenerative changes in the brain act as conclusive evidence in the diagnosis of AD during postmortem brain analysis. CSF measurement of different forms of Aβ and Tau has been incorporated in the clinical practice for the differential diagnosis of AD and other dementias. The role of immune markers in the diagnosis or prognosis of neurodegenerative diseases remains to be determined.

Immune-Based Treatment for Neurodegenerative Diseases and Alzheimer's Dementia

Since these neurodegenerative diseases have similarities in their pathophysiology—i.e., the formation of abnormal protein aggregates leading to the death of neurons—analogous immunotherapeutic approaches to stop the formation of these pathological proteins could be helpful to prevent neuronal loss and associated behavioral disorders.

Use of Omega-3 PUFA[105] and physical exercise[106] has shown some promise for treating AD-associated dementia and cognitive deficits. For example, docosahexaenoic acid (DHA) has been shown to improve brain functions and reduce AD.[105] Indeed, DHA also protects against several other risk factors for dementia, including head trauma, diabetes, and cardiovascular disease.[105] Likewise, physical exercise has been shown to elicit beneficial effects in dementia patients,[107] plausibly due to its anti-inflammatory effects mediated via a reduction in visceral fat mass and hence decreases in the release of TNF-α. Muscle fibers produce IL-6 during exercise, which stimulates the activity of glial cells, and thereby, the secretion of anti-inflammatory molecules such as

IL-1ra and IL-10. The latter inhibits the production of the pro-inflammatory cytokine TNF-α.[48]

Acetylcholinesterase inhibitors donepezil, rivastigmine, and galantamine have been shown to increase the production of antioxidants and block the release of cytokines from microglia and monocytes in the brain of AD patients, thereby eliciting anti-neuroinflammatory effects, in addition to improving neuronal transmission.[108–110] Pharmacological strategies such as Naftidrofuryl that increase the oxygen supply to brain tissue,[111] and NMDA receptor antagonists (e.g., memantine),[112] can be given alongside anti-inflammatory agents to enhance the potency of treatment, especially in mixed dementia. NSAIDs could be used to reduce neuroinflammation and treat AD. However, while the chronic NSAIDs decrease the incidence of AD, they failed to provide cognitive or clinical improvement in patients with an established diagnosis of AD. Furthermore, long-term use of NSAIDs has been reported to cause gastrointestinal ulceration and bleeding,[42] and kidney damage.[41]

Similarly, only palliative treatments are available for PD at this stage. Since dopamine is depleted in the pars compacta of the substantia nigra during PD,[83] the loss of motor functions is treated mainly with levodopa, which is converted into dopamine in the brain.[113] Other treatment approaches include dopamine agonists and monoamine oxidase B (both improve the dopaminergic pathway through independent mechanisms),[114] and anticholinergics[115] as monotherapy and as part of combination regimes. Besides, psychosocial support, proper nutrition, and healthy living could help in easing some of the symptoms of AD, PD, and associated dementia.

Future Directions for Immune-Based Research in Neurodegenerative Diseases

The efficacy of the current pharmacological approaches is limited since they target one or two of the pathways and hence are unable to modify the course of the disease significantly. Moreover, they are not without side effects. As such, it is advisable to explore new and safe therapeutic approaches that act on multiple neural and biochemical targets for the treatment of neurodegenerative diseases. For example, a pharmacological drug with anti-inflammatory activity, combined with monoamine oxidase inhibition, anti-apoptotic activity, and a neuroprotective role, could provide better prevention and treatment. An excellent case of this combined therapy regime is the intravenous administration of IgG, which has been shown to induce anti-inflammatory action *in vivo*[116] and been beneficial in the treatment of AD by inhibiting the neurotoxic effects of Aβ.[117]

More research is required on the therapeutic use of omega-3 PUFA for neurodegenerative diseases. A meta-analysis involving three randomized controlled trials of omega-3 PUFA intervention in elderly participants found no benefit of omega-3 PUFA supplementation for cognitive function and dementia in cognitively healthy older people.[118] Another study also refuted the inverse relationship between omega-3 PUFA and dementia.[119] Nevertheless, anti-inflammatory properties of omega-3 PUFA[120] could be explored further in cognitively impaired people with a longer duration of the intervention.

Precaution is advised while treating dementia with anti-inflammatory agents combined with other pharmacological drugs (e.g., anti-cholinesterase, memantine).[41,121] Physical exercise can be a safer option. Further research can help to devise a better therapeutic combination strategy for neurodegenerative diseases.

DISCUSSION

Psychiatry is a complex field with various mental disorders manifested by a diverse range of clinical symptoms, changes in molecular patterns, and diagnostic approaches that suffer from lack of complete knowledge. Past approaches of matching symptoms to a diagnosis ran into many difficulties, including the reliability of opinion and arbitrary measurement, the heterogeneous nature of psychiatric illness, and the complexity of interacting features of such disorders. The discovery of a role for immune mechanisms in the brain has now helped researchers describe mechanistic pathways associated with some of the underlying causes of psychiatric disorders. Through cross immunological and pharmacogenomic research efforts, there is now hope to achieve the goal of more accurate diagnosis and treatment of major psychiatric disorders in the near future.

A significant impediment to treating psychiatric illnesses is that current medications are often only efficacious in certain patients. A highly probable reason for this is the heterogeneity between patients who have the same psychiatric disorder yet a diverse expression of genes and alteration of different biological systems, culminating in the same condition. An example of the heterogeneity of psychiatric disorders is highlighted by a study aimed at elucidating whether there are different molecular subtypes of schizophrenia resulting in discrete pathophysiological mechanisms associated with the same symptoms.[73] Interestingly, these authors found marked variations in levels of immune components among patients.

While pharmacotherapy has been the mainstay for treatment of psychiatric disorders, it comes with the limitation of partial response and inducing tolerance to drugs. The immunomodulatory agents are used as adjunct therapies, typically administered in combination with pharmacological drug/s or other conventional strategies, such as physical exercise. However, inappropriate applications can have harmful side-effects such as inflammatory tissue damage or immunosuppression; hence caution is advised while combining therapies, until a fully proven therapy is devised.

A three way association between the high incidences of chronic inflammatory diseases (e.g. cancer, diabetes, and cardiovascular diseases), increased pro-inflammatory cytokines (TNF-α, IL-1β, and IL-6) and acute phase protein levels in systemic circulation, and comorbid depression has been reported. [122–124] This indicates that there is some mechanism for the movement of pro-inflammatory cytokines from the systemic circulation to the brain. Indeed, recent evidence has quashed the long-standing theory of the immune-privileged status of the brain and has established an active communication between the CNS and the peripheral immune system through the BBB and draining lymphatics; and that glial cells, particularly microglia, are actively involved in immunosurveillance in the CNS in conjunction with the peripheral immune cells and proteins[125] (see Figure 16.1).

In the CNS, pro-inflammatory cytokines activate macrophages in the brain and glia-driven cell-mediated immune response, triggering neuroinflammatory response.[126] Also, activated T cells migrate across the BBB and perform neuroimmune surveillance along with microglia/monocytes.[127,128] However, when stimulated by external pathogens and stress proteins, the CD4+ Th1 cells secrete pro-inflammatory cytokines such as IL-2, TNF-α, and IFN-γ in the brain, further aggravating neuroinflammation. In addition, NK cells that migrate through the BBB secrete IFN-γ and kill glial cells, implicating them in various brain disorders, including depression.[129] Moreover, B cells that cross the BBB during brain diseases such as multiple sclerosis have been linked with the development of autoimmunity in the brain.[130] However, the association of B cells of any kind with psychiatric disorders discussed in this chapter is not yet established. Taken together, this suggests that biomarkers for psychiatric disorders could be found and hence explored both in the systemic circulation as well as in the CSF.

CONCLUSION

This chapter was intended to provide insight into the immune-based biomarkers and treatments of major psychiatric disorders. However, the importance of psychological counselling and social support, as well as currently available pharmacological and non-pharmacological treatments for mental disorders, cannot be overlooked, and immunomodulatory pathways must be considered as alternative approaches until they are further developed and thoroughly tested.

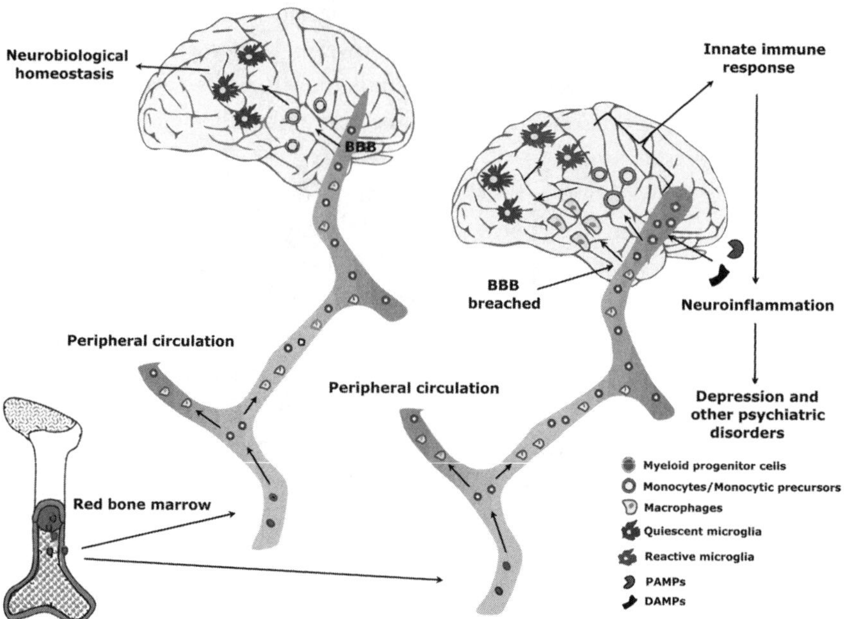

Figure 16.1 *Origin and role of microglia in the CNS.* Microglia arise during early development from progenitors in the embryonic yolk sac. During adulthood, myeloid cells in the peripheral circulation get differentiated into monocytes/monocytic precursors, which then migrate to the CNS and are subsequently transformed into quiescent forms of microglia. The latter lack phenotypic markers required for antigen presentation, however, perform phagocytosis of foreign material and are cytotoxic against infected neurons, bacteria, and viruses, thereby helping to maintain neuronal homeostasis under normal physiological conditions. Once activated in the presence of pathogen-associated molecular patterns (PAMPs) and/or damage-associated molecular patterns (DAMPs), quiescent forms of microglia undergo morphological transformation to form reactive microglia, quickly proliferate, and show MHC class I and MHC class II proteins, receptors for various cytokines, Toll-like receptors, Nod-like receptors, and antigens for T cell subsets, which are the hallmarks of macrophages in periphery and essential to mounting an innate immune response. This microglial activity is further aided by the infiltrating hematogenous macrophages, which find a way to the CNS due to a breach in the blood–brain barrier during brain injuries and pathologies. The overall effect of this immune response could be neuroinflammation and associated psychiatric disorders, such as depression.

CONFLICT OF INTEREST STATEMENT

The presented work is supported by the National Health and Medical Research Council Australia (APP 1043771 to BTB). The funders had no role in study design, data collection and analysis, decision to publish, or preparation of the manuscript.

REFERENCES

1. World Health Organization. Mental disorders affect one in four people. *WHO*, 2013; http://www.who.int/whr/2001/media_centre/press_release/en/. August 6, 2018.
2. Latoo J, Mistry M, Dunne FJ. Physical morbidity and mortality in people with mental illness. *British Journal of Medical Practitioners.* 2013;6(3):621–623.
3. World Health Organization. Comprehensive mental health action plan 2013–2020. *WHO*, 2015; http://www.who.int/mental_health/action_plan_2013/en/.
4. Kreisel T, Frank M, Licht T, et al. The role of microglia in stress-induced depression. *Brain, Behavior, and Immunity.* 2014;40:e2–e3.
5. Öngür D, Drevets WC, Price JL. Glial reduction in the subgenual prefrontal cortex in mood disorders. *Proceedings of the National Academy of Sciences.* 1998;95(22):13290–13295.
6. Mirza B, Hadberg H, Thomsen P, Moos T. The absence of reactive astrocytosis is indicative of a unique inflammatory process in Parkinson's disease. *Neuroscience.* 1999;95(2):425–432.
7. Dowlati Y, Herrmann N, Swardfager W, et al. A meta-analysis of cytokines in major depression. *Biological Psychiatry.* 2010;67(5):446–457.
8. Howren MB, Lamkin DM, Suls J. Associations of depression with C-reactive protein, IL-1, and IL-6: a meta-analysis. *Psychosomatic Medicine.* 2009;71(2):171–186.
9. Eikelenboom P, Bate C, Van Gool W, et al. Neuroinflammation in Alzheimer's disease and prion disease. *Glia.* 2002;40(2):232–239.
10. Avila J, Lucas JJ, Perez M, Hernandez F. Role of tau protein in both physiological and pathological conditions. *Physiological Reviews.* 2004;84(2):361–384.
11. Ballatore C, Lee VM-Y, Trojanowski JQ. Tau-mediated neurodegeneration in Alzheimer's disease and related disorders. *Nature Reviews. Neuroscience.* 2007;8(9):663.
12. Lee JW, Lee YK, Yuk DY, et al. Neuro-inflammation induced by lipopolysaccharide causes cognitive impairment through enhancement of beta-amyloid generation. *Journal of Neuroinflammation.* 2008;5(1):37.
13. World Health Organization. *Depression.* 2017; http://www.who.int/mediacentre/factsheets/fs369/en/. Accessed Sept. 24, 2017.
14. Belmaker R, Agam G. Major depressive disorder. *New England Journal of Medicine.* 2008(358):55–68.
15. Pendulum.org. *Bipolar Disorder—Facts and Statistics.* 2017; http://www.pendulum.org/bpfacts.html. Accessed Sept. 24, 2017.
16. van der Schot AC, Vonk R, Brans RG, et al. Influence of genes and environment on brain volumes in twin pairs concordant and discordant for bipolar disorder. *Archives of General Psychiatry.* 2009;66(2):142–151.
17. Vreeburg SA, Hoogendijk WJ, van Pelt J, et al. Major depressive disorder and hypothalamic-pituitary-adrenal axis activity: results from a large cohort study. *Archives of General Psychiatry.* 2009;66(6):617–626.
18. Watson S, Gallagher P, Ritchie JC, Ferrier IN, Young AH. Hypothalamic-pituitary-adrenal axis function in patients with bipolar disorder. *The British Journal of Psychiatry.* 2004;184(6):496–502.
19. López-Figueroa AL, Norton CS, López-Figueroa MO, et al. Serotonin 5-HT 1A, 5-HT 1B, and 5-HT 2A receptor mRNA expression in subjects with major depression, bipolar disorder, and schizophrenia. *Biological Psychiatry.* 2004;55(3):225–233.

20. Tsai S-J, Hong C-J, Hsu C-C, et al. Serotonin-2A receptor polymorphism (102T/C) in mood disorders. *Psychiatry Research.* 1999;87(2):233–237.
21. Duman RS, Heninger GR, Nestler EJ. A molecular and cellular theory of depression. *Archives of General Psychiatry.* 1997;54(7):597–606.
22. Cunha AB, Frey BN, Andreazza AC, et al. Serum brain-derived neurotrophic factor is decreased in bipolar disorder during depressive and manic episodes. *Neuroscience Letters.* 2006;398(3):215–219.
23. Malykhin N, Coupland N. Hippocampal neuroplasticity in major depressive disorder. *Neuroscience.* 2015;309:200–213.
24. Manji HK, Moore GJ, Rajkowska G, Chen G. Neuroplasticity and cellular resilience in mood disorders. *Molecular Psychiatry.* 2000;5(6):578.
25. Kreisel T, Frank MG, Licht T, et al. Dynamic microglial alterations underlie stress-induced depressive-like behavior and suppressed neurogenesis. *Molecular Psychiatry.* 2014;19(6):699–709.
26. Flynn G, Maru S, Loughlin J, Romero IA, Male D. Regulation of chemokine receptor expression in human microglia and astrocytes. *Journal of Neuroimmunology.* 2003;136(1):84–93.
27. Singhal G, Jaehne EJ, Corrigan F, Toben C, Baune BT. Inflammasomes in neuroinflammation and changes in brain function: a focused review. *Frontiers in Neuroscience.* 2014;8.
28. Patel A. Review: the role of inflammation in depression. *Psychiatria Danubina.* 2013;25(Suppl 2):S216–S223.
29. Cacci E, Claasen JH, Kokaia Z. Microglia-derived tumor necrosis factor-α exaggerates death of newborn hippocampal progenitor cells in vitro. *Journal of Neuroscience Research.* 2005;80(6):789–797.
30. Videbech P, Ravnkilde B. Hippocampal volume and depression: a meta-analysis of MRI studies. *American Journal of Psychiatry.* 2015.
31. Brietzke E, Kapczinski F. TNF-α as a molecular target in bipolar disorder. *Progress in Neuro-Psychopharmacology and Biological Psychiatry.* 2008;32(6):1355–1361.
32. Kim JS. Cytokines and adhesion molecules in stroke and related diseases. *Journal of the Neurological Sciences.* 1996;137(2):69–78.
33. Wachholz S, Eßlinger M, Plümper J, Manitz M-P, Juckel G, Friebe A. Microglia activation is associated with IFN-α induced depressive-like behavior. *Brain, Behavior, and Immunity.* 2016;55:105–113.
34. Kim Y-K, Jung H-G, Myint A-M, Kim H, Park S-H. Imbalance between pro-inflammatory and anti-inflammatory cytokines in bipolar disorder. *Journal of Affective Disorders.* 2007;104(1):91–95.
35. Marques AH, Silverman MN, Sternberg EM. Glucocorticoid dysregulations and their clinical correlates: from receptors to therapeutics. *Annals of the New York Academy of Sciences.* 2009;1179:1–18.
36. Nair A, Bonneau RH. Stress-induced elevation of glucocorticoids increases microglia proliferation through NMDA receptor activation. *Journal of Neuroimmunology.* 2006;171(1–2):72–85.
37. Koyama Y. Functional alterations of astrocytes in mental disorders: pharmacological significance as a drug target. *Frontiers in Cellular Neuroscience.* 2015;9.
38. Davis A, Gilhooley M, Agius M. Using non-steroidal anti-inflammatory drugs in the treatment of depression. *Psychiatria Danubina.* 2010;22(Suppl 1):S49–S52.

39. Nery FG, Monkul ES, Hatch JP, et al. Celecoxib as an adjunct in the treatment of depressive or mixed episodes of bipolar disorder: a double-blind, randomized, placebo-controlled study. *Human Psychopharmacology: Clinical and Experimental.* 2008;23(2):87–94.
40. Eyre HA, Air T, Proctor S, Rositano S, Baune BT. A critical review of the efficacy of non-steroidal anti-inflammatory drugs in depression. *Progress in Neuro-Psychopharmacology and Biological Psychiatry.* 2015;57:11–16.
41. Perneger TV, Whelton PK, Klag MJ. Risk of kidney failure associated with the use of acetaminophen, aspirin, and nonsteroidal antiinflammatory drugs. *New England Journal of Medicine.* 1994;331(25):1675–1679.
42. Rainsford K. Profile and mechanisms of gastrointestinal and other side effects of nonsteroidal anti-inflammatory drugs (NSAIDs). *The American Journal of Medicine.* 1999;107(6):27–35.
43. Su K-P, Huang S-Y, Chiu C-C, Shen WW. Omega-3 fatty acids in major depressive disorder: a preliminary double-blind, placebo-controlled trial. *European Neuropsychopharmacology.* 2003;13(4):267–271.
44. Stoll AL, Severus WE, Freeman MP, et al. Omega 3 fatty acids in bipolar disorder: a preliminary double-blind, placebo-controlled trial. *Archives of General Psychiatry.* 1999;56(5):407–412.
45. van Harten J. Clinical pharmacokinetics of selective serotonin reuptake inhibitors. *Clinical Pharmacokinetics.* 1993;24(3):203–220.
46. Castellano S, Ventimiglia A, Salomone S, et al. Selective serotonin reuptake inhibitors and serotonin and noradrenaline reuptake inhibitors improve cognitive function in partial responders depressed patients: results from a prospective observational cohort study. *CNS & Neurological Disorders—Drug Targets (Formerly Current Drug Targets—CNS & Neurological Disorders).* 2016;15(10):1290–1298.
47. Ernst C, Olson AK, Pinel JP, Lam RW, Christie BR. Antidepressant effects of exercise: evidence for an adult-neurogenesis hypothesis? *Journal of Psychiatry & Neuroscience.* 2006;31(2):84.
48. Gleeson M, Bishop NC, Stensel DJ, Lindley MR, Mastana SS, Nimmo MA. The anti-inflammatory effects of exercise: mechanisms and implications for the prevention and treatment of disease. *Nature Reviews Immunology.* 2011;11(9):607–615.
49. Smith LA, Cornelius V, Warnock A, Bell A, Young AH. Effectiveness of mood stabilizers and antipsychotics in the maintenance phase ofbipolar disorder: a systematic review of randomized controlled trials. *Bipolar Disorders.* 2007;9(4):394–412.
50. World Health Organization. *Schizophrenia.* 2016; http://www.who.int/mediacentre/factsheets/fs397/en/. Accessed Sep 24, 2017.
51. Tandon R, Gaebel W, Barch DM, et al. Definition and description of schizophrenia in the DSM-5. *Schizophrenia Research.* 2013;150(1):3–10.
52. Javitt DC. Glutamatergic theories of schizophrenia. *The Israel Journal of Psychiatry and Related Sciences.* 2010;47(1):4.
53. Jentsch JD, Roth RH. The neuropsychopharmacology of phencyclidine: from NMDA receptor hypofunction to the dopamine hypothesis of schizophrenia. *Neuropsychopharmacology.* 1999;20(3):201–225.
54. Rothermundt M, Arolt V, Weitzsch C, Eckhoff D, Kirchner H. Immunological dysfunction in schizophrenia: a systematic approach. *Neuropsychobiology.* 1998;37(4):186–193.

55. Drzyzga Ł, Obuchowicz E, Marcinowska A, Herman ZS. Cytokines in schizophrenia and the effects of antipsychotic drugs. *Brain, Behavior, and Immunity.* 2006;20(6):532–545.
56. Brown AS. The environment and susceptibility to schizophrenia. *Progress in Neurobiology.* 2011;93(1):23–58.
57. Gao W-J, Snyder MA. NMDA hypofunction as a convergence point for progression and symptoms of schizophrenia. *Frontiers in Cellular Neuroscience.* 2013;7:31.
58. Monji A, Kato TA, Mizoguchi Y, et al. Neuroinflammation in schizophrenia especially focused on the role of microglia. *Progress in Neuro-Psychopharmacology and Biological Psychiatry.* 2013;42:115–121.
59. Doorduin J, De Vries EF, Willemsen AT, De Groot JC, Dierckx RA, Klein HC. Neuroinflammation in schizophrenia-related psychosis: a PET study. *Journal of Nuclear Medicine.* 2009;50(11):1801–1807.
60. Naudin J, Mege J, Azorin J, Dassa D. Elevated circulating levels of IL-6 in schizophrenia. *Schizophrenia Research.* 1996;20(3):269–273.
61. Gallego J, Morell C, McNamara R, Lencz T, Malhotra A. 9. Elevated TNF-α levels in cerebrospinal fluid of patients with schizophrenia. *Schizophrenia Bulletin.* 2017;43(Suppl 1):S10.
62. Kim Y-K, Myint A-M, Lee B-H, et al. Th1, Th2 and Th3 cytokine alteration in schizophrenia. *Progress in Neuro-Psychopharmacology and Biological Psychiatry.* 2004;28(7):1129–1134.
63. Müller N, Ackenheil M, Hofschuster E, Mempel W, Eckstein R. Cellular immunity in schizophrenic patients before and during neuroleptic treatment. *Psychiatry Research.* 1991;37(2):147–160.
64. Schwarz MJ, Riedel M, Ackenheil M, Müller N. Decreased levels of soluble intercellular adhesion molecule-1 (sICAM-1) in unmedicated and medicated schizophrenic patients. *Biological Psychiatry.* 2000;47(1):29–33.
65. Sperner-Unterweger B, Whitworth A, Kemmler G, et al. T-cell subsets in schizophrenia: a comparison between drug-naive first episode patients and chronic schizophrenic patients. *Schizophrenia Research.* 1999;38(1):61–70.
66. Hakobyan S, Boyajyan A, Sim RB. Classical pathway complement activity in schizophrenia. *Neuroscience Letters.* 2005;374(1):35–37.
67. Toro C, Deakin J. Adult neurogenesis and schizophrenia: a window on abnormal early brain development? *Schizophrenia Research.* 2007;90(1):1–14.
68. Pocklington AJ, O'Donovan M, Owen MJ. The synapse in schizophrenia. *European Journal of Neuroscience.* 2014;39(7):1059–1067.
69. Fourgeaud L, Boulanger LM. Synapse remodeling, compliments of the complement system. *Cell.* 2007;131(6):1034–1036.
70. Rutkowski MJ, Sughrue ME, Kane AJ, Mills SA, Fang S, Parsa AT. Complement and the central nervous system: emerging roles in development, protection and regeneration. *Immunology and Cell Biology.* 2010;88(8):781–786.
71. Dong C, Davis RJ, Flavell RA. MAP kinases in the immune response. *Annual Review of Immunology.* 2002;20(1):55–72.
72. Kyosseva SV, Elbein AD, Griffin WST, Mrak RE, Lyon M, Karson CN. Mitogen-activated protein kinases in schizophrenia. *Biological Psychiatry.* 1999;46(5):689–696.

73. Schwarz E, Izmailov R, Spain M, et al. Validation of a blood-based laboratory test to aid in the confirmation of a diagnosis of schizophrenia. *Biomarker Insights.* 2010;5:39.
74. Correll CU, Leucht S, Kane JM. Lower risk for tardive dyskinesia associated with second-generation antipsychotics: a systematic review of 1-year studies. *American Journal of Psychiatry.* 2004;161(3):414–425.
75. Sommer IE, de Witte L, Begemann M, Kahn RS. Nonsteroidal anti-inflammatory drugs in schizophrenia: ready for practice or a good start? A meta-analysis. *The Journal of Clinical Psychiatry.* 2012;73(4):414–419.
76. Horrobin DF. Omega-3 fatty acid for schizophrenia. *American Journal of Psychiatry.* 2003;160(1):188–189.
77. Malchow B, Reich-Erkelenz D, Oertel-Knöchel V, et al. The effects of physical exercise in schizophrenia and affective disorders. *European Archives of Psychiatry and Clinical Neuroscience.* 2013;263(6):451–467.
78. Dawson TM, Dawson VL. Molecular pathways of neurodegeneration in Parkinson's disease. *Science.* 2003;302(5646):819–822.
79. Palop JJ, Mucke L. Amyloid-[beta]-induced neuronal dysfunction in Alzheimer's disease: from synapses toward neural networks. *Nature Neuroscience.* 2010;13(7):812–818.
80. Babic T. The cholinergic hypothesis of Alzheimer's disease: a review of progress. *Journal of Neurology, Neurosurgery & Psychiatry.* 1999;67(4):558.
81. Agostinho P, Cunha AR, Oliveira C. Neuroinflammation, oxidative stress and the pathogenesis of Alzheimer's disease. *Current Pharmaceutical Design.* 2010;16(25):2766–2778.
82. Kamboh M, Demirci F, Wang X, et al. Genome-wide association study of Alzheimer's disease. *Translational Psychiatry.* 2012;2(5):e117.
83. Sveinbjornsdottir S. The clinical symptoms of Parkinson's disease. *Journal of Neurochemistry.* 2016;139(S1):318–324.
84. Hauser DN, Hastings TG. Mitochondrial dysfunction and oxidative stress in Parkinson's disease and monogenic Parkinsonism. *Neurobiology of Disease.* 2013;51:35–42.
85. Surmeier DJ. Calcium, ageing, and neuronal vulnerability in Parkinson's disease. *The Lancet Neurology.* 2007;6(10):933–938.
86. American Psychiatric Association. *Diagnostic and Statistical Manual of Mental Disorders,* Fifth Edition. Arlington, VA, American Psychiatric Association; 2013.
87. Shimomura T, Mori E, Yamashita H, et al. Cognitive loss in dementia with Lewy bodies and Alzheimer disease. *Archives of Neurology.* 1998;55(12):1547–1552.
88. McKhann GM, Knopman DS, Chertkow H, et al. The diagnosis of dementia due to Alzheimer's disease: recommendations from the National Institute on Aging-Alzheimer's Association workgroups on diagnostic guidelines for Alzheimer's disease. *Alzheimer's & Dementia.* 2011;7(3):263–269.
89. Stefaniak J, O'Brien J. Imaging of neuroinflammation in dementia: a review. *Journal of Neurology, Neurosurgery, and Psychiatry.* 2016;87(1):21–28.
90. Wenk GL. Neuropathologic changes in Alzheimer's disease. *Journal of Clinical Psychiatry.* 2003;64:7–10.
91. Rubio-Perez JM, Morillas-Ruiz JM. A review: inflammatory process in Alzheimer's disease, role of cytokines. *The Scientific World Journal.* 2012;2012.

92. Liu L, Chan C. The role of inflammasome in Alzheimer's disease. *Ageing Research Reviews.* 2014;15:6–15.
93. Fillit H, Ding W, Buee L, et al. Elevated circulating tumor necrosis factor levels in Alzheimer's disease. *Neuroscience Letters.* 1991;129(2):318–320.
94. Cacabelos R, Barquero M, Garcia P, Alvarez X, Varela de Seijas E. Cerebrospinal fluid interleukin-1 beta (IL-1 beta) in Alzheimer's disease and neurological disorders. *Methods and Findings in Experimental and Clinical Pharmacology.* 1991;13(7):455–458.
95. Yasojima K, Schwab C, McGeer EG, McGeer PL. Up-regulated production and activation of the complement system in Alzheimer's disease brain. *The American Journal of Pathology.* 1999;154(3):927–936.
96. Janeway CA, Travers P, Walport M, Shlomchik MJ. The complement system and innate immunity. 2001.
97. Hsuchou H, Kastin AJ, Mishra PK, Pan W. C-reactive protein increases BBB permeability: implications for obesity and neuroinflammation. *Cellular Physiology and Biochemistry.* 2012;30(5):1109–1119.
98. McGeer PL, McGeer E, Yasojima K. Alzheimer disease and neuroinflammation. *Advances in Dementia Research*: Springer, Vienna; 2000:53–57.
99. Zhang W, Wang T, Pei Z, et al. Aggregated α-synuclein activates microglia: a process leading to disease progression in Parkinson's disease. *The FASEB Journal.* 2005;19(6):533–542.
100. Banati RB, Gehrmann J, Schubert P, Kreutzberg GW. Cytotoxicity of microglia. *Glia.* 1993;7(1):111–118.
101. Hostenbach S, Cambron M, D'haeseleer M, Kooijman R, De Keyser J. Astrocyte loss and astrogliosis in neuroinflammatory disorders. *Neuroscience Letters.* 2014;565:39–41.
102. Aizenstein HJ, Nebes RD, Saxton JA, et al. Frequent amyloid deposition without significant cognitive impairment among the elderly. *Archives of Neurology.* 2008;65(11):1509–1517.
103. Lee H-g, Perry G, Moreira PI, et al. Tau phosphorylation in Alzheimer's disease: pathogen or protector? *Trends in Molecular Medicine.* 2005;11(4):164–169.
104. Kitazawa M, Oddo S, Yamasaki TR, Green KN, LaFerla FM. Lipopolysaccharide-induced inflammation exacerbates tau pathology by a cyclin-dependent kinase 5-mediated pathway in a transgenic model of Alzheimer's disease. *Journal of Neuroscience.* 2005;25(39):8843–8853.
105. Cole GM, Ma Q-L, Frautschy SA. Omega-3 fatty acids and dementia. *Prostaglandins, Leukotrienes and Essential Fatty Acids.* 2009;81(2):213–221.
106. Lautenschlager NT, Cox KL, Flicker L, et al. Effect of physical activity on cognitive function in older adults at risk for Alzheimer disease: a randomized trial. *Journal of the American Medical Association.* 2008;300(9):1027–1037.
107. Ahlskog JE, Geda YE, Graff-Radford NR, Petersen RC. Physical exercise as a preventive or disease-modifying treatment of dementia and brain aging. Mayo Clinic Proceedings. 2011;86(9):876–884.
108. Tabet N. Acetylcholinesterase inhibitors for Alzheimer's disease: anti-inflammatories in acetylcholine clothing! *Age and Ageing.* 2006;35(4):336–338.

109. Loveman E, Green C, Kirby J, et al. The clinical and cost-effectiveness of donepezil, rivastigmine, galantamine and memantine for Alzheimer's disease. 2006.
110. Simonsen AH, McGuire J, Hansson O, et al. Novel panel of cerebrospinal fluid biomarkers for the prediction of progression to Alzheimer dementia in patients with mild cognitive impairment. *Archives of Neurology.* 2007;64(3):366–370.
111. Emeriau J-P, et al. Efficacy of naftidrofuryl in patients with vascular or mixed dementia: results of a multicenter, double-blind trial. *Clinical Therapeutics.* 2000;22(7):834–844.
112. Areosa SA, Sherriff F, McShane R. Memantine for dementia. *Cochrane Database of Systematic Reviews.* 2005;3:CD003154.
113. Block G, Liss C, Reines S, Irr J, Nibbelink D. Comparison of immediate-release and controlled release carbidopa/levodopa in Parkinson's disease. *European Neurology.* 1997;37(1):23–27.
114. Lyytinen J, Kaakkola S, Ahtila S, Tuomainen P, Teräväinen H. Simultaneous MAO-B and COMT inhibition in L-dopa-treated patients with Parkinson's disease. *Movement Disorders.* 1997;12(4):497–505.
115. Katzenschlager R, Sampaio C, Costa J, Lees A. Anticholinergics for symptomatic management of Parkinson's disease. *The Cochrane Library.* 2002.
116. Nimmerjahn F, Ravetch JV. Anti-inflammatory actions of intravenous immunoglobulin. *Annual Review of Immunology.* 2008;26:513–533.
117. Dodel R, Du Y, Depboylu C, et al. Intravenous immunoglobulins containing antibodies against β-amyloid for the treatment of Alzheimer's disease. *Journal of Neurology, Neurosurgery & Psychiatry.* 2004;75(10):1472–1474.
118. Sydenham E, Dangour AD, Lim W-S. Omega 3 fatty acid for the prevention of cognitive decline and dementia. *Sao Paulo Medical Journal.* 2012;130(6):419.
119. Laurin D, Verreault R, Lindsay J, Dewailly É, Holub BJ. Omega-3 fatty acids and risk of cognitive impairment and dementia. *Journal of Alzheimer's Disease.* 2003;5(4):315–322.
120. Simopoulos AP. Omega-3 fatty acids in inflammation and autoimmune diseases. *Journal of the American College of Nutrition.* 2002;21(6):495–505.
121. Kos T, Popik P. A comparison of the predictive therapeutic and undesired side-effects of the NMDA receptor antagonist, memantine, in mice. *Behavioural Pharmacology.* 2005;16(3):155–161.
122. Il'yasova D, Colbert LH, Harris TB, et al. Circulating levels of inflammatory markers and cancer risk in the Health, Aging and Body Composition cohort. *Cancer Epidemiology Biomarkers & Prevention.* 2005;14(10):2413–2418.
123. De Rekeneire N, Peila R, Ding J, et al. Diabetes, hyperglycemia, and inflammation in older individuals: the Health, Aging and Body Composition study. *Diabetes Care.* 2006;29(8):1902–1908.
124. Volpato S, Guralnik JM, Ferrucci L, et al. Cardiovascular disease, interleukin-6, and risk of mortality in older women: the Women's Health and Aging Study. *Circulation.* 2001;103(7):947–953.
125. Capuron L, Miller AH. Immune system to brain signaling: neuropsychopharmacol ogical implications. *Pharmacology & Therapeutics.* 2011;130(2):226–238.
126. Dinarello CA. Proinflammatory cytokines. *CHEST Journal.* 2000;118(2):503–508.

127. Hickey W, Hsu B, Kimura H. T-lymphocyte entry into the central nervous system. *Journal of Neuroscience Research.* 1991;28(2):254–260.
128. Engelhardt B. Molecular mechanisms involved in T cell migration across the blood–brain barrier. *Journal of Neural Transmission.* 2006;113(4):477–485.
129. Poli A, Kmiecik J, Domingues O, et al. NK cells in central nervous system disorders. *The Journal of Immunology.* 2013;190(11):5355–5362.
130. von Büdingen H-C, Kuo TC, Sirota M, et al. B cell exchange across the blood–brain barrier in multiple sclerosis. *The Journal of Clinical Investigation.* 2012;122(12):4533.

INDEX

Tables and figures are indicated by an italic t and f following the paragraph number